Power

Second Edition

EDN Series for Design Engineers

Brown	*Power Supply Cookbook, Second Edition*
Dostál	*Operational Amplifiers, Second Edition*
Dye	*Radio Frequency Transistors: Principles and Practical Applications, Second Edition*
Gates Energy Products	*Rechargeable Batteries Applications Handbook*
Hickman	*Electronic Circuits, Systems and Standards: The Best of EDN*
Marston	*Newnes Electronic Circuits Pocket Book*
Marston	*Integrated Circuit and Waveform Generator Handbook*
Marston	*Diode, Transistor and FET Circuits Manual*
Pease	*Troubleshooting Analog Circuits*
Sinclair	*Passive Components*
Williams	*Analog Circuit Design: Art, Science and Personalities*

Power Supply Cookbook

Second Edition

Marty Brown

Boston Oxford Johannesburg Melbourne New Delhi

Newnes is an imprint of Butterworth–Heinemann.

A member of the Reed Elsevier group

Recognizing the importance of preserving what has been written, Butterworth–Heinemann prints its books on acid-free paper whenever possible.

Butterworth–Heinemann supports the efforts of American Forests and the Global ReLeaf program in its campaign for the betterment of trees, forests, and our environment.

Library of Congress Cataloging-in-Publication Data
Brown, Marty.
Power supply cookbook / Marty Brown.—2nd ed.
p. cm.
Includes bibliographical references and index.
ISBN 0-7506-7329-X
1. Electric power supplies to apparatus—Design and construction. 2. Power electronics. 3. Electronic apparatus and appliances—power supply. I. Title.
TK7868.P6 B76 2001
621.381'044—dc21 00-050054

British Library Cataloguing-in-Publication Data
A catalogue record for this book is available from the British Library.

The publisher offers special discounts on bulk orders of this book.
For information, please contact:

Manager of Special Sales
Butterworth–Heinemann
225 Wildwood Avenue
Woburn, MA 01801–2041
Tel: 781-904-2500
Fax: 781-904-2620

For information on all Newnes publications available, contact our World Wide Web home page at: http://www.newnespress.com

10 9 8 7 6 5 4 3 2 1

Printed in the United States of America

Contents

Preface

Power Supply Cookbook was written by a practicing design engineer for practicing design engineers. Through designing power supplies for many years, along with a variety of electronic products ranging from industrial control to satellite systems, I have acquired a great appreciation for the "systems-level" development process and the trade-offs associated with them. Many of the approaches I use involve issues outside the immediate design of the power supply and their impact on the design.

Power Supply Cookbook, Second Edition has been updated with the latest advances in the field of efficient power conversion. Efficiencies of between 80 to 95 percent are now possible using these new techniques. The major losses within the switching power supply and the modern techniques to reduce them are discussed at length. These include: synchronous rectification, lossless snubbers, and active clamps. The information on methods of control, noise control, and optimum printed circuit board layout has also been updated.

As with the previous edition, the "cookbook" approach taken in *Power Supply Cookbook, Second Edition* facilitates information finding for both the novice and seasoned engineer. The information is organized so that the reader need only read the material for the degree of in-depth knowledge he or she wishes to acquire. Because of the enclosed design flow, the typical power supply can be designed schematically in less than 8 hours, which can cut weeks from the expected design period.

The purpose of this book is not to advance the bastions of academia, but to offer the tried and true design approaches implemented by many engineers in the power field. It offers advice and examples which can be immediately applied to the reader's own designs.

Introduction

This book is an invaluable adjunct to those engineers wanting to better understand power supply operation in order to effectively implement the computer-aided design (CAD) tools available. The broad implementation and success of CAD tools, along with the internationalization of the world's design resources, has led to competition that has shortened the typical product design cycle from more than a year to a matter of months. As a result, it is important for design engineers to locate and apply just the right amount of information without a long learning period.

Power Supply Cookbook, Second Edition is organized in a rather unique manner and, if followed correctly, can greatly shorten the amount of time needed to design a power supply. By presenting intuitive descriptions of the power supply system's operation along with commonly used circuit approaches, it is designed to help anyone with a working electronics knowledge to design a very complex switching power supply quickly.

I developed the concept for *Power Supply Cookbook* after having spent many hours working with design engineers on their power supply designs and, subsequently, my own designs.

The "Cookbook" Method of Organization

Power Supply Cookbook, Second Edition follows the same tried and true "cookbook" organization as its predecessor. This easy-to-use format helps readers quickly locate the power supply design sections they need without reading the book from start to finish. Additionally, the text follows the design flow that a seasoned power supply designer would follow. Circuit sections are designed in a way that provides information needed by subsequent circuit sections. Coverage of more complicated design areas, such as magnetics and feedback loops, is presented in a step-by-step format to help designers reduce the opportunity for mistakes.

The results of the calculations in this book lead to a conservative ("middle of the road") design. The results are "calculated estimates" that can be adjusted one way or another to enhance a performance or a physical property of the power supply. These compromises are discussed in the appropriate sections of the text.

For best results, the new reader should follow this flow:

A. Read Chapter 1 on the role of the power supply within the system and design program. This chapter provides the reader with insight as to the role of the power supply within the overall system, and develops the power supply design specification.

B. Read the introduction sections for the type of power supply you wish to develop (linear, pulsewidth modulated [PWM] switching, or high-efficiency).

C. Follow the order of the design "flowchart" and refer to the appropriate section within the book. Within each section, read the basic operation of that subcircuit. Then choose a design implementation that would best fit your requirements from the selection of common industry design approaches.

D. Calculate the component values and ratings from the design equations using your particular set of operating conditions.
E. "Paste" the resulting subcircuit into the main schematic and proceed to the next subcircuit to be designed.
F. At the end of the "paper design" (estimated 8 to 12 hours), read the section on PCB layout and begin building the first prototype.
G. Debug and test the prototype.
H. Finalize the physical and electrical design in preparation for production release.

The appendices are provided for those technical areas that are common among the various power supply technologies. They also present more detail for those designers who wish a deeper understanding of the subjects. The material on the design of basic PWM switching power supplies should be followed for all switching power supply designs. Chapter 4 describes how one can further enhance the overall efficiency of the power supply being designed.

In short, this book is written for working engineers by a working engineer. I hope you find it infinitely useful.

1. The Role of the Power Supply within the System and Design Program

The power supply assumes a very unique role within a typical system. In many respects, it is the mother of the system. It gives the system life by providing consistent and repeatable power to its circuits. It defends the system against the harsh world outside the confines of the enclosure and protects its wards by not letting them do harm to themselves. If the supply experiences a failure within itself, it must fail gracefully and not allow the failure to reach the system.

Alas, mothers are taken for granted, and their important functions are not appreciated. The power system is routinely left until late in the design program for two main reasons. First, nobody wants to touch it because everybody wants to design more exciting circuits and rarely do engineers have a background in power systems. Secondly, bench supplies provide all the necessary power during the system debugging stage and it is not until the product is at the integration stage that one says "Oops, we forgot to design the power supply!" All too frequently, the designer assigned to the power supply has very little experience in power supply design and has very little time to learn before the product is scheduled to enter production.

This type of situation can lead to the "millstone effect" which in simple terms means "You designed it, you fix it (*forever*)." No wonder no one wants to touch it and, when asked, disavows any knowledge of having ever designed a power supply.

1.1 Getting Started. This Journey Starts with the First Question

In order to produce a good design, many questions must be asked prior to the beginning of the design process. The earlier they are asked the better off you are. These questions also avoid many problems later in the design program due to lack of communication and forethought. The basic questions to be asked include the following.

From the marketing department

1. From what power source must the system draw its power? There are different design approaches for each power system and one can also get information as to what adverse operating conditions are experienced for each.

2. What safety and radio frequency interference and electromagnetic interference (RFI/EMI) regulations must the system meet to be able to be sold into the target market? This would affect not only the electrical design but also the physical design.
3. What is the maintenance philosophy of the system? This dictates what sort of protection schemes and physical design would match the application.
4. What are the environmental conditions in which the product must operate? These are temperature range, ambient RF levels, dust, dirt, shock, vibration, and any other physical considerations.
5. What type of graceful degradation of product performance is desired when portions of the product fail? This would determine the type of power busing scheme and power sequencing that may be necessary within the system.

From the designers of the other areas of the product

1. What are the technologies of the integrated circuits that are being used within the design of the system? One cannot protect something, if one doesn't know how it breaks.
2. What are the "best guess" maximum and minimum limits of the load current and are there any intermittent characteristics in its current demand such as those presented by motors, video monitors, pulsed loads, and so forth? Always add 50 percent more to what is told to you since these estimates always turn out to be low. Also what are the maximum excursions in supply voltage that the designer feels that the circuit can withstand. This dictates the design approaches of the cross-regulation of the outputs, and feedback compensation in order to provide the needs of the loads.
3. Are there any circuits that are particularly noise-sensitive? These include analog-to-digital and digital-to-analog converters, video monitors, etc. This may dictate that the supply has additional filtering or may need to be synchronized to the sensitive circuit.
4. Are there any special requirements of power sequencing that are necessary for each respective circuit to operate reliably?
5. How much physical space and what shape is allocated for the power supply within the enclosure? It is always too small, so start negotiating for your fair share.
6. Are there any special interfaces required of the power supply? This would be any power-down interrupts, etc., that may be required by any of the product's circuits.

This inquisitiveness also sets the stage for the beginning of the design by defining the environment in which the power supply must operate. This then forms the basis of the design specification of the power supply.

1.2 Power System Organization

The organization of the power system within the final product should complement the product philosophy. The goal of the power system is to distribute power effectively to each section of the entire product and to do it in a fashion that meets the needs of each subsection within the product. To accomplish this, one or more power system organization can be used within the product.

For products that are composed of one functional "module" that is inseparable during the product's life, such as a cellular telephone, CRT monitor, RF receiver, etc., an integrated power system is the traditional system organization. Here, the product has one main power supply which is completely self-contained and outputs directly to the product's circuits. An *integrated power system* may actually have more than one power supply within it if one of the load circuits has power demand or sequencing requirements which cannot be accommodated by the main power supply without compromising its operation.

For those products that have many diverse modules that can be reconfigured over the life of the product, such as PCB card cage systems and cellular telephone ground stations, etc., then the *distributed power system* is more appropriate. This type of system typically has one main "bulk" power supply that provides power to a bus which is distributed throughout the entire product. The power needs of any one module within the system are provided by smaller, *board-level* regulators. Here, voltage drops experienced across connectors and wiring within the system do not bother the circuits.

The integrated power system is inherently more efficient (less losses). The distributed system has two or more power supplies in series, where the overall power system efficiency is the product of the efficiencies of the two power supplies. So, for example, two 80 percent efficient power supplies in series produces an overall system efficiency of 64 percent.

The typical power system can usually end up being a combination of the two systems and can use switching and linear power supplies.

The engineer's motto to life is "Life is a tradeoff" and it comes into play here. It is impossible to design a power supply system that meets *all* the requirements that are initially set out by the other engineers and management and keep it within cost, space, and weight limits. The typical initial requirement of a power supply is to provide infinitely adaptable functions, deliver kilowatts within zero space, and cost no money. Obviously, some compromise is in order.

1.3 Selecting the Appropriate Power Supply Technology

Once the power supply system organization has been established, the designer then needs to select the technology of each of the power supplies within the system. At the early stage of the design program, this process may be iterative between reorganizing the system and the choice of power supply technologies. The important issues that influence this stage of the design are:

1. Cost.
2. Weight and space.
3. How much heat can be generated within the product.
4. The input power source(s).
5. The noise tolerance of the load circuits.
6. Battery life (if the product is to be portable).
7. The number of output voltages required and their particular characteristics.
8. The time to market the product.

The three major power supply technologies that can be considered within a power supply system are:

1. Linear regulators.
2. Pulsewidth modulated (PWM) switching power supplies.
3. High efficiency resonant technology switching power supplies.

Each of these technologies excels in one or more of the system considerations mentioned above and must be weighed against the other considerations to determine the optimum mixture of technologies that meet the needs of the final product. The power supply industry has chosen to utilize each of the technologies within certain areas of product applications as detailed in the following.

Linear

Linear regulators are used predominantly in ground-based equipments where the generation of heat and low efficiency are not of major concern and also where low cost and a short design period are desired. They are very popular as board-level regulators in distributed power systems where the distributed voltage is less than 40 VDC. For off-line (plug into the wall) products, a power supply stage ahead of the linear regulator must be provided for safety in order to produce dielectric isolation from the ac power line. Linear regulators can only produce output voltages lower than their input voltages and each linear regulator can produce only one output voltage. Each linear regulator has an average efficiency of between 35 and 50 percent. The losses are dissipated as heat.

PWM switching power supplies

PWM switching power supplies are much more efficient and flexible in their use than linear regulators. One commonly finds them used within portable products, aircraft and automotive products, small instruments, off-line applications, and generally those applications where high efficiency and multiple output voltages are required. Their weight is much less than that of linear regulators since they require less heatsinking for the same output ratings. They do, however, cost more to produce and require more engineering development time.

High efficiency resonant technology switching power supplies

This variation on the basic PWM switching power supply finds its place in applications where still lighter weight and smaller size are desired, and most importantly, where a reduced amount of radiated noise (interference) is desired. The common products where these power supplies are utilized are aircraft avionics, spacecraft electronics, and lightweight portable equipment and modules. The drawbacks are that this power supply technology requires the greatest amount of engineering design time and usually costs more than the other two technologies.

The trends within the industry are away from linear regulators (except for board-level regulators) towards PWM switching power supplies. Resonant and quasi-resonant switching power supplies are emerging slowly as the technology matures and their designs are made easier. To help in the selection, Table 1–1 summarizes some of the trade-offs made during the selection process.

Table 1–1 Comparison of the Four Power Supply Technologies

	Linear Regulator	PWM Switching Regulator	Resonant Transition Switching Regulator	Quasi-Resonant Switching Regulator
Cost	Low	High	High	Highest
Mass	High	Low-medium	Low-medium	Low-medium
RF Noise	None	High	Medium	Medium
Efficiency	35–50%	70–85%	78–92%	78–92%
Multiple outputs	No	Yes	Yes	Yes
Development time to production	1 week	8 person-months[a]	10 person-months[a]	10 person-months[a]
		5 person-months[b]	8 person-months	8 person-months[b]

[a] Based upon a reasonable level of experience and facilities.

[b] With the use of this book.

1.4 Developing the Power System Design Specification

Before actually designing the power system, the designer should develop the power system design specification. The design specification acts as the performance goal that the ultimate power supply must meet in order for the entire product to meet its overall performance specification. Once developed, it should be viewed as a semi-firm document and should only be changed after the needs of the product formally change.

When developing the design specification, the power supply designer must keep in mind what is a reasonable requirement and what is an idealistic requirement. Engineers not experienced in power supply design often will produce requirements on the power supply that either will cost an unnecessary fortune and take up too much space or will be impossible to meet with the present state of the technology. Here the power supply designer should press the other engineers, managers, and marketers for compromises that will prompt them to review their requirements to decide what they can actually live with.

The power system specification will be based upon the questions that should previously have been asked of the other departments involved in defining and designing the product. Some of the requirements can be anticipated to grow, such as the current needed by various subsystems within the product. Always add 25 to 50 percent to the output current capabilities of the power supply during the design process to accommodate this inevitable event. Also, the space allocated to the power system and its cost will almost always be less than what will be finally required. Some negotiations will be in order. Since the power system is a support function within the product, its design will always be modified in reaction to design issues within the other sections of the product. This will always make the power supply design the last circuit to be released for production. Recognizing and addressing these potential trouble areas early in the design period will help avoid delays later in the program.

To develop a good design specification, the designer should understand the meaning of the terms used within the power supply field. These are measurable power supply parameters with a common set of test conditions that the actual design affects. These parameters are the following.

Input voltage

$V_{in(nom)}$ The input voltage at which the product expects to operate for >99 percent of its life.

$V_{in(low)}$ The lowest anticipated operational input voltage (brown-out).

$V_{in(hi)}$ The highest anticipated operational average input voltage.

Line Frequency(s) dc, 50, 60, or 400 Hz, etc.

Include any adverse operating conditions that may require the supply to operate outside the conventional specifications such as:

Dropout A period of time over which the input line voltage completely disappears (the specification is typically 8 mS for 60 Hz ac off-line applications).

Surge A defined period of time where the input voltage will exceed the $V_{in(hi)}$ specification that the unit must survive and during which it may need to operate.

Transients These are very high voltage "spikes" (+/–) that are characteristic of the input power system.

Emergency operation Any operation required of the product during any adverse operating periods. This may be because the product's function is so critical for the survival of the operator of the unit, that it must operate to just short of its own destruction.

Input current

$I_{in(max)}$ This is the maximum average input current. Its maximum limit may be specified by a safety regulatory agency.

Output voltage(s)

$V_{out(rated)}$ The nominal output voltage (ideal).

$V_{out(min)}$ The output voltage below which the load should be inhibited or turned off.

$V_{out(max)}$ The maximum output voltage under which normal operation of the load circuits can operate.

$V_{out(abs)}$ The voltage at which the loads reach their destructive limits.

Ripple voltage (switching power supplies) This is measured in peak-to-peak volts, and its frequency and level should be acceptable to the load circuits.

Output current

$I_{out(rated)}$ The maximum average current that will be drawn from an output.

$I_{out(min)}$ The minimum current that will be drawn from the output during normal operation.

I_{sc} The maximum current limit that should be delivered into a short-circuited load.

Describe any unusual load demand characteristics related to any output. These consist of intermittent loads such as motors, CRTs, etc., and also any loads that may be removed from or added to the system as part of an overall system architecture, such as probes, handsets, and the like.

Dynamic load response time: This is the amount of time it requires the power supply to recover to within load regulation limits in response to a step change in the load.

Line regulation: Percentage change in the output voltage(s) in response to a change in the input voltage.

$$\text{Line Reg.} = \frac{V_{o(\text{hi-in})} - V_{o(\text{lo-in})}}{V_{o(\text{nom-in})}} \cdot 100(\%) \tag{1.0}$$

Load regulation: Percentage change in the output voltage(s) in response to a change in load current from one-half rated to rated load current.

$$\text{Load Reg.} = \frac{V_{o(\text{full-load})} - V_{o(\text{half-load})}}{V_{o(\text{rated-load})}} \cdot 100(\%) \tag{1.1}$$

Overall efficiency: This will determine how much heat will be generated within the product and whether any heatsinking will be needed in the physical design.

$$\text{Effic.} = \frac{P_{\text{out}}}{P_{\text{in}}} \cdot 100(\%) \tag{1.3}$$

Protections

- Input fusing limits.
- Overcurrent foldback on the outputs.
- Overvoltage trip protection limits.
- Undervoltage lockout on the input power line.
- Any graceful degradation features and repair philosophy after system failure.

Operating and Storage Ambient Temperature Ranges Outside the Product

Safety regulatory agency issues

- Dielectric withstanding voltage (hipot).
- Insulation resistance.
- Enclosure considerations (interlocks, insulation class, shock, marking, etc.).

RFI/EMI (Radiofrequency and electromagnetic interference) which regulatory agency specifications the product must meet.

- Conducted EMI: line filtering.
- Radiated RFI: physical layout and enclosures.

Special functionalities required of the power supply. These include any power-on resets and power-fail signals needed by any microcomputers in the system, remote turn-off, output voltage or current programming, power sequencing, status signals, etc.

This now forms a very good basis from which to begin a power supply design. This specification is now at a point that it can dictate which design paths must be pursued in order to meet the above specifications and will help to guide the designer during the design process.

1.5 A Generalized Design Approach to Power Supplies: Introducing the Building-block Approach to Power Supply Design

All power supply engineers follow a general pattern of steps in the design of power supplies. If the pattern is followed, each step actually sets the foundation for subsequent design steps and will guide the designer through a path of least resistance to the desired result. This text presents an approach that consists of two facets: first it breaks the power supply into distinct blocks that can be designed in a modular fashion; secondly, it prescribes the order in which the blocks are to be designed in order to ease their "pasting" together. The reader is further helped by the inclusion of typical industry design approaches for each block of various applications used by power supply designers in the field. Each block includes the associated design equations from which the component values can be quickly calculated. The result is a coherent, logical design flow in which the unknowns are minimized. The approach is organized such that the typical inexperienced designer can produce a "professional" grade power supply schematic in under 8 working hours, which is about 40 percent of the entire design process. The physical design, such as breadboarding techniques, low-noise printed circuit board (PCB) layouts, transformer winding techniques, etc., are shown through example. The physical factors always present a problem, not only to the inexperienced designer, but to the experienced designer as well. It is hoped that these practical examples will keep the problems to a minimum. All power supplies, regardless of whether they are linear or switching, follow a general design flow. The linear power supplies, though, because of the maturity of the technology and the level of integration offered by the semiconductor manufacturers, will be presented mainly via examples. The design flow of the switching power supplies, which are much more complicated, will be covered in more detail in the respective chapters dealing with the selected power supply technology. The generalized approach is as follows.

1. Select the appropriate technology and topology for your application.
2. Perform "black box" approximations knowing only the design specification requirements. This results in estimates of semiconductor power losses, peak currents and voltages. It may also indicate to the designer that the chosen topology is inappropriate and a different choice is necessary. It also allows the designer to order any semiconductor samples that may be required during the breadboarding phase of the program.
3. Design the power supply schematically, guided by the design flowcharts.
4. Build the breadboard using the techniques outlined in the physical layout and construction sections in the text.
5. TEST, TEST, TEST! Test the power supply against the requirements stated in the design specification. If they do not meet the requirements, some design modifications may be necessary. Make "baseline" measurements so

that you can measure any subsequent changes in the power supply's performance. Conduct tests with the final product connected to the supply to check for unwanted interactions. And by all means, begin to measure items related to safety and RFI/EMI prior to submitting the final product to the approval bodies.
6. Finalize the physical design. This would include physical packaging within the product, heatsink design, and the PCB design.
7. Submit the final product for approval body safety and RFI/EMI testing and approval. Some modifications are usually required, but if you have done your homework in the previous design stages, these can be minor.
8. Production Release!

It all sounds simple, but the legendary and cursed philosopher, Murphy, runs wild through the field of power supply design, so expect many a visit from this unwelcome guest.

1.6 A Comment about Power Supply Design Software

There is an abundance of software-based power supply design tools, particularly for PWM switching power supply designs. Many of these software packages were written by the semiconductor manufacturers for their own highly integrated switching power supply integrated circuits (ICs). Many of these ICs include the power devices as well as the control circuitry. These types of software packages should only be used with the targeted products and not for general power supply designs. The designs presented by these manufacturers are optimized for minimum cost, weight, and design time, and the arrangements of any external components are unique to that IC.

There are several generalized switching power supply design software packages available primarily from circuit simulator companies. Caution should be practiced in reviewing all software-based switching power supply design tools. Designers should compare the results from the software to those obtained manually by executing the appropriate design equations. Such a comparison will enable designers to determine whether the programmer and his or her company really understands the issues surrounding switching power supply design. Remember, most of the digital world thinks that designing switching power supplies is just a matter of copying schematics.

The software packages may also obscure the amount of latitude a designer has during a power supply design. By making the program as broad in its application as possible, the results may be very conservative. To the seasoned designer, this is only a first step. He or she knows how to "push" the result to enhance the power supply's performance in a certain area. All generally applied equations and software results should be viewed as calculated estimates. In short, the software may then lead the designer to a result that works but is not optimum for the system.

1.7 Basic Test Equipment Needed

Power supplies, especially switching power supplies, require the designer to view parameters not commonly encountered in the other fields of electronics. Aside from ac and dc voltage, the designer must also look at ac and dc current measurements and waveforms, and RF spectrum analysis. Although the vision of large capital expenditures flashes through your mind when this is mentioned, the basic equipment can be obtained for under US $3000. The equipment can be classified as necessary and optional, but somewhere along the line, all the equipment will have to be used whether one buys the items or rents them.

Necessary test equipment

1. A 100 MHz or higher bandwidth, time-based oscilloscope. The bandwidth is especially needed for switching power supply design. A digital oscilloscope may miss important transients on some of the key waveforms, so evaluate any digital oscilloscope carefully.
2. 10 : 1 voltage probes for the oscilloscope.
3. A dc/ac volt and ampere multimeter. A true RMS reading meter is optional.
4. An ac and/or dc current probe for the oscilloscope. Especially needed for switching power supply design. Some appropriate models are Tektronics P6021 or P6022 and A6302 or A6303, or better.
5. A bench-top power supply that can simulate the input power source. This will be a large dc power supply with voltage and current ratings in excess of what is needed. For off-line power supplies, use a variac with a current rating in excess of what is needed.

Note: Please isolate all test equipment from earth ground when testing.

Optional test equipment

1. Spectrum analyzer. This can be used to view the RFI and EMI performance of the power supply prior to submission to a regulatory agency. It would be too costly to set up a full testing laboratory, so I would recommend using an third-party testing house.
2. A true RMS wattmeter for conveniently measuring efficiency and power factor. This is needed for off-line power supplies.

2. An Introduction to the Linear Regulator

The linear regulator is the original form of the regulating power supply. It relies upon the variable conductivity of an active electronic device to drop voltage from an input voltage to a regulated output voltage. In accomplishing this, the linear regulator wastes a lot of power in the form of heat, and therefore gets hot. It is, though, a very electrically "quiet" power supply.

The linear power supply finds a very strong niche within applications where its inefficiency is not important. These include wall-powered, ground-base equipment where forced air cooling is not a problem; and also those applications in which the instrument is so sensitive to electrical noise that it requires an electrically "quiet" power supply—these products might include audio and video amplifiers, RF receivers, and so forth. Linear regulators are also popular as local, board-level regulators. Here only a few watts are needed by the board, so the few watts of loss can be accommodated by a simple heatsink. If dielectric isolation is desired from an ac input power source it is provided by an ac transformer or bulk power supply.

In general, the linear regulator is quite useful for those power supply applications requiring less than 10 W of output power. Above 10 W, the heatsink required becomes so large and expensive that a switching power supply becomes more attractive.

2.1 Basic Linear Regulator Operation

All power supplies work under the same basic principle, whether the supply is a linear or a more complicated switching supply. All power supplies have at their heart a closed negative feedback loop. This feedback loop does nothing more than hold the output voltage at a constant value. Figure 2–1 shows the major parts of a series-pass linear regulator.

Linear regulators are step-down regulators only; that is, the input voltage source must be higher than the desired output voltage. There are two types of linear regulators: the *shunt regulator* and the *series-pass regulator*. The shunt regulator is a voltage regulator that is placed in parallel with the load. An unregulated current source is connected to a higher voltage source, the shunt regulator draws output current to maintain a constant voltage across the load given a variable input voltage and load current. A common example of this is a Zener diode regulator. The series-pass linear regulator is more efficient than the shunt regulator and uses an active semiconductor as the series-pass unit, between the input source and the load.

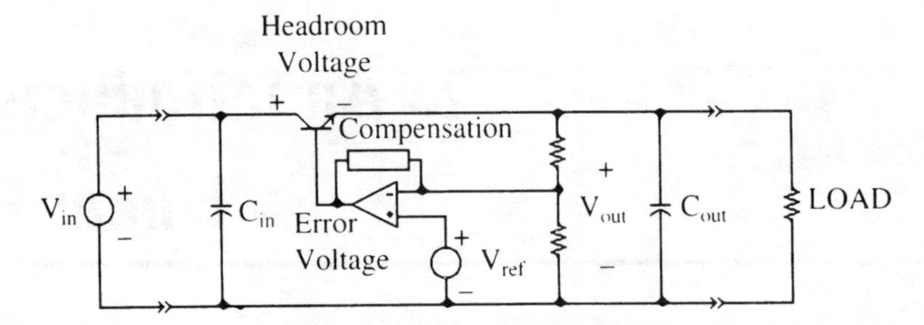

Figure 2–1 The basic linear regulator.

The series-pass unit operates in the linear mode, which means that the unit is not designed to operate in the full on or off mode but instead operates in a degree of "partially on." The negative feedback loop determines the degree of conductivity the pass unit should assume to maintain the output voltage.

The heart of the negative feedback loop is a high-gain operational amplifier called a *voltage error amplifier.* Its purpose is to continuously compare the difference between a very stable voltage reference and the output voltage. If the output differs by mere millivolts, then a correction to the pass unit's conductivity is made. A stable voltage reference is placed on the noninverting input and is usually lower than the output voltage. The output voltage is divided down to the level of the voltage reference. This divided output voltage is placed into the inverting input of the operational amplifier. So at the rated output voltage, the center node of the output voltage divider is identical to the reference voltage.

The gain of the error amplifier produces a voltage that represents the greatly amplified difference between the reference and the output voltage (error voltage). The error voltage directly controls the conductivity of the pass unit thus maintaining the rated output voltage. If the load increases, the output voltage will fall. This will then increase the amplifier's output, thus providing more current to the load. Similarly, if the load decreases, the output voltage will rise, thus making the error amplifier respond by decreasing pass unit current to the load.

The speed by which the error amplifier responds to any changes on the output and how accurately the output voltage is maintained depends on the error amplifier's *feedback loop compensation.* The feedback compensation is controlled by the placement of elements within the voltage divider and between the negative input and the output of the error amplifier. Its design dictates how much gain at dc is exhibited, which dictates how accurate output voltage will be. It also dictates how much gain at a higher frequency and bandwidth the amplifier exhibits, which dictates the time it takes to respond to output load changes or *transient response time.*

The operation of a linear regulator is very simple. The very same circuitry exists in the heart of all regulators, including the more complicated switching regulators. The voltage feedback loop performs the ultimate function of the power supply—the maintaining of the output voltage.

2.2 General Linear Regulator Considerations

The majority of linear regulator applications today are board-level, low-power applications that are easily satisfied through the use of highly integrated 3-

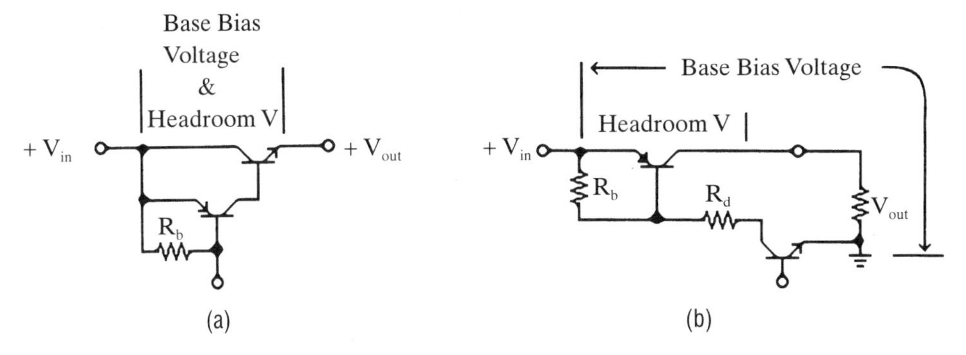

Figure 2–2 The pass unit's influence on the dropout voltage: (a) NPN pass unit; (b) PNP pass unit (low dropout).

terminal regulator integrated circuits. Occasionally, though, the application calls for either a higher output current or greater functionality than the 3-terminal regulators can provide.

There are design considerations that are common to both approaches and those that are only applicable to the nonintegrated, custom designs. These considerations define the operating boundary conditions that the final design will meet, and the relevant ones must be calculated for each design. Unfortunately, many engineers neglect them and have trouble over the entire specified operating range of the product after production.

The first consideration is the *headroom voltage*. The headroom voltage is the actual voltage drop between the input voltage and the output voltage during operation. This enters predominantly into the later design process, but it should be considered first, just to see whether the linear supply is appropriate for the needs of the system. First, more than 95 percent of all the power lost within the linear regulator is lost across this voltage drop. This headroom loss is found by

$$P_{HR} = (V_{in(max)} - V_{out}) I_{load(rated)} \tag{2.1}$$

If the system cannot handle the heat dissipated by this loss at its maximum specified ambient operating temperature, then another design approach should be taken. This loss determines how large a heatsink the linear regulator must have on the pass unit.

A quick estimated thermal analysis will reveal to the designer whether the linear regulator will have enough thermal margin to meet the needs of the product at its highest specified operating ambient temperature. One can find such a thermal analysis in Appendix A.

The second major consideration is the minimum *dropout voltage* of a particular topology of linear regulator. This voltage is the minimum headroom voltage that can be experienced by the linear regulator, below which it falls out of regulation. This is predicated only by how the pass transistors derive their drive bias current and voltage. The common positive linear regulator utilizes an NPN bipolar power transistor (see Figure 2–2a). To generate the needed base-emitter voltage for the pass transistor's operation, this voltage must be derived from its own collector-emitter voltage. For the NPN pass units, this is the actual minimum headroom voltage. This dictates that the headroom voltage cannot get any lower than the base-emitter voltage (~0.65 VDC) of the NPN pass unit plus the drop across any base drive devices (transistors and resistors). For the three terminal regulators such as the MC78XX series, this voltage is 1.8 to 2.5 VDC. For custom designs using NPN pass transistors for positive outputs, the dropout voltage may be higher. For applications where the input voltage may come even closer than 1.8–2.5 VDC to the output voltage, a *low dropout regulator* is recommended. This topology utilizes a PNP pass transistor, which now derives its base-emitter voltage from the output voltage instead of the headroom or input voltage (see Figure 2–2b). This allows the regulator to have a dropout voltage of 0.6 VDC minimum. P-Channel MOSFETs can also be used in this function and can exhibit dropout voltages close to zero volts.

The dropout voltage becomes a driving issue when the input to the linear regulator during normal operation is allowed to fall close to the output voltage. If operating from an ac wall transformer, this would occur at brown-out conditions (minimum ac voltages). The low dropout regulator (e.g., LM29XX) would allow the regulator to operate to a lower ac input voltage. Low dropout regulators are also widely used as *post regulators* on the output of switching power supplies. Within switching regulators, the efficiency is of great concern, so the headroom drop needs to be kept to a minimum. Here, the low dropout regulator will save several W of loss over a conventional NPN-based linear regulator. If the application will never see headroom voltages less than 2.5 V, then use the conventional linear regulators (e.g., MC78XX).

Another consideration is the type of pass unit to be used. From a headroom loss standpoint, it makes absolutely no difference whether a bipolar power transistor or a power MOSFET is used. The difference comes in the drive circuitry. If the headroom voltage is high, the controller (usually a ground-oriented circuit) must pull current from the input or output voltage to ground. For a single bipolar pass transistor this current is

$$I_B = I_{Load}/h_{FE} \tag{2.2}$$

The power lost just in driving the bipolar pass transistor is

$$P_{drive} = V_{in(max)} \cdot I_B \quad or \quad V_{out} \cdot I_B \tag{2.3}$$

This drive loss can become significant. A driver transistor can be added to the pass transistor to increase the effective gain of the pass unit and thus decrease the drive current, or a power MOSFET can be used as a pass unit that uses magnitudes less dc drive current than the bipolar power transistor. Unfortunately, the MOSFET requires up to 10 VDC to drive the gate. This can drastically increase the dropout voltage. In the vast majority of linear regulator applications, there is little difference in operation between a buffered pass unit and a MOSFET insofar as efficiency is concerned. Bipolar transistors are much less expensive than power MOSFETs and have less propensity to oscillate.

The linear regulator is a mature technology and therefore can usually be accommodated by the integrated solutions provided by the semiconductor manufacturers. For applications beyond the limits of these integrated linear regulators alone, usually adding more components around the IC will satisfy the requirement. Otherwise, a completely custom approach would need to be utilized. These various approaches are overviewed in the design examples in the following section.

2.3 Linear Power Supply Design Examples

Linear regulators can be designed to meet a variety of cost and functional needs. The design examples that follow illustrate that linear regulator designs can

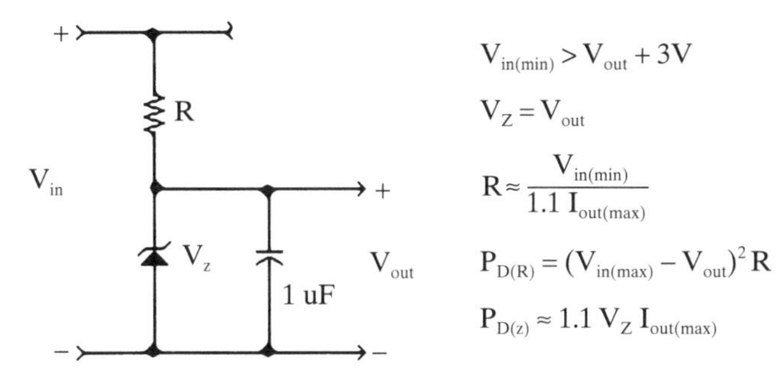

Figure 2–3 A Zener shunt regulator.

range from the very elementary to the more complex. Designs for enhanced 3-terminal regulator designs will be abbreviated, since the integrated circuit datasheets usually contain great detail. Due to the relatively large power loss of linear regulators, the thermal considerations typically represent a significant problem. Some thermal analysis and design is done in the examples. For further insight on this please refer to Appendix A.

2.3.1 Elementary Discrete Linear Regulator Designs

These types of linear regulators were commonly built before the advent of operational amplifiers and they can save money in consumer designs. Some of their drawbacks include drift with temperature and limited load current range.

The Zener shunt regulator

This type of regulator is typically used for very local voltage regulation for less than 200 mW of a load. A series resistance is placed between a higher voltage and is used to limit the current to the load and Zener diode. The Zener diode compensates for the variation in load current. The Zener voltage will drift with temperature. The drift characteristics are given in many Zener diode datasheets. Its load regulation is adequate for most supply specifications for integrated circuits. It also has a higher loss than the series-pass type of linear regulator, since its loss is set for the maximum load current, which for any load remains less than that value. A Zener shunt regulator can be seen in Figure 2–3.

The one-transistor series-pass linear regulator

By adding a transistor to the basic Zener regulator, one can take advantage of the gain that the bipolar transistor offers. The transistor is hooked up as an emitter follower, which can now provide a much higher current to the load, and the Zener current can be lowered. Here the transistor acts as a rudimentary error amplifier (refer to Figure 2–4). When the load current increases, it places a higher voltage into the base, which increases its conductivity, thus restoring the voltage to its original level. The transistor can be sized to meet the demands of the load and the headroom loss. It can be a TO-92 transistor for those loads up to 0.25 W or a TO-220 for heavier loads (depending on heatsinking).

2.3.2 Basic Three-Terminal Regulator Designs

Three-terminal regulators are used in the majority of board-level regulator applications. They excel in cost and ease of use for these applications. They can also, with care, be used as the basis or higher functionality linear regulators.

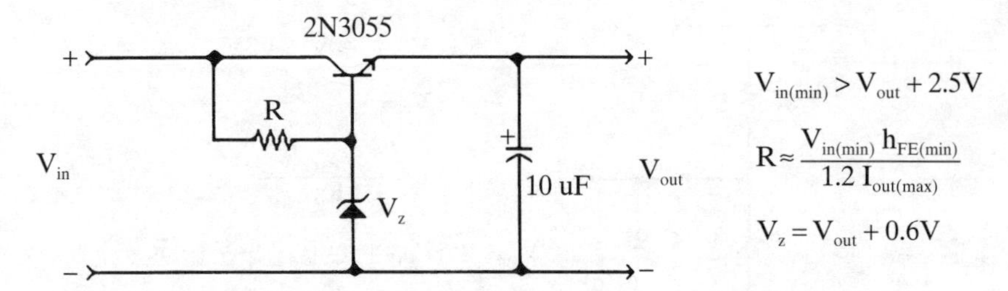

Figure 2–4 A discrete bipolar series-pass regulator.

The most often ignored consideration is the overcurrent limiting method used in 3-terminal regulators. They typically use an overtemperature cutoff on the die of the regulator which is typically between +150°C and +165°C. If the load current is passed through the 3-terminal regulator, and if the heatsink is too large, the regulator may fail due to overcurrent (bondwire, IC traces, etc.). If the heatsink is too small, then one may not be able to get enough power from the regulator. Another consideration is if the load current is being conducted by an external pass-unit the overtemperature cutoff will be nonfunctional, and another method of overcurrent protection will be needed.

2.3.2.1 The Basic Three-Terminal Positive Regulator Design

This example will illustrate the design considerations that should be undertaken with each 3-terminal regulator design. Many designers view only the electrical specifications of the regulators and forget the *thermal derating* of the part. At high headroom voltages, and at high ambient operating temperatures, the regulator can only deliver a fraction of its full-rated performance. Actually, in the majority of the 3-terminal applications, the heatsink determines the regulator's maximum output current. The manufacturer's electrical ratings can be viewed as having the part bolted onto a large piece of metal and placed in an ocean. Any application not employing those unorthodox components must operate at a lower level. The following example illustrates a typical recommended design procedure.

Design Example 1. Using Three-Terminal Regulators

Specification	Input:	12 VDC (max)
		8.5 VDC (min)
	Output:	5.0 VDC
		0.1–0.25 Amp
	Temperature:	–40–+50°C

Note: The 1N4001 is required for discharging the 100 μF capacitor when the system is turned off.

Thermal Design (refer also to Appendix A)

Given in data sheet: $R_{\theta JC} = 5°C/W$

$R_{\theta JA} = 65°C/W$

$T_{j(max)} = 150°C$

$$P_{D(max)} = (V_{in(max)} - V_{out}) \cdot I_{load(max)}$$
$$= (12 - 5\,V)(0.25\,A) = 1.75\,W \text{ (headroom loss)}$$

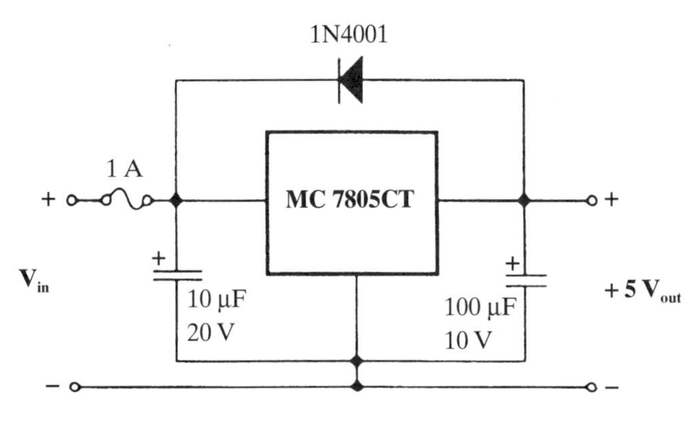

Figure 2–5 A 3-terminal regulator.

Without a heatsink the junction temperature will be:

$$T_j = P_D \cdot R_{\theta JA} + T_{A(max)} = (1.75\,W)(65°C/W) + 50$$
$$= 163.75°C.$$

A small "clip-on" style heatsink is required to bring the junction temperature down to below its maximum ratings.
Refer to Appendix A for aid in the selection of heatsinks.

Selecting the heatsink—Thermalloy P/N 6073B

Given in heatsink data: $R_{\theta SA} = 14°C/W$
Using a silicon insulator $R_{\theta CS} = 65°C/W$

The new worst case junction temperature is now:

$$T_{j(max)} = P_D(R_{\theta JC} + R_{\theta CS} + R_{\theta SA}) + T_A$$
$$= (1.75W)(5\,°C/W + 65\,°C/W + 14\,°C/W) + 50\,°C$$
$$= 84.4\,°C$$

2.3.2.2 Three-Terminal Regulator Design Variations

The following design examples illustrate how 3-terminal regulator integrated circuits can form the basis of higher-current, more complicated designs. Care must be taken, though, because all of the examples render the overtemperature protection feature of the 3-terminal regulators useless. Any overcurrent protection must now be added externally to the integrated circuit.

The current-boosted regulator
The design shown in Figure 2–6 adds just a resistor and a transistor to the 3-terminal regulator to yield a linear regulator that can provide more current to the load. The current-boosted positive regulator is shown, but the same equations hold for the boosted negative regulator. For the negative regulators, the power transistor changes from a PNP to an NPN. Beware, there is no overcurrent or overtemperature protection in this particular design.

The current-boosted 3-terminal regulator with overcurrent protection
This design adds the overcurrent protection externally to the IC. It employs the base-emitter (0.6 V) junction of a transistor to accomplish the overcurrent

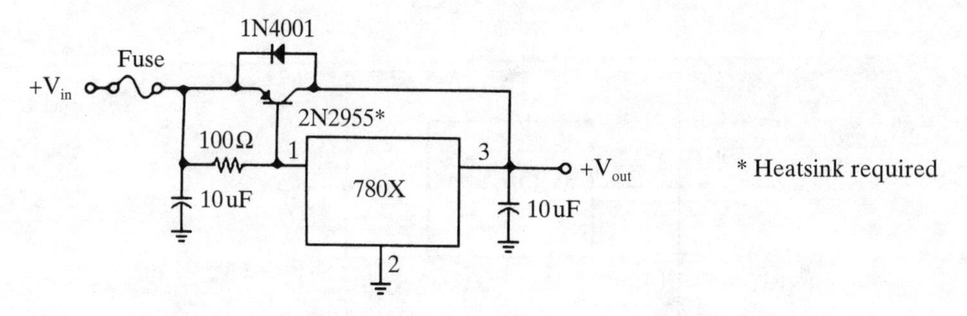

Figure 2–6 Current-boosted 3-terminal regulator without overcurrent protection.

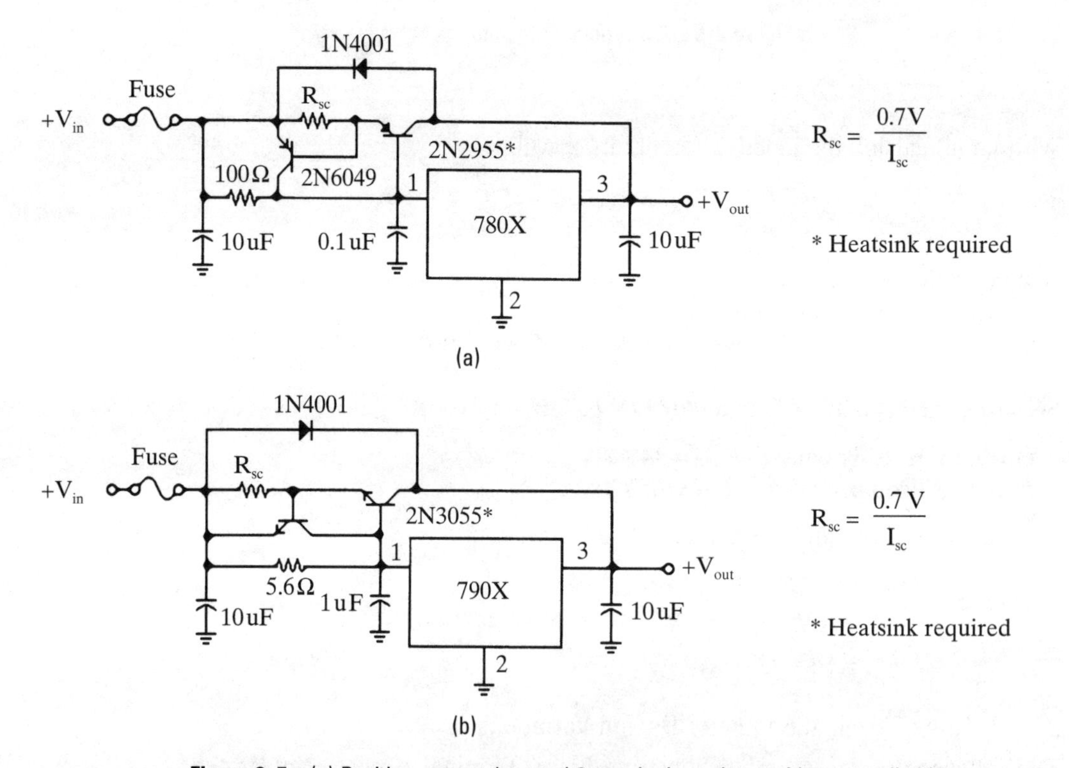

Figure 2–7 (a) Positive current-boosted 3-terminal regulator with current limiting. (b) Negative current-boosted 3-terminal regulator with current limiting.

threshold and gain of the overcurrent stage. For the negative voltage version of this, all the external transistors change from NPN to PNP and vice versa. These can be seen in Figures 2–7a and b.

2.3.3 Floating Linear Regulators

A floating linear regulator is one way of achieving high-voltage linear regulation. Its philosophy is one in which the regulator controller section and the series-pass transistor "float" on the input voltage. The output voltage regulation is accomplished by sensing the ground, which appears as a negative voltage when referenced to the output voltage. The output voltage serves as the "floating ground" for the controller and the power for the controller and series-pass transistor is drawn from the headroom voltage (the input-to-output difference) or is provided by an auxiliary isolated power supply.

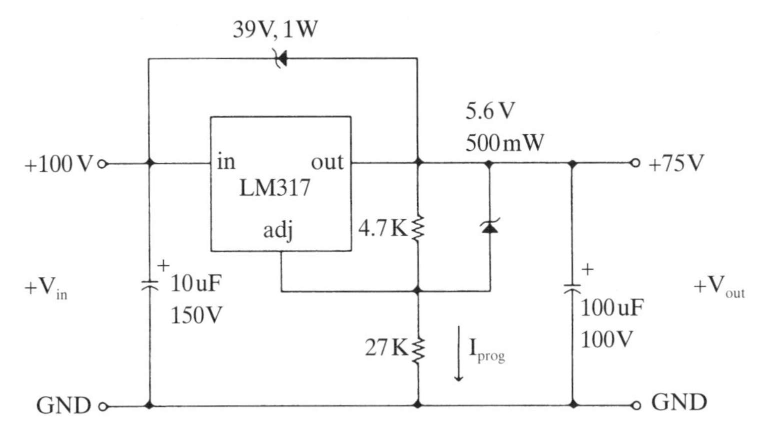

Figure 2–8 A high voltage floating linear regulator.

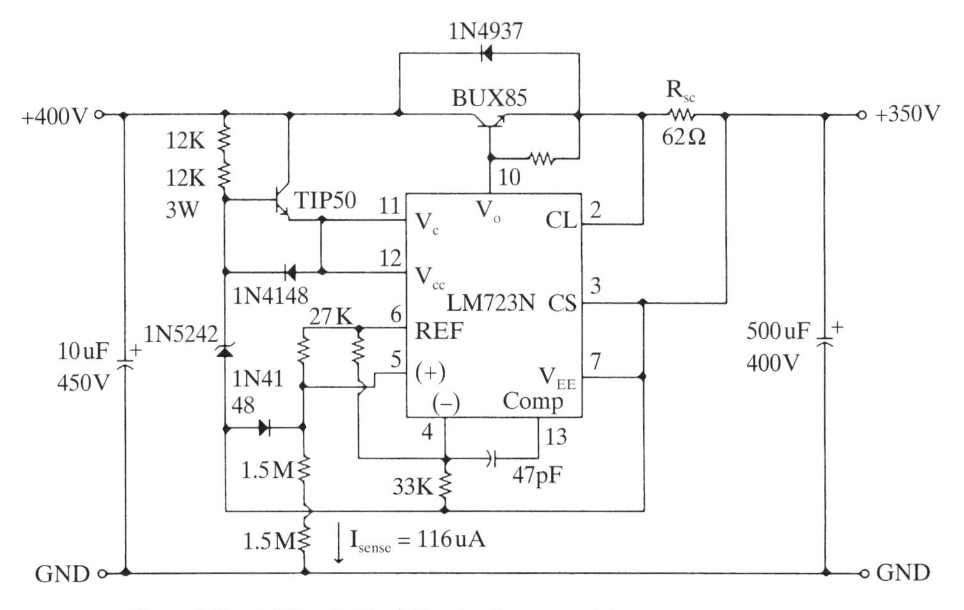

Figure 2–9 A 350 volt, 10 mA floating linear regulator.

The power transistor still needs to have a breakdown voltage rating greater than the input voltage, since at start-up, it must see the entire input voltage across it. Other methods such as a *bootstrap Zener diode* can also be used in order to shunt the voltage around the pass transistor, but only when the input voltage itself is switched on and off to activate the power supply. Also, caution must be taken to ensure that any controller input or output pin never goes negative with respect to the floating ground of the IC. Protection diodes are usually used for this purpose. One last caution is the little-known breakdown voltage of common resistors. If the output voltage exceeds 200 V, more than one sensing resistor must be placed in series in order to avoid the 250 V breakdown characteristic of 1/4 W resistors.

A common low-voltage positive floating regulator is the LM317 (the negative regulator complementary part is the LM337). The MC1723 can also be used to create a floating linear regulator, but care must be taken to protect the IC against the high voltage.

The first example shows how an LM317 can be modified to create a 70 V linear regulator from a 100 V input voltage. Several design restrictions must be strictly followed, for example, the operational headroom voltage must not exceed the voltage rating of the bootstrap Zener diode or regulation will be lost. Also the use of the protection diode on the error amplifier is mandatory. This regulator can be seen in Figure 2–8.

The second example illustrates a 350 V floating linear regulator that can provide up to 10 mA of load current from a 400 to 450 V unregulated source. The TIP50 provides the bias supply for the controller, which must withstand the full input voltage during start-up and power supply foldback. The controller is "grounded" on the output voltage and the minimum headroom voltage is 15 V. To readjust the output voltage, one changes the value of the two series resistors in the voltage sensing branch and this is set by

$$R_{sense} = (V_{out} + 4.0\text{V}) / I_{sense} \tag{2.4}$$

Floating linear regulators are particularly suited for high-output voltage regulation, but may be used anywhere. This regulator can be seen in Figure 2–9.

3. Pulsewidth Modulated Switching Power Supplies

Although pulsewidth modulated (PWM) switching power supplies have been around for a long time, it wasn't until the mid-1970s that they became more accepted and broadly applied. Switching power supplies offer many advantages over linear regulators.

Switching power supplies are more efficient and are smaller in size than linear regulators of similar ratings. They are, however, more difficult to design and radiate more electromagnetic interference (EMI).

Today, there are two ways to approach the design of switching power supplies. The design of board-level, dc/dc (dc-in, dc-out) switching power supplies can be copied directly from the semiconductor maker's datasheet and can use standard components from other manufacturers. However, if any of the requirements fall outside the standardized approaches, then it becomes a custom design, and is much more complex.

This book is organized so that the massive process of designing a custom switching power supply is broken down into smaller, more understandable pieces. Each piece is then explained in "non-power engineer" terms, and commonly accepted design approaches are illustrated with the relevant design equations. The intent is for the reader to read the section, choose the best design approach to meet his or her needs, use his or her particular system parameters, and produce a subcircuit that can be inserted into a larger power supply design. The design order is the way that seasoned power engineers use to approach their designs, and has proven to provide answers before the questions have arisen.

3.1 The Fundamentals of PWM Switching Power Supplies

The operation of switching power supplies can be relatively easy to understand. Unlike linear regulators which operate the power transistor in the linear mode, the PWM switching power supply operates the power transistors in both the saturated and cutoff states. In these states, the volt-ampere product across the power transistor is always kept low (saturated, low-*V*/high-*I*; and cutoff, Hi-*V*/No-*I*). This EI product within the power device is the loss within all the power semiconductors.

This more efficient operation of the PWM switching power supply is done by "chopping" the direct current (dc) input voltage into pulses whose amplitude is the magnitude of the input voltage and whose duty cycle is controlled by a switching regulator controller. Once the input voltage is converted to an ac rectangular waveform, the amplitude can be stepped up or down by a transformer. Additional output voltages can be derived by adding secondaries to the transformer. Ultimately these ac waveforms are then filtered to provide the dc output voltages.

The controller, whose main purpose is to maintain a regulated output voltage, operates very much like a linear style controller. That is, the functional blocks, voltage reference, and error amplifier are arranged identical to the linear regulator's. The difference is, the output of the error amplifier (the error voltage) is then placed into a voltage-to-pulsewidth converter stage prior to driving the power switches.

There are two major operational types of switching power supplies: the *forward-mode* converter and the *boost-mode* converter. Although their arrangements of parts are subtly different, their operation is very different and each has advantages in certain areas of application.

3.1.1 The Forward-mode Converter

Forward-mode regulators form a large family of switching power supply topologies. They can be recognized by an L-C filter just after the power switch or after the output rectifier on the secondary of a transformer. A simple form of the forward-mode regulator can be seen in Figure 3–1. This is called the *buck regulator*.

Its operation can be seen as analogous to a mechanical flywheel and a one-piston engine. The L-C filter, like the flywheel, stores energy between the power pulses of the driver. The input to the L-C filter (*choke input filter*) is the chopped input voltage. The L-C filter volt-time averages this duty-cycle modulated input voltage waveform. The L-C filtering function can be approximated by

$$V_{out} \approx V_{in} \cdot \text{duty cycle}. \tag{3.1}$$

The output voltage is maintained by the controller by varying the duty cycle. The buck converter is also known as a *step-down converter*, since its output must be less than the input voltage.

The operation of the buck regulator can be seen by breaking its operation into two periods (refer to Figure 3–2). When the switch is turned on, the input voltage is presented to the input of the L-C filter. The inductor current ramps linearly upward and is described as

$$i_{L(on)} = \frac{(V_{in} - V_{out})t_{on}}{L_o} + i_{init}. \tag{3.2}$$

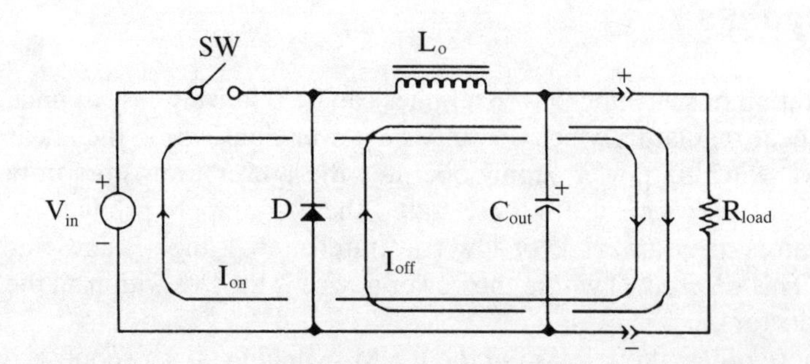

Figure 3–1 A basic forward-mode converter (buck converter shown).

The energy stored within the inductor during this period is

$$E_{\text{stored}} = \left(\frac{1}{2}\right) L_o \left(i_{\text{pk}} - i_{\text{min}}\right)^2 \tag{3.3}$$

This input energy is stored by the flux contained within the core material of the inductor.

When the power switch is turned off, the input voltage to the inductor wants to fly below ground and the diode (D), called a *catch diode*, becomes forward biased. This continues to conduct the current that was formerly flowing through the power switch and some of the stored energy is discharged to the load. This forms a local current loop that includes the diode, inductor, and the load. The current through the inductor is described during this period by

$$i_{\text{L(off)}} = i_{\text{pk}} - \frac{V_{\text{out}} \cdot t_{\text{off}}}{L_o} \tag{3.4}$$

The current waveform, this time, is a negative linear ramp whose slope is $-V_{\text{out}}/L$. When the power switch once again turns on, the diode snaps off and the current now flows through the input power source and the power switch. The inductor's current (i_{min}) just prior to the switch being turned on, becomes the initial current the power switch must then initially pass.

The dc output load current value falls between the peak and the minimum current values. In typical applications, the peak inductor current is about 150 percent of the dc load current and the minimum current is about 50 percent.

The advantages of forward-mode converters are: they exhibit lower output peak-to-peak ripple voltages than do boost-mode converters, and they can provide much higher levels of output power. Forward-mode converters can provide up to kilowatts of power.

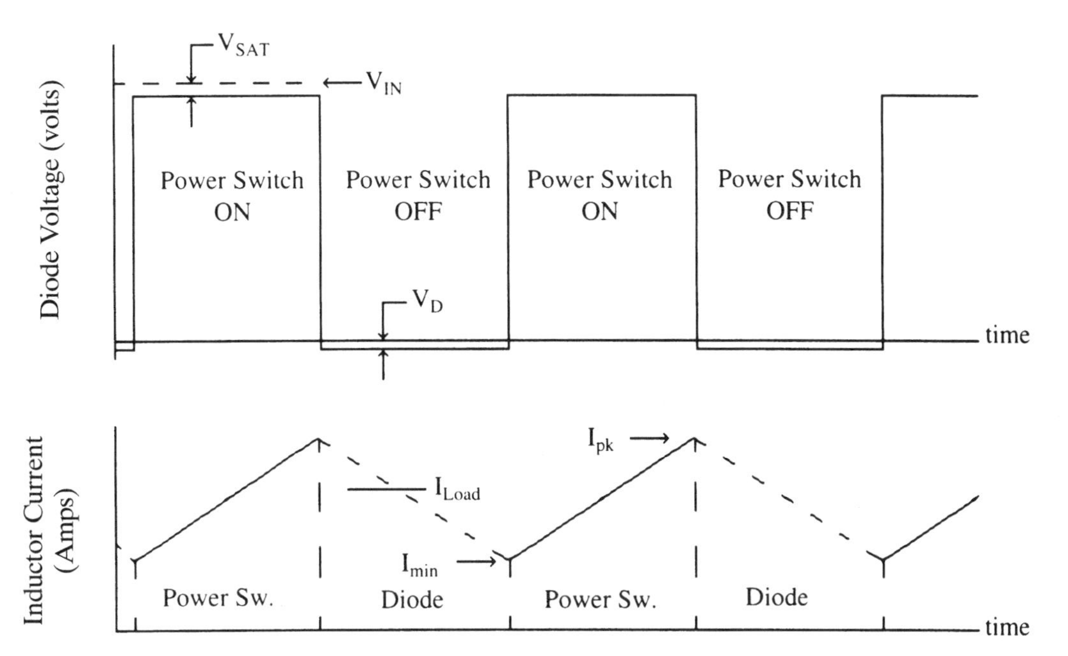

Figure 3–2 The voltage and current waveforms for a forward-mode converter (buck converter).

A transformer can be placed between the power switch and the L-C filter which serves as a voltage step-up or step-down of the input voltage. These topologies form a family of converters called *transformer-isolated forward converters* (refer to Figures 3–14 through 3–17). The transformer offers some distinct advantages, such as providing a dielectric barrier from the input to the output and the ability to add additional outputs, and makes the output voltages independent from the level of the input voltage.

3.1.2 The Boost-mode Converter

The second family of converters are the boost-mode converters. The most elementary boost-mode (or boost-derived) converter can be seen in Figure 3–3. It is called a *boost converter.*

As one can notice, the boost-mode converter has the same parts as the forward-mode converter, but they have been rearranged. This new arrangement causes the converter to operate in a completely different fashion than the forward-mode converter. This time, when the power switch is turned on, a current loop is created that only includes the inductor, the power switch, and the input voltage source. The diode is reverse-biased during this period. The inductor's current waveform (Figure 3–4) is also a positive linear ramp and is described by

$$i_L(t_{on}) = \frac{V_{in} \cdot t_{on}}{L} \tag{3.5}$$

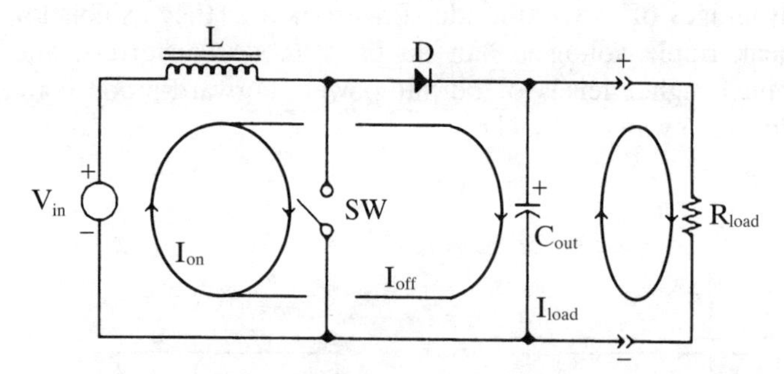

Figure 3–3 A basic boost-mode converter (boost converter shown).

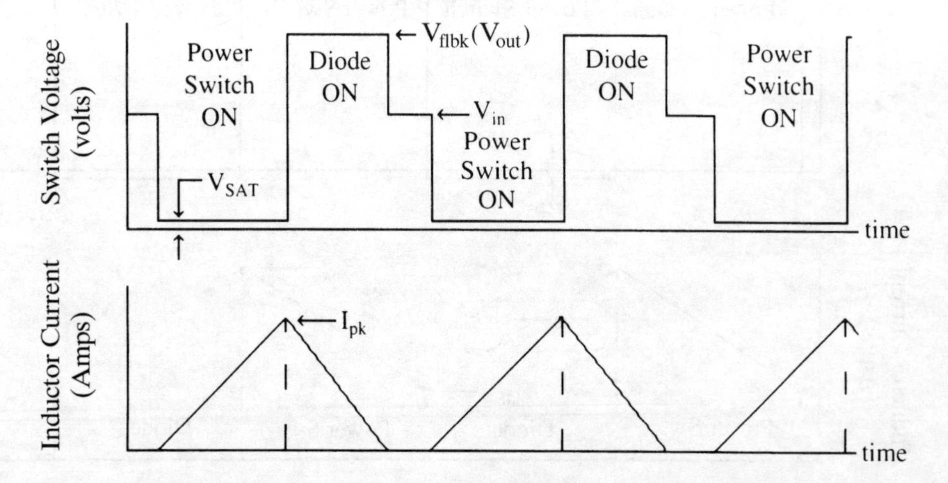

Figure 3–4 Waveforms for a discontinuous-mode boost converter.

Energy is stored in the flux within the inductor's core material. When the power switch is turned off, the inductor's voltage, "flies back" above the input voltage. Quickly, the diode becomes forward biased when the inductor's voltage exceeds the output voltage. The inductor voltage is then clamped at the value of the output voltage. This voltage level is referred to as the *flyback voltage* and is the value of the output voltage plus one diode forward voltage drop. The inductor current during the power switch's off period is described by

$$i_L(t_{off}) = I_{pk(on)} - \frac{(V_{out} - V_{in}) \cdot t_{off}}{L} \tag{3.6}$$

When the core's flux is completely emptied prior to the next cycle, it is referred to as the *discontinuous-mode* of operation. This is seen in the inductor current and voltage waveforms in Figure 3–4. When the core does not completely empty itself, a residual amount of energy remains in the core. This is called the *continuous mode* of operation and can be seen in Figure 3–5. The majority of boost-mode converters operate in the discontinuous mode since there are some intrinsic instability problems when operating in the continuous mode.

The energy stored within the inductor of a discontinuous-mode boost converter is described by

$$E_{stored} = \left(\frac{1}{2}\right) L (i_{pk})^2 \tag{3.7}$$

The energy delivered per second (joules/second or watts) must be sufficient to meet the continuous power demands of the load. This means that the energy stored during the ON time of the power switch must have a high enough I_{pk} to satisfy equation 3.8:

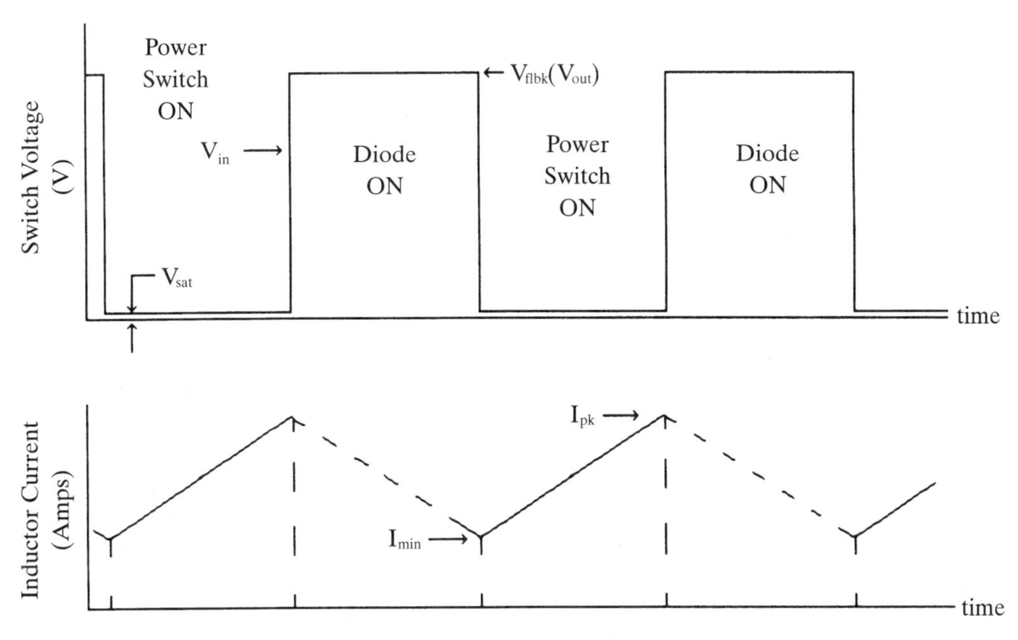

Figure 3–5 Waveforms for a continuous-mode boost converter.

$$P_{load} < P_{out} = f_{op}\left[\frac{1}{2}(L)(I_{pk})^2\right] \quad (3.8)$$

where f_{op} is the frequency of operation of the converter.

The *boost converter* shown in Figure 3–3 can only be used as a step-up converter. That is, the output voltage must be higher than the highest value of input voltage input. If the inductor is replaced by a transformer, as seen in Figure 3–15, a topology called a *flyback converter* is made. The flyback voltage and current, as seen by the power switch, are similar to those in the boost converter, but are affected by the turns ratio of the transformer. The flyback voltage, which is still $V_{out} + V_{diode}$ on the secondary, is scaled by the turns ratio of the transformer when viewed from the power switch. The transformer also provides a dielectric barrier from the input to the output, and additional output voltages can be derived from the same transformer. The outputs also become independent from the level of the input voltage, thus giving the flyback topology the highest input dynamic range of all the topologies.

Due to the higher peak currents within boost-mode converters, they can only be used in applications of 150 W or less. They have the least parts of all the topologies and are therefore very popular in the low to medium power applications.

3.2 The Building-block Approach to PWM Switching Power Supply Design

PWM switching power supplies lend themselves quite nicely to an organized approach to their design. They are more complicated and therefore can be partitioned into more functional blocks of an elementary nature. The switching power supply designer, whether knowingly or unknowingly, does its design in a functional block approach and this approach will be presented.

Executing the design in a specific order and fashion makes the design flow much easier by predetermining information needed for subsequent portions. In general, PWM and resonant switching power supply designs start with the overall considerations, next comes the design of the power sections, then the designs proceed through the control and ancillary functions, and finally is the testing and perfecting stage. It all begins with a well-defined design specification as outlined in Section 1.3. By first defining the operation of the supply and the environment in which it must operate, the design flow takes the form as seen in Figure 3–6. Some initial decisions must be made at the beginning, namely which topology of switching power supply to use.

Once the topology is selected, the design path is determined and the design may proceed. By proceeding through the block diagram in Figure 3–7, in the order indicated by the design flow in Figure 3–6, the design will progress relatively quickly. The first-time designer may be able to produce a very good "paper design" (or schematic) within 8 working hours, if he or she has a good library of data books. Each functional block in Figure 3–7 will have a choice of typical design approaches for that block. The designer determines from his or her requirements which approach is most appropriate for the needs of the supply. Then, by executing the design equations and using the parameters supplied by the design specification, the block can be designed within a matter of minutes.

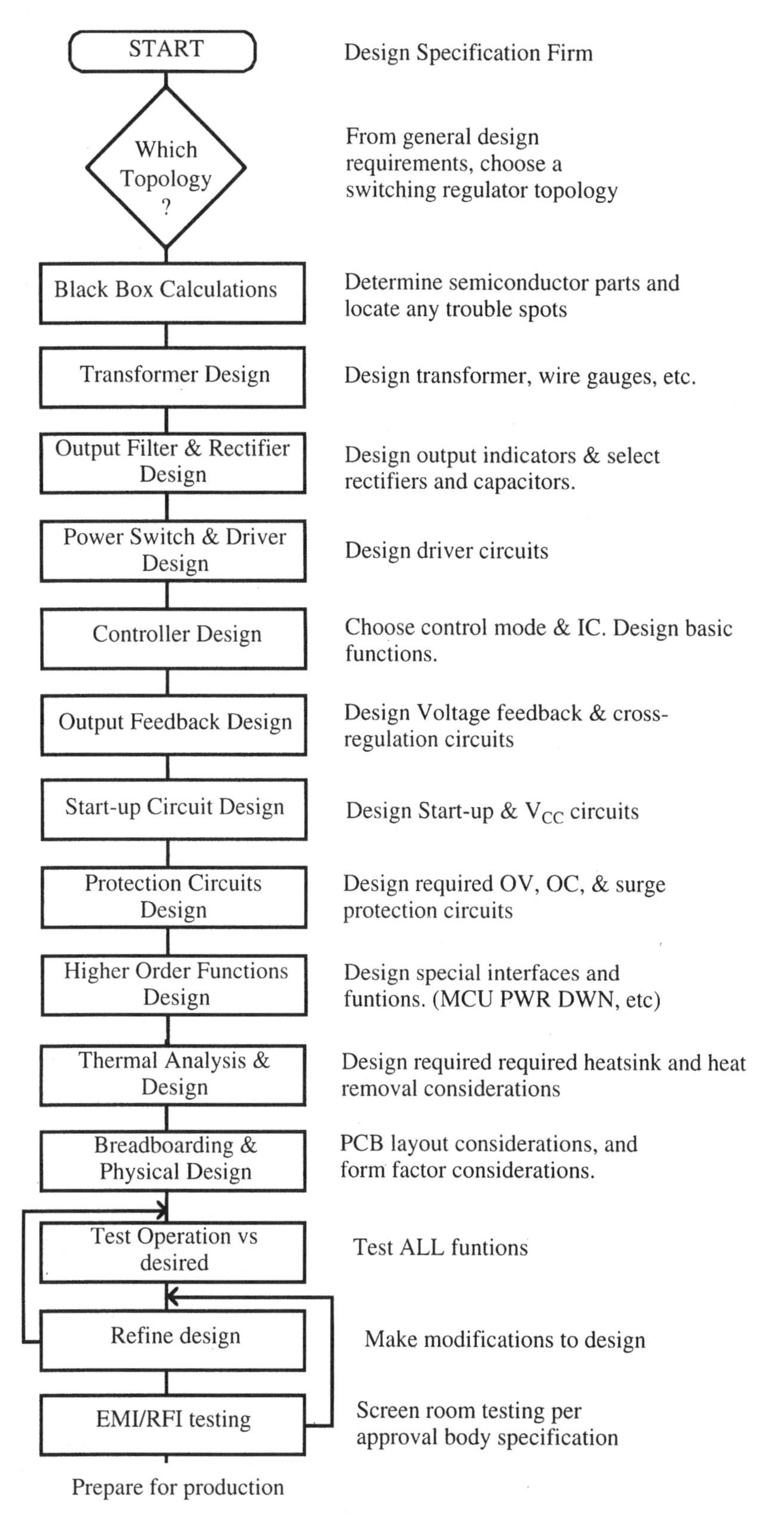

Figure 3–6 Design flow for a PWM switching power supply.

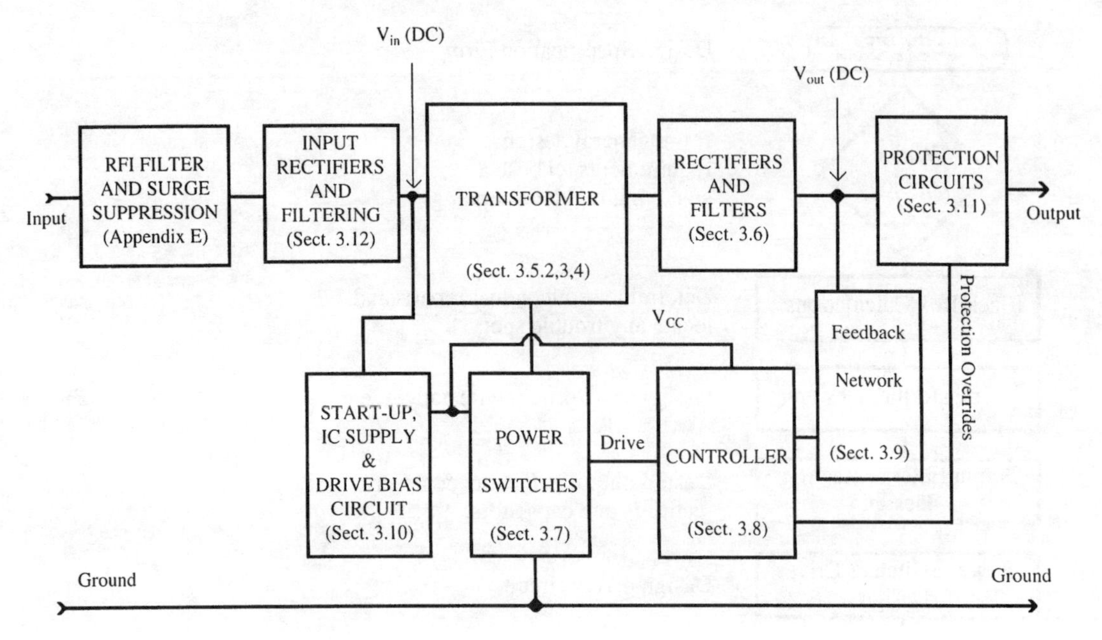

Figure 3–7 The functional block diagram for a PWM switching power supply.

3.3 Which Topology of PWM Switching Power Supply to Use?

A major decision that must be considered at the beginning of a switching power supply design is which basic topology to use. The term *topology* refers to the arrangement of the power components within the switching power supply design. This arrangement has a large bearing on which environment the supply can operate in safely and how much power the power supply can provide to the loads. This is the point in the design process where the major cost versus performance tradeoffs are made. Each topology has its relative merit. One topology may have a low parts cost, but can only provide a limited amount of power; another has ample power capability, but costs more, and so on. More than one topology will work for any application, but one of the choices may provide the best performance at the needed cost. A summary of the relative merits of the various topologies is given in Table 3–1.

The major factors that determine the optimum choice of topology are:

1. Is transformer isolation needed from input to output?
2. How much input voltage appears across the primary winding of the transformer or inductor?
3. What is the peak current through the power switches?
4. What is the maximum operating voltage across the power switches?

The nontransformer isolated topologies are used for *board-level* converters. These are in distributed power systems where an intermediate bus voltage is distributed throughout the system and each board within the system has its own power supplies. The bus voltage is always a “safe” level that is not deemed lethal to the operator of the equipment, therefore dielectric isolation is optional. I still heartily recommend transformer isolation in most applications. The added cost

Table 3–1 Comparison of the PWM Switching Regulator Topologies

Topology	Power Range (W)	$V_{in(dc)}$ Range	In/Out Isolation	Typical Efficiency (%)	Relative Parts Cost
Buck	0–1000	5–40	No	78	1.0
Boost	0–150	5–40	No	80	1.0
Buck-boost	0–150	5–40	No	80	1.0
1T forward	0–150	5–500	Yes	78	1.4
Flyback	0–150	5–500	Yes	80	1.2
Push-pull	100–1000	50–1000	Yes	75	2.0
Half-bridge	100–500	50–1000	Yes	75	2.2
Full-bridge	400–2000+	50–1000	Yes	73	2.5

is minimal compared to the added level of protection for the load. Transformer isolation is mandatory for all switching power supplies having a dc input voltage of 40 VDC or higher.

The amount of voltage appearing across the primary of the transformer is indicative of how much peak current is flowing through the power switches. Switching power supplies are constant power circuits. That is, the lower the primary voltage, the higher the peak currents in order to provide the needed output power. For power transistors and MOSFETs in TO-220 packages and smaller, a maximum peak current limit of 20 A is recommended. Above 20 A, the failure modes of the power switches become very erratic and the power devices are difficult to protect. By using another topology, the peak current can be reduced.

The higher the maximum voltage the power switches experience, the greater the likelihood that they will exceed their safe operating areas (SOA). Voltage spikes are very common within switching power supplies, and the opportunity of these spikes exceeding the avalanche voltage rating of the power switch becomes more likely. For transformer isolated topologies, the industry has settled into certain topologies that they use within the different ranges of applications. This is shown in Figure 3–8.

The *flyback* topology (see Figure 3–12) is the favorite below 100 to 150 W because of its low parts count (hence cost) and intrinsically better efficiency. But because its peak currents are much higher than the forward-mode converters, it reaches the SOA limits of the power switches at a relatively low output power. Between an output power of 150 and 500 W the *half-bridge* (see Figure 3–15) becomes the favorite. The parts cost more but they are still reasonable. The half-bridge converter only places one-half of the input voltage across the primary winding and therefore exhibits fairly high peak currents. It therefore is only used to 500 W or less. Above 500 W and into many kilowatts, the *full-bridge* (see Figure 3–16) topology is used. This requires four power switches, two of which have floating drive circuits, and is the most costly to implement, but at these output power levels the added cost is necessary. The *push-pull* (see Figure 3–14) can also be used in this region, but it suffers from a potentially severe failure mode called *core imbalance*. This is where the flux within the transformer will operate non-symmetrically about a "zero" balance point. This will cause the

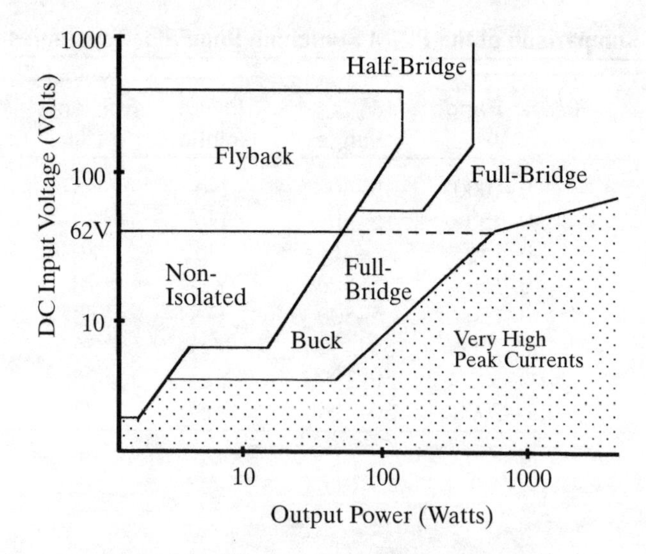

Figure 3–8 Where various topologies are used.

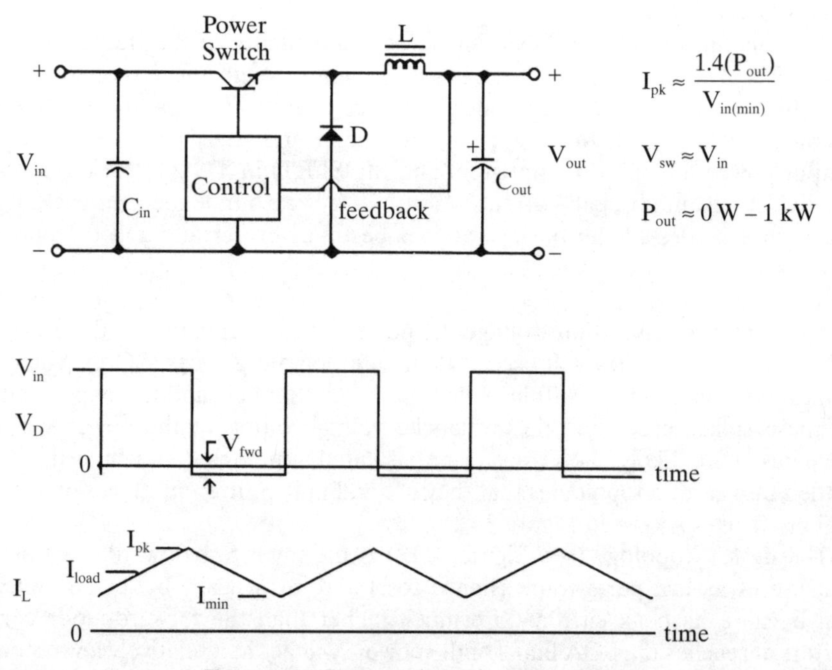

Figure 3–9 The buck (step-down) converter.

transformer to saturate in the direction of one power switch and burn it out within nanoseconds when step changes occur in the load. Pulse-to-pulse current-protected voltage-mode control or current-mode control techniques must be used to avoid this problem.

By reviewing Table 3–1 and Figure 3–8, you can develop a good idea as to where your application fits. The appropriate topology then can be chosen. Figures 3–9 through 3–16 show the basic PWM switching power supply topologies along with their salient waveforms and some estimated important parameters.

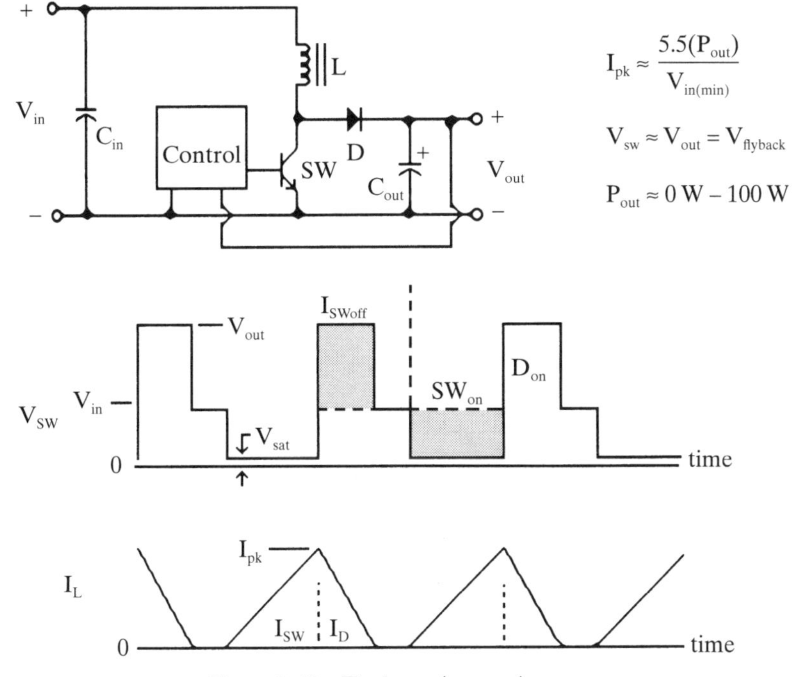

Figure 3–10 The boost (step-up) converter.

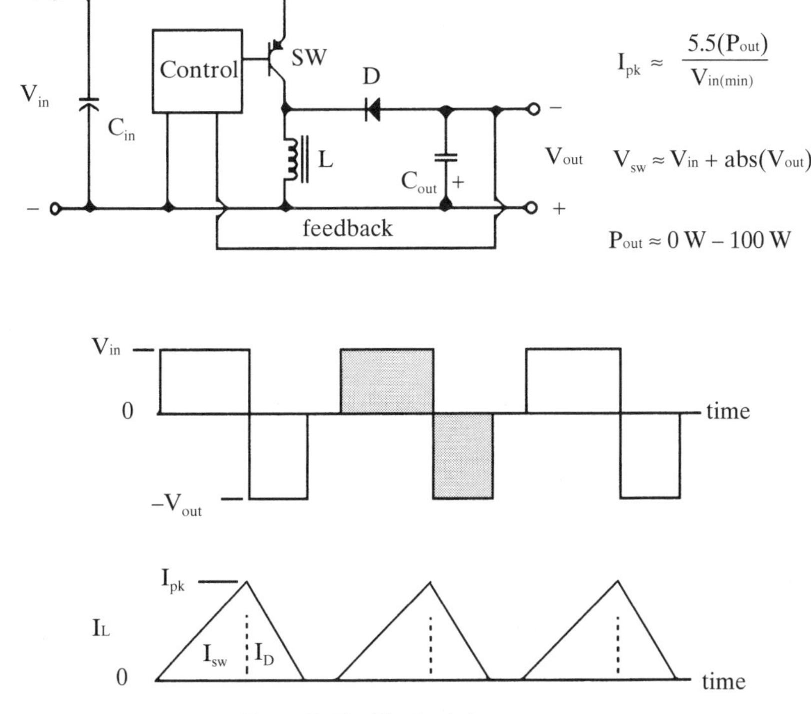

Figure 3–11 The buck-boost converter.

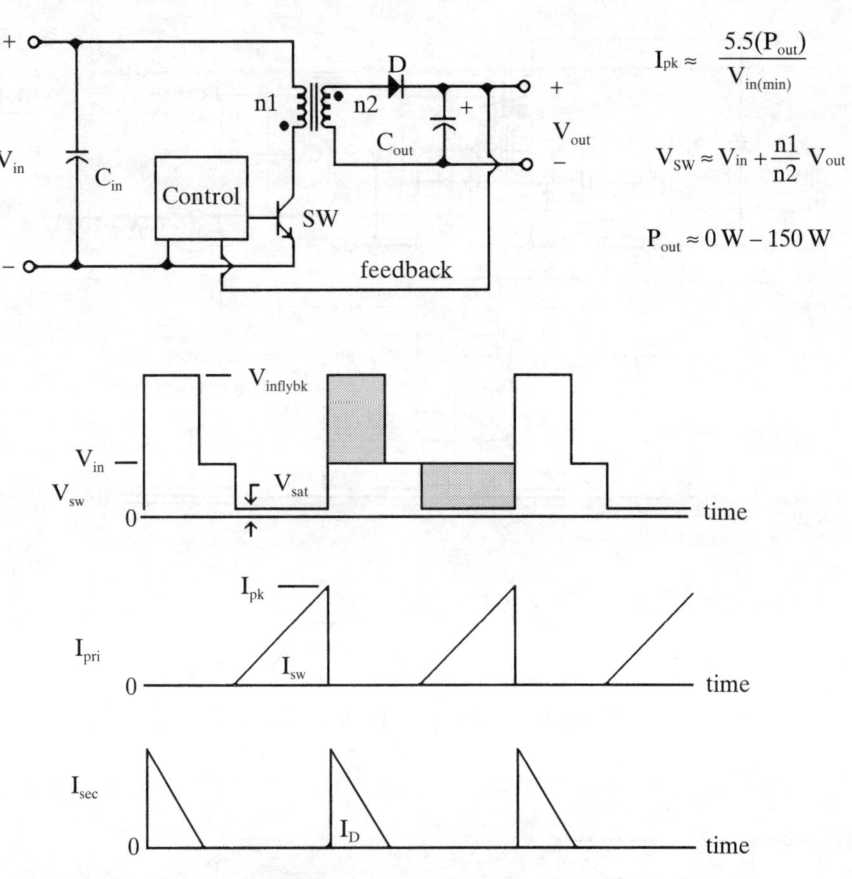

Figure 3–12 The flyback converter.

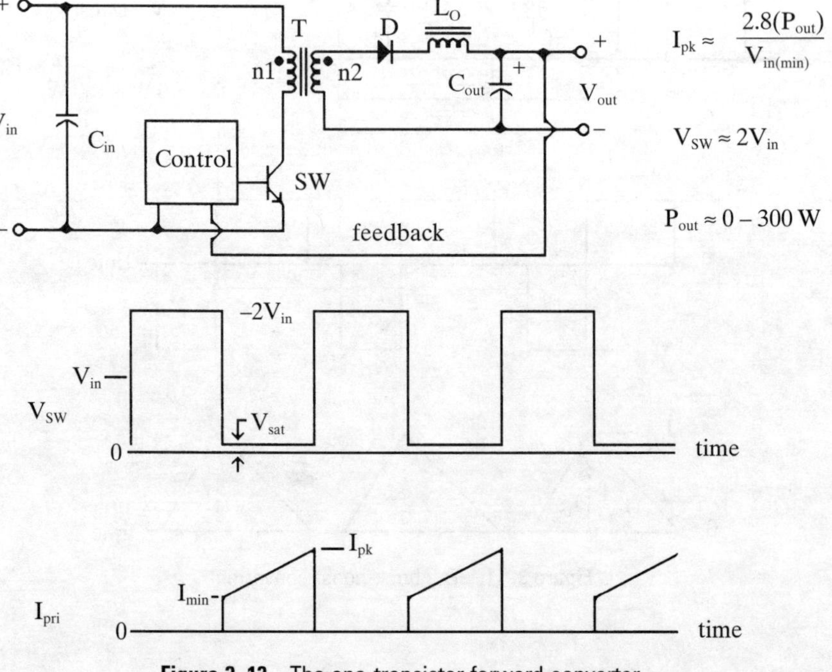

Figure 3–13 The one-transistor forward converter.

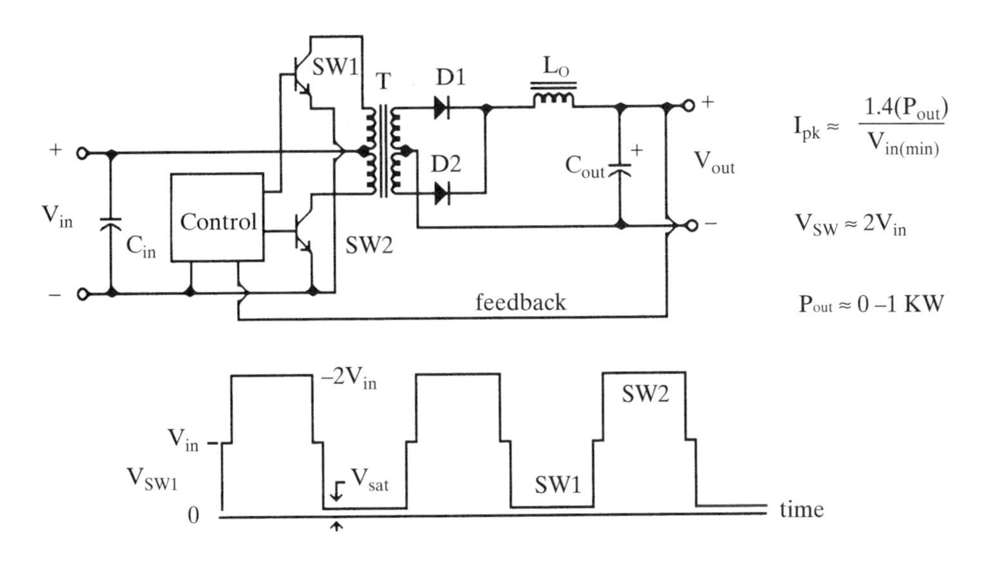

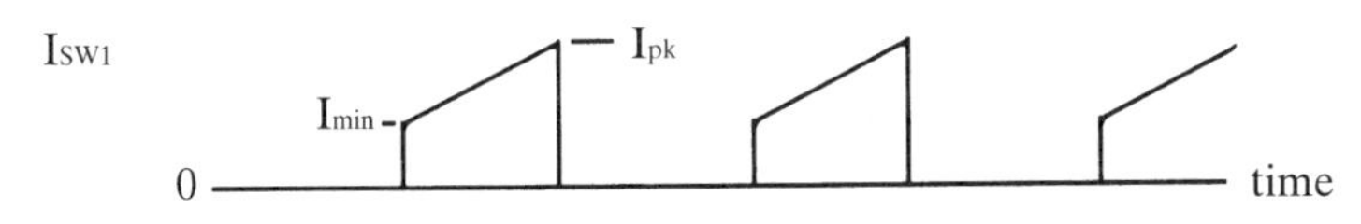

Figure 3–14 The push-pull converter.

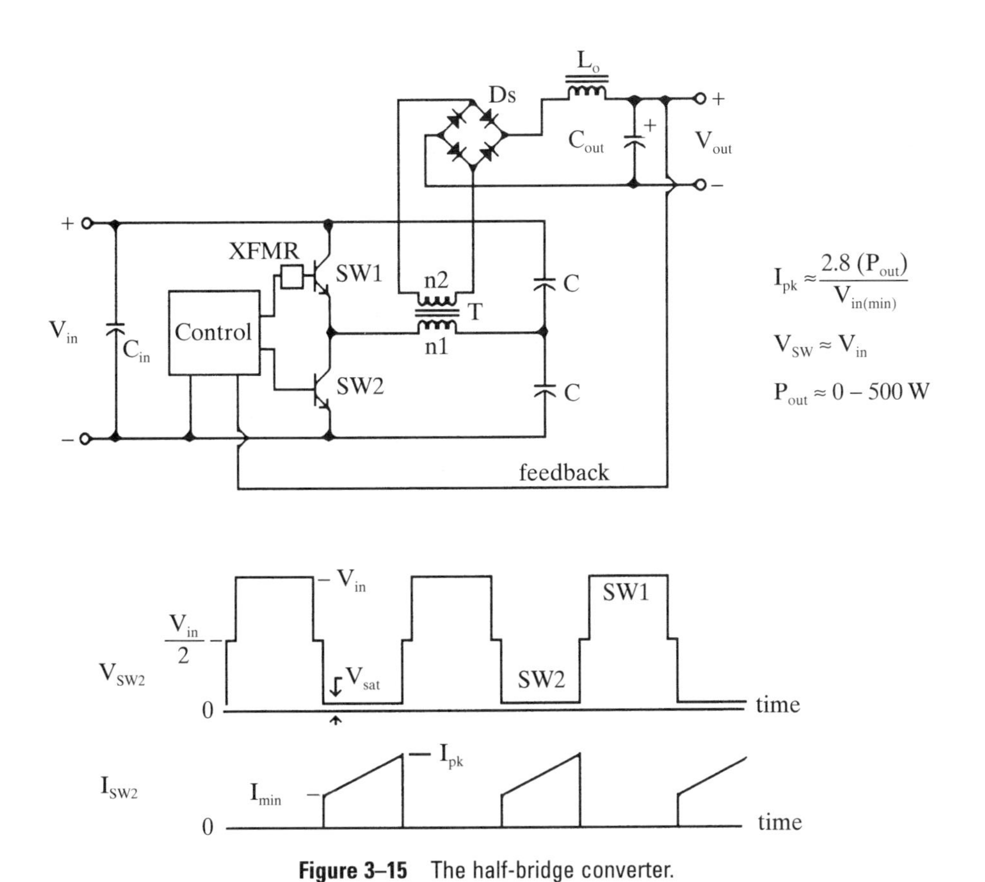

Figure 3–15 The half-bridge converter.

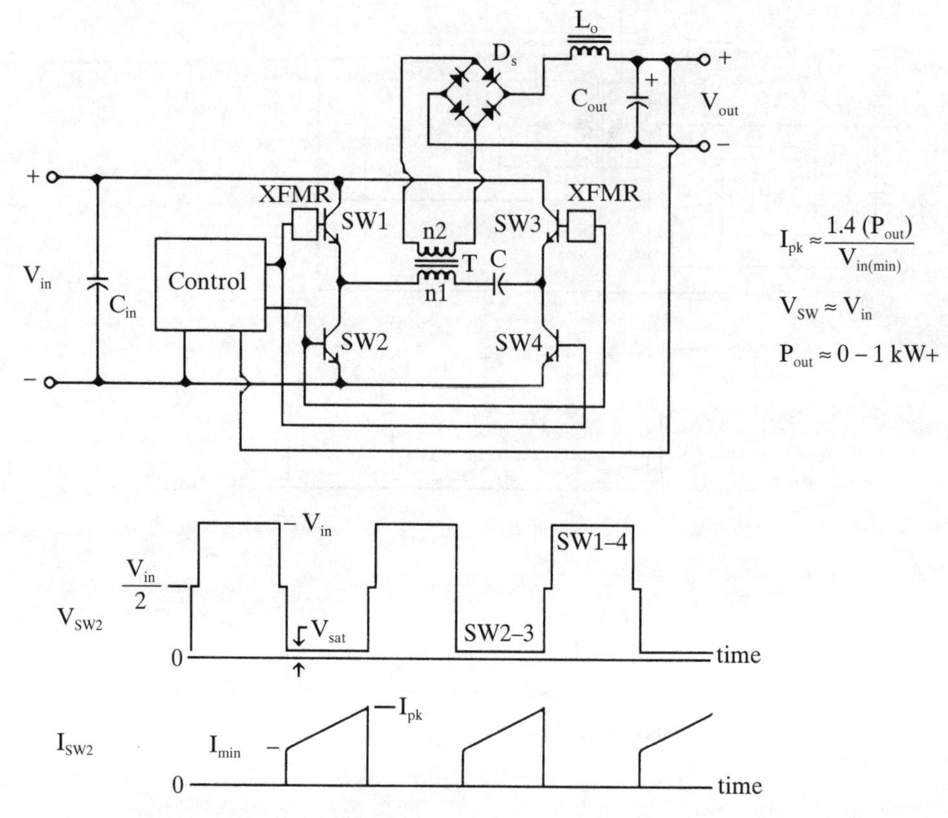

Figure 3–16 The full-bridge converter.

3.4 The "Black Box" Considerations for Switching Power Supplies

This preliminary step in the design phase, predetermines some of the major parameters the switching power supply will exhibit. This allows the designer to determine whether his or her choice of topology is correct and also allows the designer to order components for the breadboard, long before they are required. It also results in knowing some of the particularly important parameters needed later during the design. For determining the "black box" estimates, the designer need only know the environmental parameters from the design specification and then treat the future power supply as a black box; that is, a box with only input and output lines emerging from it (refer to Figure 3–17).

The estimates are done as follows:

1. Output power

$$P_{\text{out}} = \sum_{m=1}^{n} \left(V_{\text{out}(m)} \cdot I_{\text{out}(m)}\right) \tag{3.9}$$

2. Input power

$$P_{\text{in}} \cong \frac{P_{\text{out}}}{\text{effic(est)}} \tag{3.10}$$

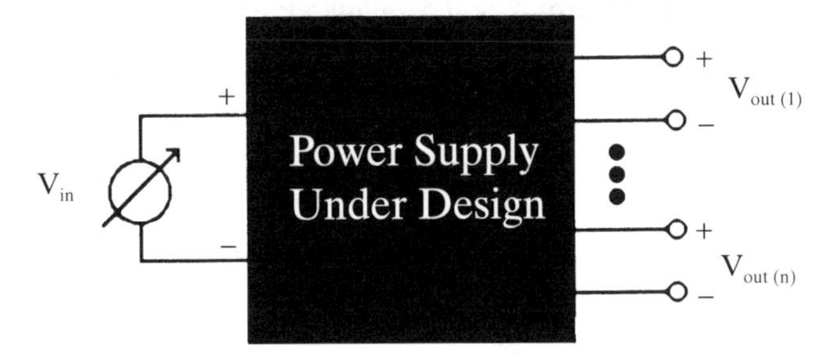

Figure 3–17 The SMPS treated like a black box.

3. Average input currents

$$I_{\text{in(av)(nom)}} = \frac{P_{\text{in}}}{V_{\text{in(nom)}}} \tag{3.11}$$

where "nom" is the operational value of the input voltage during normal operation.

Solve for the highest input average input current that occurs at the lowest specified input voltage. This value allows the designer to determine the size of the wire used in the primary winding of the transformer or inductor.

4. The input peak current

This is completely determined by the topology previously chosen.

$$I_{\text{pk}} = \frac{k \cdot P_{\text{out}}}{V_{\text{in(min)}}} \tag{3.12}$$

where k =: 1.4 for the Buck, Half-forward, and Full-Bridge.
=: 2.8 for the Half-bridge, IT Forward.
=: 5.5 for the Boost, Buck-boost, and Flyback.

The value of the peak current is useful in the design of the flyback-mode inductors and transformers. For the forward-mode supplies, it is just a curiosity at this time.

5. Selecting the power switches and rectifiers

Each topology has predictable voltage and current stresses for the power switches and rectifiers. These estimates have about a 90 percent confidence factor. Selecting the power devices at this stage in the design cycle can save precious time later in the program by not having to wait for parts. Table 3–2 contains equations that may be conservative in nature, but will work in the application.

6. Estimating component losses (optional)

The relative losses among the various sections within a PWM switching power supply are somewhat predictable from experience. These loss proportions are of course affected by the design paths pursued by the designer, but at this stage only a "good guess" estimate is desired. Table 3–3 shows the "typical" efficiency for each major topology and the percentage of loss that occurs between the power switch stage and the output rectifier stage.

Table 3–2 Estimating the Significant Minimum Parameters of the Power Semiconductors

	Bipolar Power Switch		MOSFET Power Switch		Rectifier(s)	
Topology	V_{CEO}	I_C	V_{DSS}	I_D	V_R	I_F
Buck	V_{in}	I_{out}	V_{in}	I_{out}	V_{in}	I_{out}
Boost	V_{out}	$\frac{2P_{out}}{V_{in(min)}}$	V_{out}	$\frac{2P_{out}}{V_{in(min)}}$	V_{out}	I_{out}
Buck/boost	$V_{in} - V_{out}$	$\frac{2P_{out}}{V_{in(min)}}$	$V_{in} - V_{out}$	$\frac{2P_{out}}{V_{in(min)}}$	$V_{in} - V_{out}$	I_{out}
Flyback	$1.7V_{in(max)}$	$\frac{2P_{out}}{V_{in(min)}}$	$1.5V_{in(max)}$	$\frac{2P_{out}}{V_{in(min)}}$	$10V_{out}$	I_{out}
One Transistor Forward	$2V_{in}$	$\frac{1.5P_{out}}{V_{in(min)}}$	$2V_{in}$	$\frac{1.5P_{out}}{V_{in(min)}}$	$3V_{out}$	I_{out}
Push-pull	$2V_{in}$	$\frac{1.2P_{out}}{V_{in(min)}}$	$2V_{in}$	$\frac{1.2P_{out}}{V_{in(min)}}$	$2V_{out}$	I_{out}
Half-bridge	V_{in}	$\frac{2P_{out}}{V_{in(min)}}$	V_{in}	$\frac{2P_{out}}{V_{in(min)}}$	$2V_{out}$	I_{out}
Full-bridge	V_{in}	$\frac{1.2P_{out}}{V_{in(min)}}$	V_{in}	$\frac{1.2P_{out}}{V_{in(min)}}$	$2V_{out}$	I_{out}

To determine each section's respective loss for your power supply topology, use the following

$$P_{loss(ckt)} = P_{in} \cdot (1 - \text{effic}) \cdot P_{(\%)} \tag{3.13}$$

where $P(\%)$ is the typical estimated percentage loss of the desired section with respect to the total loss of the power supply (see Table 3–3).

When the topology has multiple power switches, then multiply the $P_{loss(ckt)}$ by

0.5 (50%) for the push-pull and half-bridge
0.25 (25%) for the full-bridge.

To estimate the loss within each output rectifier within a multiple output switching power supply, use the ratio of the desired output's power compared to the total output power as in Equation 3.14.

$$P_{r(n)} \approx P_{in} \cdot (1 - \text{effic}) \cdot P_{(\%)} \cdot \left(\frac{P_{out(n)}}{P_{out}} \right) \tag{3.14}$$

These loss figures can be used in determining the appropriate packages needed for the semiconductors. That is, do some of the semiconductors need

Table 3–3 The "Black Box" Estimates for Losses within the Various Topologies

Topology	Power Switch Type: Bipolar	Power Switch Type: MOS	Overall Estimated Efficiency (%)	Estimated Percent of Total Loss ($P_{(\%)}$): Power Switch and Drive (%)	Output Rectifier (%)	Magnetics (%)	Miscellaneous (%)
Buck	×		72	42	48	5	5
		×	76	35	55	5	5
Boost	×		74	55	35	5	5
		×	77	48	42	5	5
Buck-boost	×		74	55	35	5	5
		×	77	48	42	5	5
Flyback	×		75	44	46	5	5
		×	78	33	57	5	5
Half-forward	×		74	44	46	5	5
		×	77	33	57	5	5
Push-pull	×		69	50	40	5	5
		×	72	40	50	5	5
Half-bridge	×		69	48	42	5	5
		×	72	40	50	5	5
Full-bridge	×		65	50	40	5	5
		×	70	40	50	5	5

to be attached to a heatsink? Also, a guess at how much heatsinking is required can be made.

This completes the black box estimates portion of the design program. As one can see, some very useful information can be determined by performing these estimates.

3.5 Design of the Magnetic Elements

The design of the magnetic elements forms the backbone of a good switching power supply design. Their proper electrical and physical design have a large affect on the reliable operation of every switching power supply design. Entire books have been written devoted to their design and theory, but in keeping with the premise behind this publication, I have chosen to take a different approach. Since switching power supplies are a specialized narrow application of magnetic elements, the design process for the magnetic elements can be greatly focused and simplified. This yields the quickest working design without having to understand the subtleties involved with each facet of the design. For more detail in the operation of the core materials, refer to Appendix D.

3.5.1 The Generalized Design Flow of the Magnetic Elements

Part of the confusion in designing the magnetic elements is the inexactness involved in their design. At best, the results of calculations should be considered a "calculated estimate." Pushing the result of a calculation in one direction or another will still yield a working design, but may offer advantages in some other facet of the power supply design such as core size, power supply input dynamic range, etc. These trade-offs will be discussed at the appropriate points in the text.

The generalized design flow for the magnetic elements is as follows.

1. Select the core material appropriate for the application and for the frequency of operation (Appendix D).
2. Select the desirable core style that will meet the needs of the application and of any regulator agency the power supply should meet (Appendix D).
3. Determine the size of the core needed to provide the required output power of the power supply (Section 3.5.2).
4. Determine whether an airgap is needed and calculate the number of turns needed for each winding. Then determine whether the accuracy of the output voltages meets the needs of the requirements and whether the winding actually fit into the selected core size (Sections 3.5.3 through 3.5.7).
5. Wind the magnetic component using the described physical winding techniques given in Section 3.5.9.
6. During the prototype stage, verify its operation with respect to the level of voltage spikes, cross regulation, output accuracy and ripple, and RFI, etc., and make corrections where necessary.

If the above procedure is executed, the initial design of any magnetic component should take less than 30 minutes. Please refer to the end of Appendix D for the listing of the core manufacturer's websites where the needed information can be downloaded.

3.5.2 Determining the Size of the Magnetic Core

Each manufacturer uses different core sizing procedures. Some use graphs, others simply state how much power each core can handle for a particular application, and some use cryptic equations that are confused by the mixture of unrelated units. The following two procedures are generalized approaches for estimating the initial core size.

The following methods take two forms. The first is a simple table of size versus power which yields a good "guess" if you do not have many or complicated windings. The second is a calculated approach which has "fudge factors" that involve the number of windings, and whether the component must meet a regulatory agency requirement. The results of both should be considered estimates, so samples of the resulting core size should be requested along with samples of the next larger size. This will eliminate the need to go back later if the windings grow too large for the core.

There are five major factors in selection of the size of a core for an application:

Factor:	**Parameter Affected:**
Output power	A_c (core cross sectional area)
Bipolar or unipolar flux	A_c (core cross sectional area)
Input voltage	A_w (window area)
Number of windings	A_w (window area)
Winding configuration	A_w (window area)

Other considerations enter into the operation of the magnetic element, such as the desired range of leakage inductance for each winding, electrostatic shielding, etc. (see Appendix D). These factors affect the size and cost of the final magnetic element.

Core Sizing Method 1

Use Table 3–4 to determine the range of power into which your application fits. Consult a core manufacturer's data book for cores that fit your application and select the closest or next larger size of core. Order those samples (refer to Table 3–4).

Core Sizing Method 2

This method begins with the assumption of a one-winding transformer. Each winding and each additional regulatory requirement only tend to make the winding area and hence the core to grow. The influences of each of the factors are multiplied together to arrive at a combined "scaling factor." The scaling factor is then used to scale the size of the basic one-winding inductor.

The first step is to determine the size of the one-winding inductor. This is found by using Equation 3.15.

$$W_a A_c = \frac{0.68 \cdot P_{out} \cdot d_w \cdot 10^3}{B_{max} \cdot f} \qquad (3.15)$$

where: d_w is the area of the wire used in the primary winding in circular mils or square inches (refer to the wire table in Appendix E).
B_{max} is the peak operating flux density (G).
f is the frequency of operation.
P_{out} is the total output power of the supply.

For the MKS (pure metric) system the following equation is used:

$$W_a A_c = \frac{0.68 \cdot P_{out} \cdot d_w}{B_{max} \cdot f} \qquad (3.16)$$

where: d_w is the area of the wire used in the primary winding in cm^2 (refer to the wire table in Appendix E).
B_{max} is the peak operating flux density (Teslas).
f is the frequency of operation (Hz).
P_{out} is the total output power of the supply (W).

Next the scaling influences need to be identified and a combined scaling factor calculated. These scaling factors are found in Table 3–5.

Combine the individual scaling factors as follows:

$$K_{net} = K_a \cdot K_b \ldots \text{etc.} \qquad (3.17)$$

The estimated size for the final transformer is obtained from:

$$W_a A_c' = K_{net} \cdot W_a A_c \qquad (3.18)$$

Table 3–4 Approximate Core Size versus Output Power

Output Power (W)	MPP Toroid Diameter (inch(mm))	E–E, E–L, etc. Core (each side) (inch(mm))
<5	0.65 (16)	0.5 (11)
<25	0.80 (20)	1.1 (30)
<50	1.1 (30)	1.1 (30)
<100	1.5 (38)	1.8 (47)
<250	2.0 (51)	2.4 (60)

Table 3–5 Transformer Growth Scaling Factors

Consideration	Scaling Factor	
Flyback transformer	1.1	
One secondary	1.2	} or
Two or more secondaries	1.3	} or
Isolated secondaries	1.4	
UL or CSA approval	1.1	} or
IEC approval	1.2	
Faraday shield	1.1	

In the U.S., the result will be in inches^4; for the metric system the results will be in meters^4. The relationship to convert between the two numbering system results is

$$\text{meters}^4 = 2.402 \times 10^6 \text{ inches}^4 \tag{3.19a}$$

$$\text{inches}^4 = 4.162 \times 10^{-7} \text{ meters}^4 \tag{3.19b}$$

Some of the core manufacturer's data sheets specify a core parameter W_aA_c that can be compared with the results of the above calculations. The core with the closest or next higher value should be chosen.

If you desire to use the core manufacturer's own method of determining core size, then feel free to do so. This phase of the transformer design is a gross estimation.

3.5.3 Designing the Forward-mode Transformer

There are two main functions the forward-mode transformer performs: the first is to provide a dielectric isolation barrier from the input to the output; and the second is to step up or step down the pulsewidth modulated ac input voltage signal. The design flow becomes a step-by-step design procedure.

The forward-mode transformer stores no energy other than a small amount within the magnetization of the core material itself. There are only two major considerations which are important during the gross schematic design of the transformer:

1. The peak operating flux density (B_{max}) should never closely approach or enter saturation over the power supply's entire operating range.
2. Do the resulting windings provide an accurate enough output voltage to meet the design specifications?

Other considerations are important during the physical winding process and involve winding losses, leakage inductance, shielding, and physical space. They do not need to be considered now.

The first step is to determine the number of turns needed for the primary winding. For this, the parameters from the core data sheet of the particular core and core material are used. Also, the minimum level of flux density already should have been determined (refer to Appendix D). The equation for determining the number of turns for the primary winding in the CGS System (U.S.) is

$$N_{pri} = \frac{V_{in(nom)} \cdot 10^8}{4 \cdot f \cdot B_{max} \cdot A_c} \tag{3.20a}$$

where A_c is the effective core cross sectional area of the desired core (cm^2).
$V_{in(nom)}$ is the typical operating input voltage (V).
B_{max} is the maximum operating flux density (gauss (webers/cm^2)).

In the MKS System (Europe) system this is

$$N_{pri} = \frac{V_{in(nom)}}{4 \cdot f \cdot B_{max} \cdot A_c} \quad (3.20b)$$

where A_c is the effective core cross sectional area of the desired core (m^2).
$V_{in(nom)}$ is the typical operating input voltage (V).
B_{max} is the maximum operating flux density (teslas (webers/m^2)).

Some core companies use still a third metric system, milliTeslas (mT), and millimeters. These companies tend to be Japanese.

$$N_{pri} = \frac{V_{in(nom)} \cdot 10^9}{4 \cdot f \cdot B_{max} \cdot A_c} \quad (3.20c)$$

where A_c is the effective core cross sectional area of the desired core (mm^2).
$V_{in(nom)}$ is the typical operating input voltage (V).
B_{max} is the maximum operating flux density (mT (webers/mm^2)).

This number of turns now serves as the reference winding upon which all the other windings will be determined.

Next, one determines the number of turns of the highest power output secondary winding. The voltage drop across the output rectifiers cannot be ignored. The equation to determine the number of turns for this winding is

$$N_{sec} = \frac{1.1(V_{out} + V_{fwd})}{N_{pri}(V_{in(min)})DC_{max}} \quad (3.21)$$

where V_{fwd} is the forward voltage drop of the anticipated output rectifier.
DC_{max} is the maximum expected duty cycle (0.95 is good).
$V_{in(min)}$ is the minimum expected input voltage.

This equation can be solved for the needed secondary turns at the lowest anticipated input voltage. At any input voltage lower than that, the regulator will fall out of regulation.

The next step is to determine the number of turns for the other secondaries based on the turns of the first secondary winding. The starting point is

$$N_{sec(n)} = \frac{(V_{out(n)} + V_D)N_{sec(1)}}{(V_{out(1)} + V_{D1})} \quad (3.22)$$

where $V_{out(n)}$ is the additional output voltage.
V_D is the anticipated rectifier's forward voltage drop.

The result of this calculation will always yield a noninteger number, but many cores can only accept an integer number of turns. Therefore, one must round off the results to the closest integer. This results in an error in the eventual output voltages of these additional outputs. One must now check to see if these errors are too much for the application. Using the original output's "volts per turn" value, calculate the new output voltages using the rounded-off turns on the additional windings. If the error in any of the output voltages is too great, first consider changing the technology of the rectifier—to one that has a higher or lower forward voltage drop. If that is insufficient, add a turn to the original output winding, recalculate the new value for the additional turns, and recheck that the error is acceptable. If the result is still unacceptable, go back to the primary and add turns and recalculate the new secondaries. Remember that adding turns on the primary winding is moving in the safe direction of lower flux density and adding turns to the secondary makes the power supply fall out of regulation at a lower input voltage. This iterative process should continue until all the outputs have an acceptable amount of error from the "ideal" output voltage. The designer will have to accept some error in the output voltages as a fact of life.

The next step is to decide how the secondary windings should be arranged. That is, whether it is desired to have isolated secondaries, center-tapped or non-center-tapped, or have "autotransformer" style secondaries that share windings of lower voltage outputs (refer to Figure 3–18).

Remember that the amount of current flowing through each half of a center-tapped winding is one-half the continuous current and therefore needs only one-half the wire cross-sectional area. Also, autotransfomer windings have more than one output current flowing though the lower winding, so adjustments to the wire gauges are necessary.

Lastly in the "paper design," the designer should verify that the turns and their respective wire gauges fit inside the area reserved for windings. This is done by summing the products of each winding times its respective wire cross-sectional area and verifying that the sum is less than the winding area specified for the core or bobbin.

$$W_a \approx k \cdot \sum (N_i \cdot A_{w(i)}) \quad (3.23)$$

where k is the between 1.2 and 1.4 for winding inefficiencies and insulating tape.

The final design step is the method for physically building the transformer; for this refer to Section 3.5.8.

3.5.4 Designing the Flyback Transformer

The flyback transformer operates differently from the forward-mode transformer. Instead of both windings conducting current simultaneously, energy is stored within the core material by the primary winding and then transferred to the secondary circuit when the primary turns off. Accordingly, the classic transformer relationships of reflected impedances and primary-to-secondary turns ratios do not apply directly. Now the voltage, time, and energy become the main concerns.

To start the design, the peak current should have been estimated during the proceeding "black box" estimation phase. The core style and material should have also been selected (refer to Appendix D). This time an air gap is necessary for reliable operation.

One begins by realizing that the primary winding behaves like an elementary inductor during the power switch's on-time and obeys the relationship in Equation 3.24:

$$I_{pk} = \frac{V_{in} \cdot T_{on}}{L_{pri}} \quad (3.24)$$

Rearranging for L_{pri} and substituting $T_{on} = \partial_{(max)}/f$, one can solve for the maximum primary inductance using known power supply operating parameters using Equation 3.25:

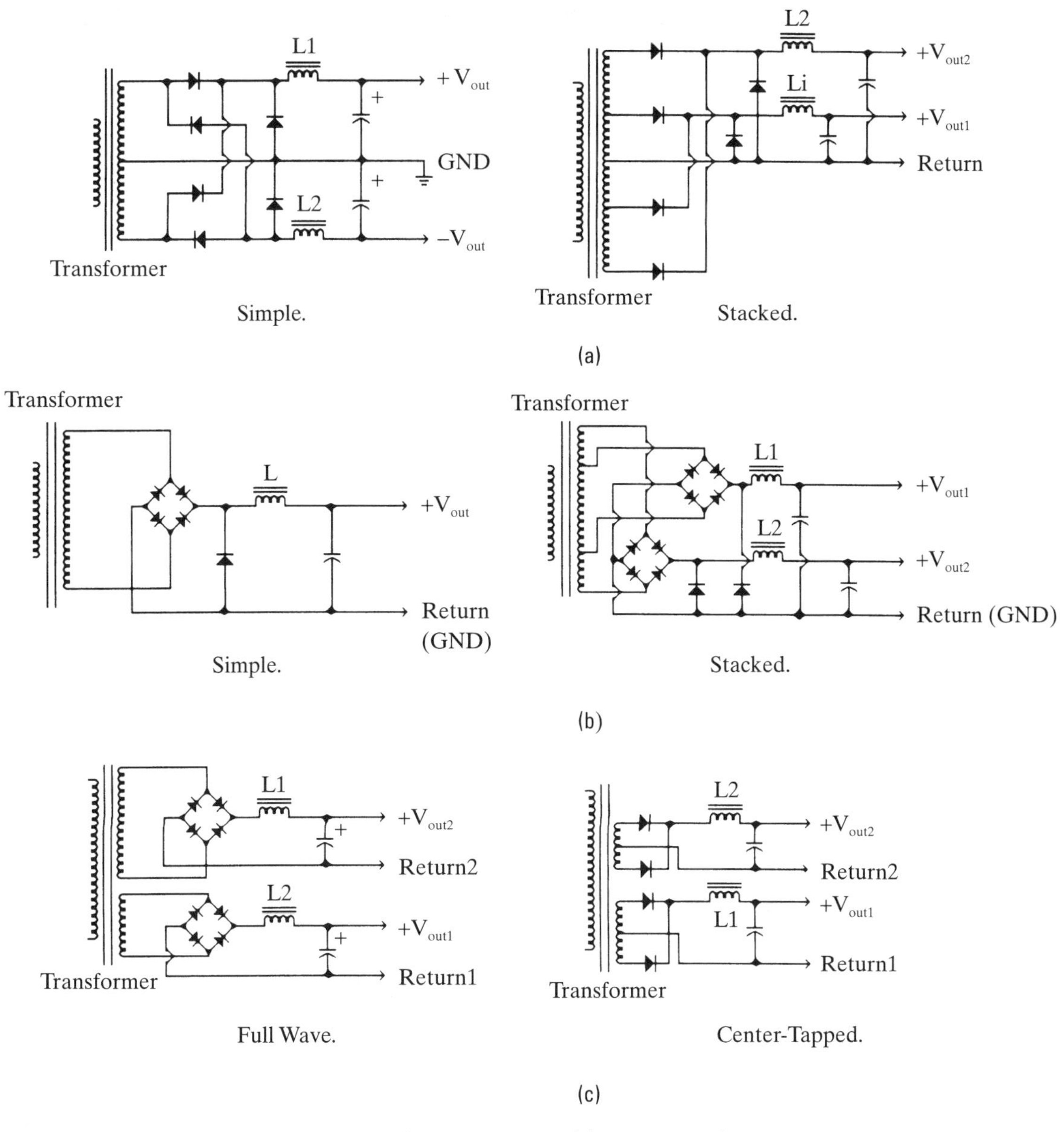

Figure 3–18 Forward-mode secondary winding arrangements: (a) center-tapped secondaries; (b) full-wave secondaries; (c) isolated secondaries.

$$L_{pri} = \frac{V_{in(min)} \cdot \partial_{max}}{I_{pk} \cdot f} \quad (3.25)$$

where $\partial_{(max)}$ is the maximum duty cycle (usually 50 percent or 0.5).

This inductance value should be considered the maximum value since any higher value will cause the power supply to fall out of regulation above the minimum operational input voltage.

The energy entering the core during each on time of the power switch is

$$E_{stored} = \frac{L_{pri} \cdot (I_{pk})^2}{2} \quad (3.26)$$

To verify that the maximum continuous output power capability of the transformer can satisfy the maximum power requirements of the load, one calculates

$$P_{in(core)} = \frac{1}{2} L_{pri} (I_{pk})^2 \cdot f_{op} > P_{out} \tag{3.27}$$

An air-gap is required for all unipolar flux drive applications such as this. One method of achieving this is shown in Equation 3.28a (CGS system (U.S.)).

$$l_{gap} \approx \frac{0.4\pi L_{pri} I_{pk}^2 \cdot 10^8}{A_c B_{max}^2} \text{ cm} \tag{3.28a}$$

where A_c is the effective core cross-sectional area (cm^2).
B_{max} is the maximum operational flux density (gauss (webers/cm^2)).

In the MKS (Europe) system:

$$l_{gap} \approx \frac{0.4\pi L_{pri} I_{pk}^2}{A_c B_{max}^2} \text{ meters} \tag{3.28b}$$

where A_c is the effective core cross-sectional area (meters^2).
B_{max} is the maximum operational flux density (T (webers/m^2)).

This is an estimated air-gap length. The designer should use the closest value of air-gap available within the standard core part number.

The core manufacturer provides a parameter called A_L for each airgap length. This parameter is the inductance of the core when 1,000 turns is placed on it (U.S.). To determine the number of turns required to achieve the desired value of inductance one solves Equation 3.29:

$$N_{pri} = 1{,}000 \sqrt{\frac{L_{pri}}{A_L}} \tag{3.29}$$

where L_{pri} is the value of the primary inductance in mH.

If the A_L value is not given for a particular core/air-gap combination, Equation 3.30 can be used. Do not mix CGS and MKS units (gauss and centimeters with teslas and meters).

$$N_{pri} = \frac{B_{max} l_{gap(actual)}}{0.4\pi I_{pk}} \tag{3.30}$$

This value of N_{pri} represents the maximum value of primary inductance that can place the required energy in the core within the time period available at the lowest expected input voltage. Any future modifications to this number should be in the downward direction.

Now determine the turns of the highest power secondary winding by using Equation 3.31.

$$N_{sec} = \frac{N_{pri}(V_{out} + V_{fwd})(1 - \partial_{(max)})}{V_{in(min)} \cdot \partial_{(max)}} \tag{3.31}$$

where ∂_{max} is the maximum on-time duty cycle (usually 50 percent).

The result of Equation 3.31 should be viewed as the maximum number of turns, since any more turns would increase the secondary's inductance and take longer to empty the core's energy. This calculation will always result in a non-integer number of turns. Many cores do not support fractional turns, so round this number to the closest fractional turn that the core will allow.

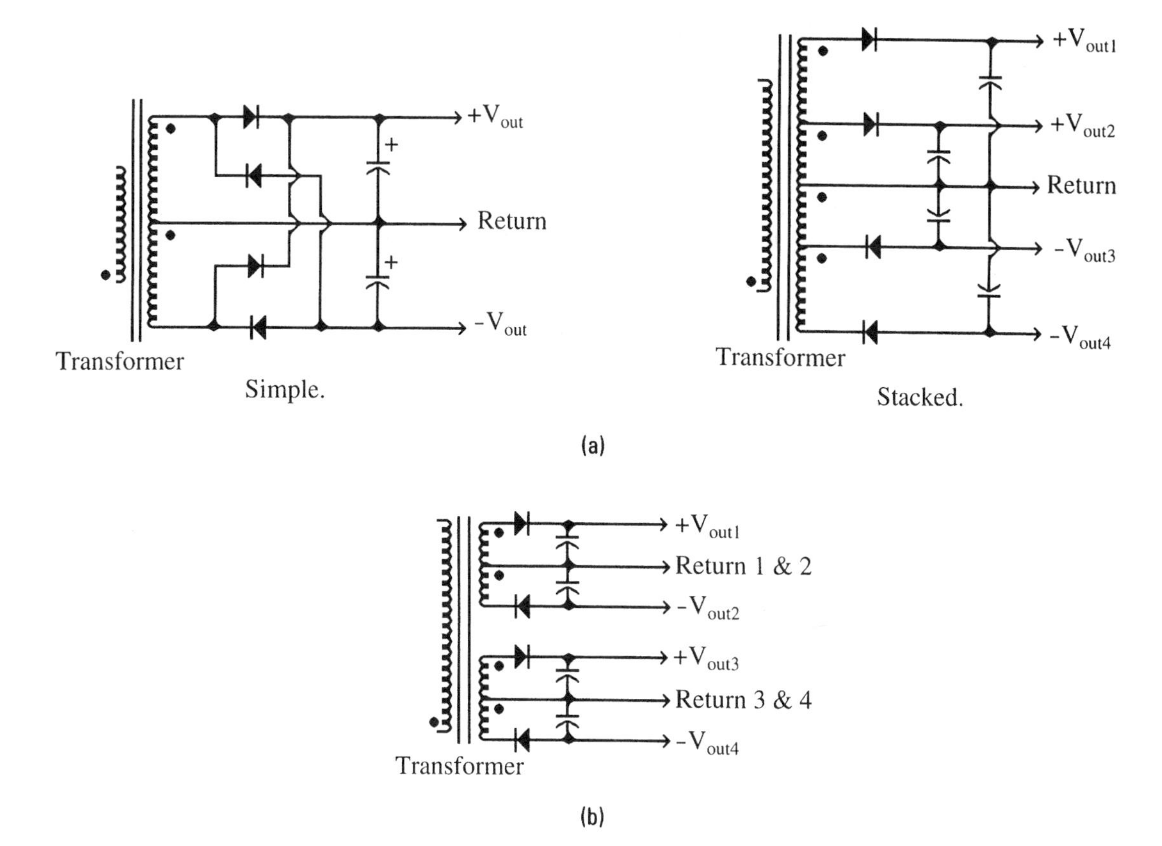

Figure 3–19 Flyback transformer secondary arrangements: (a) center-tapped secondaries; (b) isolated secondaries.

To determine the turns for any additional windings, use the procedure given for the forward-mode transformer (refer to Equation 3.22). Once again, if the resulting errors in the output voltages from the "ideal" values of output voltage are excessive, an iterative process will have to be under taken. Start by removing one turn from the original secondary winding and recalculating the output voltages (including the forward voltage drops of the rectifiers). Eventually, some error will have to be accepted in the output voltages.

The arrangement of the secondary should now be considered. The designer may either want to choose autotransformer style secondaries (i.e., shared turns with lower voltage windings) or isolated windings. Since flyback secondaries are half-wave rectified, noncenter-tapped winding or full-wave rectifier bridges are not possible (refer to Figure 3–19). When the desired secondary winding arrangement is decided, the windings must be checked to see if they still fit within the core's winding area. This is done in the same fashion as seen in Equation 3.23.

The physical design of the transformer is critical in flyback converters. If they are not properly designed physically, excessive voltage spikes could be generated that would adversely affect the reliable operation of the semiconductor components (refer to Section 3.5.8).

3.5.5 Designing the Forward-mode Filter Choke

The forward-mode filter choke is the inductor filter on every forward-mode converter's output. Its purpose is to store energy for the load during the periods when the power switch(es) are turned off. Its electrical function is to integrate the rectangular switching pulses into dc.

Its design is somewhat simple. First, the core should be chosen. Usually, a mopermalloy toroid is used for this application. This is because this material is self-gapped. Gapped ferrite cores can also be used with no difficulties. If a gapped ferrite core is desired follow the core sizing procedure in Section 3.5.2 for a simple one-winding inductor. The following method will demonstrate how to design a choke using the mopermalloy toroid.

The first step is to determine the minimum inductance needed for the output, using Equation 3.32:

$$L_{\text{min}} = \frac{[V_{\text{in(max)}} - V_{\text{out}}] \cdot T_{\text{off(est)}}}{1.4 \cdot I_{\text{out(min)}}} \tag{3.32}$$

where $V_{\text{in(max)}}$ is the highest peak voltage following the output rectifier of that particular output.
V_{out} is the output voltage.
$T_{\text{off(est)}}$ is the estimated on time of power switches at the highest input voltage (30 percent of $1/f_{\text{op}}$ is a good guess).
$I_{\text{out(min)}}$ is the lightest expected load current for that output.

This value represents the minimum value for the inductance, below which the core will empty of flux at the minimum rated load current for that output.

For mopermalloy toroids, the method for estimating the needed core size is done by calculating an intermediate value of the energy being stored in the core, using Equation 3.33:

$$E_{\text{L}} = L \cdot I^2_{\text{out(av)}} \tag{3.33}$$

Refer to Figure 3–20 and locate this value on the x-axis and move vertically until the first curve is intersected. Then move horizontally and read the part number. Then referring to that part number's data sheet and reading the value of AL for the core, the designer can calculate the number of turns needed by using Equation 3.29. In general, the higher the average current flowing through the choke, the lower the recommended permeability of the MPP core material.

Next the designer must check whether the turns will fit into the window or winding area of the toroid. The percent of occupied window area is determined by

$$\%\ \text{window} = \frac{N \cdot A_{\text{wire}}}{A_{\text{window}}}(100) \tag{3.34}$$

where A_{wire} is the cross-sectional area of the wire (refer to the wire table in Appendix F) (in^2 or m^2).
A_{window} is the available wire area of the toroid (window area) (in^2 or m^2).
N is the number of turns.

If this value is greater than 40 to 50 percent, then too much of the window is taken up by the wire. This is because a winding shuttle must fit through that remaining hole and it cannot pass through in less than 50 percent of the available window area. The solution is to go to the next larger core size or to drop the wire gauge by one size. The latter will increase the temperature of the inductor through added resistive losses.

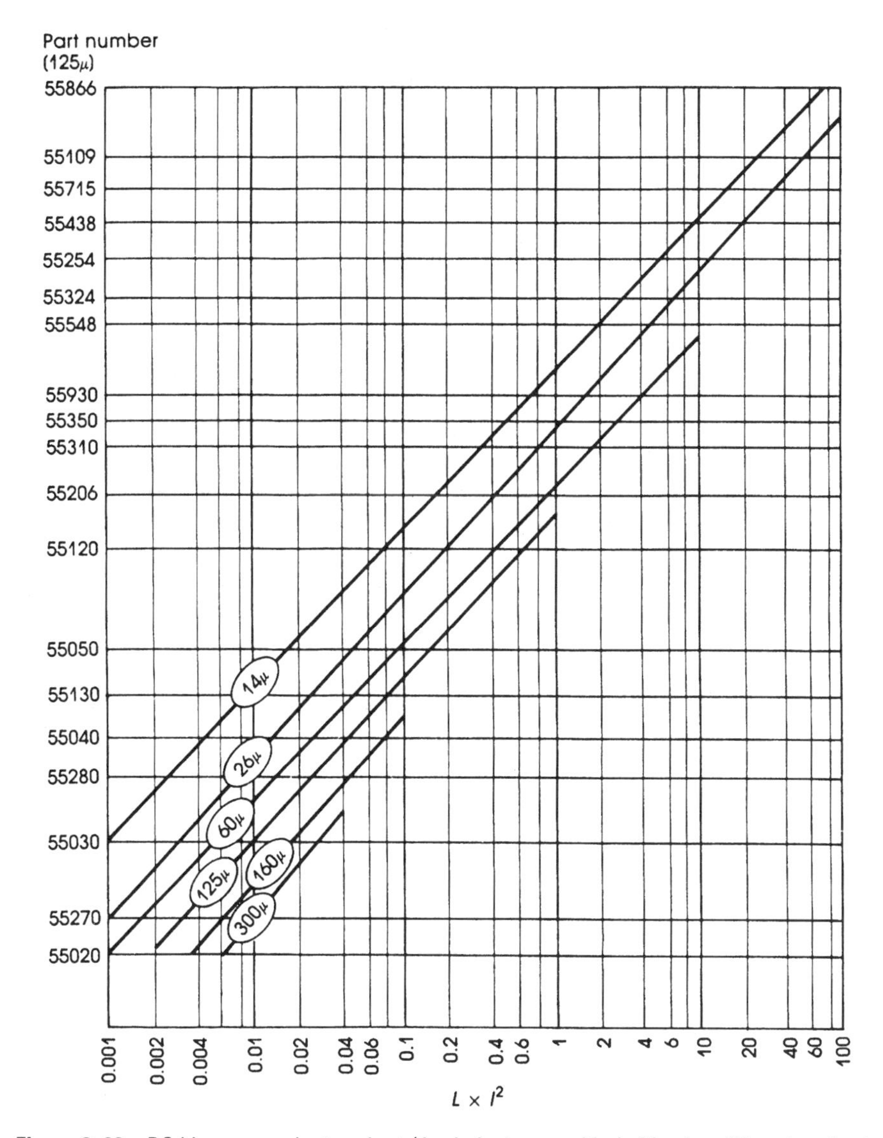

Figure 3–20 DC bias core selector chart (L = inductance with dc bias in millihenries; I = dc current in amperes) (Courtesy of Magnetics, Inc.).

Lastly, if the supply operates at a high frequency and the current through the filter is high, then the one might consider using *litz wire*, which is a wire made up of much smaller gauge wires to reduce the skin effect. Litz wire has an overall larger diameter than single-strand wire with the same overall cross-sectional area.

3.5.6 Designing the Mutually Coupled Forward-mode Filter Choke

Within multiple-output forward converters, it is possible to easily combine the filter chokes of complementary outputs (i.e., +/– 5V, etc.) together on the same core (see Figure 3–21). This offers several advantages; it saves space, vastly improves cross-regulation of those outputs, and exhibits superior ripple voltage levels on both outputs.

First the core style and material should be selected. This is done in an identical fashion as the single-output filter choke. Either a mopermalloy (MPP) toroid (refer to Section 3.5.5) or a ferrite bobbin core (refer to Section 3.5.2)

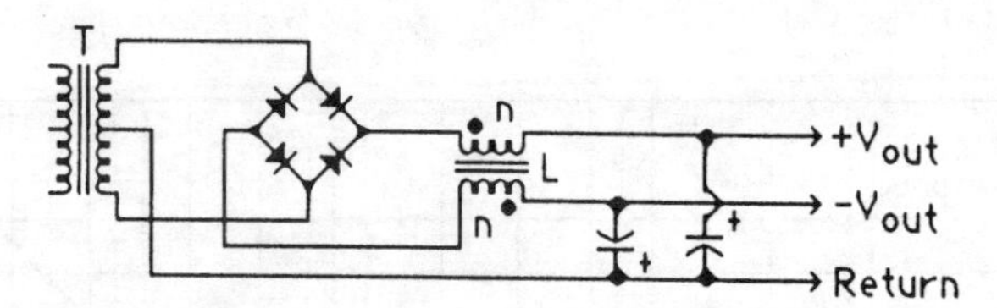

Figure 3–21 The mutually coupled output filter choke.

can be used. For the MPP toroid core, determine the size of the core needed, add the two load currents together, use the appropriate wire gauge for this current, and use Equation 3.33 for the toroid. The necessary core winding area will be almost twice as large as for the single-winding choke.

The number of turns for both windings is determined by calculating the minimum inductance and the number of turns for the lowest current output. This is done using Equations 3.32 and 3.29. The other winding will have the identical number of turns.

These windings should be *bifilar* wound, which means that the wires are twisted together prior to winding them on the core or bobbin. This guarantees that the turns will be identical, which is critical for its operation. A twist pitch of about three twists per linear inch is good for #22 AWG and a lower twist pitch works for larger gauge wires.

The fluxes contributed by both windings should add within the core. Since the winding *sense* (i.e., wound in the same direction) and the voltage polarities are both opposite in complementary output, then the polarity dots should be on opposite sides of the inductor (see Figure 3–20). If the windings are mistakenly hooked up in the wrong polarity, the windings will fight one another and the overall operation of the power supply will be impaired.

More windings could be added to the mutually coupled filter choke core, but I highly recommend against this temptation. If the windings are not exact (to the turn), the supply will loose approximately 1 percent in efficiency for each turn in error on each output. Instead, use a mutually coupled filter choke for each complementary set of outputs, and use the output cross-sensing technique described in Section 3.9.

3.5.7 Designing the dc Filter Choke

The DC filter choke is used for ripple voltage and current reduction on the output of the switching power supply immediately following the existing filters. It is also used as an EMI filter on switching power supplies that have a single power line on the input, such as battery and distributed power systems.

The dc filter inductor has dc current flowing through the inductor with a small ac signal summed onto it. With such a large dc current flowing through the choke, an air-gap is needed. The common choice of core is the MPP toroid core. These cores have distributed air-gaps within the core material and come in various permeabilities. The rule of thumb is that the higher the dc currents flowing through the inductor, the lower the permeability that should be used.

Actually the design of the dc EMI filter inductor is an easy matter. The core manufacturer provides a graph entitled “Normal Magnetization Curves” for the MPP cores, as seen in Figure 3–22. A permeability of 60 or below is recommended.

The first step is to determine the wire size needed. This can be done by knowing the average dc current flowing through the inductor, then referring to

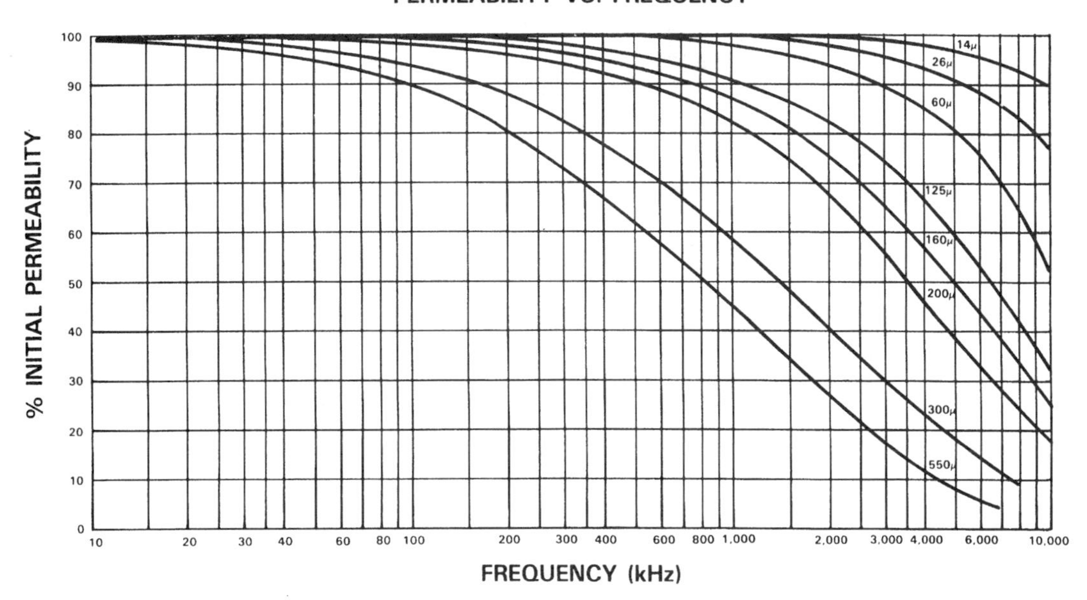

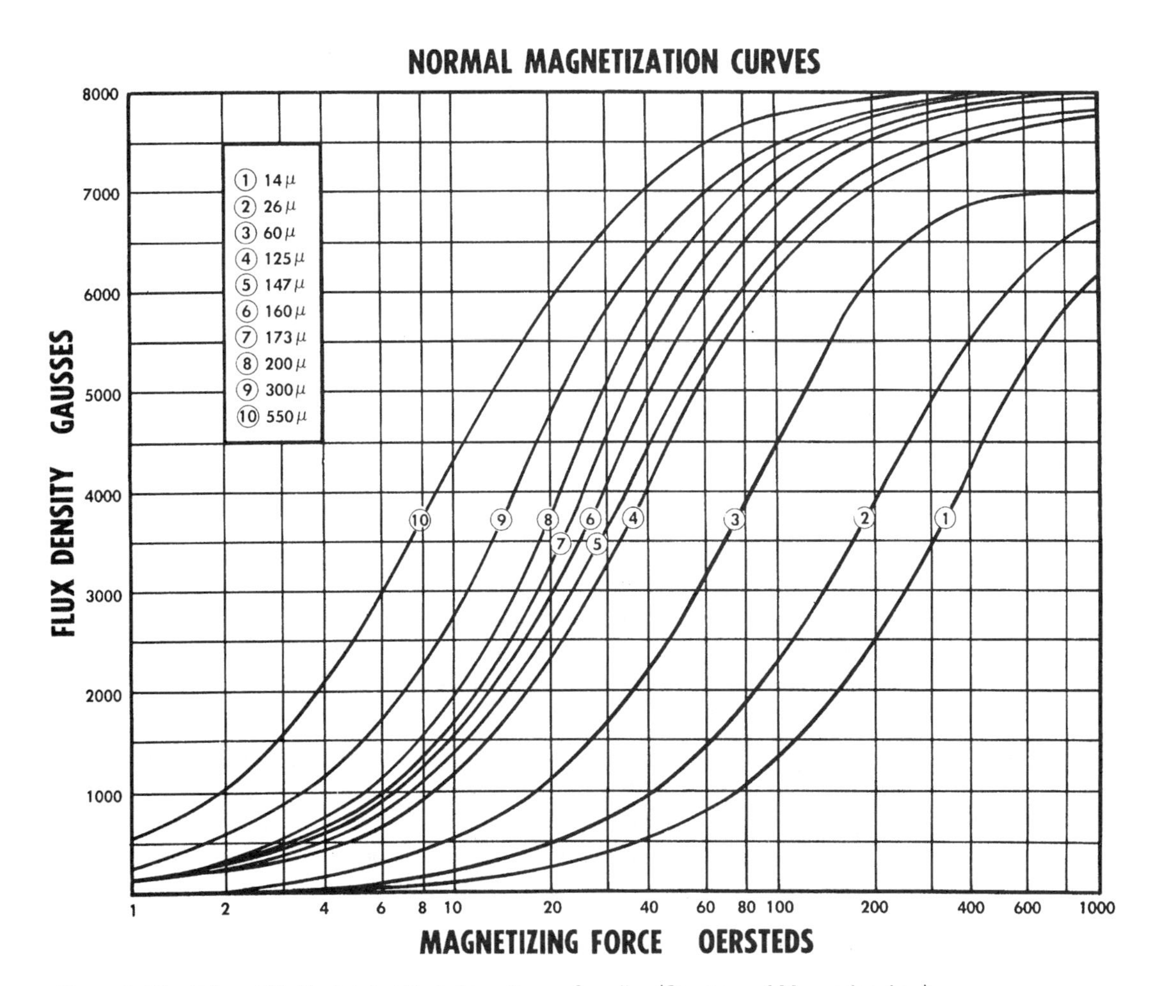

Figure 3–22 Pulsewidth Modulated Switching Power Supplies (Courtesy of Magnetics, Inc.).

the wire table (see Appendix F) and finding the wire gauge that can pass this current. Litz wire is not needed since the amount of ac passing through the inductor is negligible.

Next refer to the normal magnetization curves and select a value of H (magnetizing force, oersteds (Oe)) that is below the point on the curves where the permeability starts to drop due to the saturation of the core material. From the chart in figure 3–22 a value of 20 Oe is a good value. Choosing a permeability of 60 yields a reasonably low value of flux density.

Next comes an iterative process. A winding factor of 50 percent or less is good for manufacturability; assume that approximately 10 turns are going to be placed on the core. Calculate the wire area by multiplying the chosen wire cross-sectional area by 10. Then refer to the core data sheets and find a core that is at least double this value.

For the initial choice of core, calculate the actual number of turns needed by using Equation 3.35.

$$N = \frac{H \cdot l}{0.4 \cdot \pi \cdot I_{av}} \tag{3.35}$$

where H is the chosen magnetizing force (Oe)
l is the magnetic path length of the core (cm or meters)
I_{av} is the average current flowing through the inductor (A)

Check the winding factor once again to see if the window fillage is less than 50 percent. If it is larger go to the next larger core, and if it is less than 30 percent go to the next smaller core. Then recalculate the new turns for that core. Litz wire is not needed for this application, since the amount of ac current flowing through the winding is very small.

3.5.8 Base and Gate Drive Transformers

The purpose of gate or base drive transformers is to provide isolation between the controller section and a floating power switch. Their design is relatively easy, but it is important to the reliable operation of the switching power supply.

There are several important factors to consider during the design of the gate or base drive transformer:

1. The dielectric isolation of the transformer should be at least able to withstand twice the input voltage. Although this transformer is not tested during the HIPOT test, failure of its insulation would cause a catastrophic failure in the control circuitry in the event of a power switch failure.
2. The turns ratio is typically 1:1, but if other turns ratios are used, the output voltage should not cause an avalanche failure of the power switch.
3. Winding techniques that exhibit good primary-to-secondary coupling should be used. Any degradation in good coupling causes the isolated power switches to switch slower than the grounded power switches.

The design of the base or gate drive transformer is similar to the forward-mode power transformer. For unipolar drivers (as seen in Figure 3–23a), coupling capacitors should be used between the driver and the transformer, and on the output between the transformer and the power switch. These coupling capacitors should be at least 10 times the value of the gate-to-source capacitance of the chosen power MOSFET. This is because these capacitors form a capacitive voltage divider with the gate-to-source capacitance and will reduce the gate's drive voltage. For bipolar driver outputs (as in Figure 3–23b), the input coupling capacitor can be omitted.

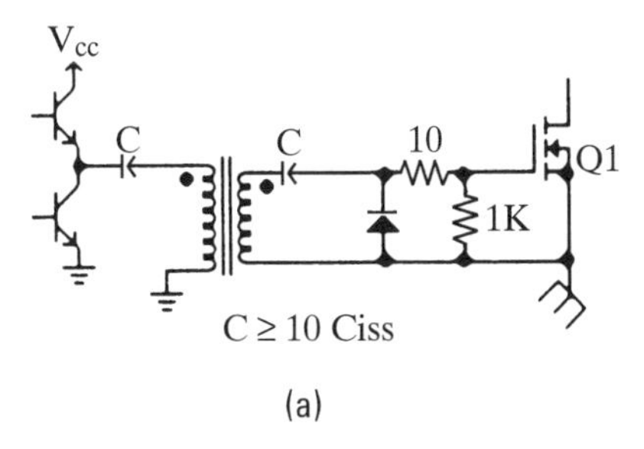

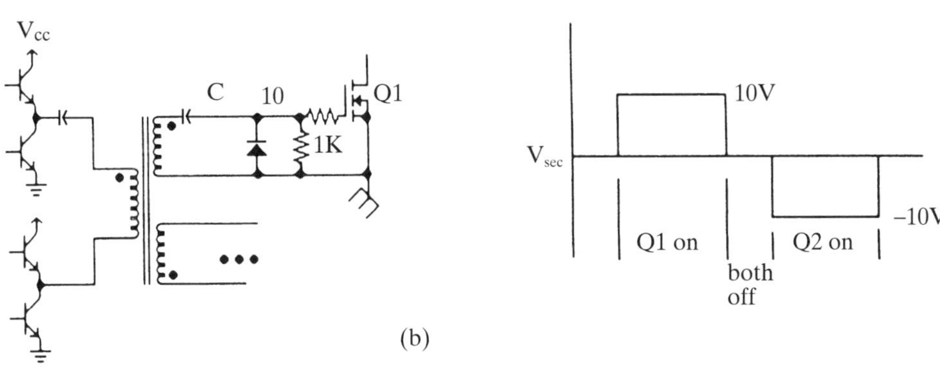

Figure 3–23 Examples of transformer-coupled base and gate drives: (a) single MOSFET drive circuit; (b) dual MOSFET drive.

A dc restoration circuit is needed following the output coupling capacitor to make the drive voltage referenced to the power switch's common. The supply voltage of the driver should be well bypassed so that its voltage does not "droop" during the drive pulse.

Remember that a forward-mode transformer reflects the impedance from one side to the other. This means that if drivers are single-ended on the primary side (i.e., active turn-on, passive turn-off) the power switch will still have a slow turn-off. If totem-pole outputs are driving the primary are used, the power switch's response will be fast.

A ferrite toroid or E core can be used for a drive transformer. No gap is needed since the input coupling capacitor guarantees that the core will operate in a bipolar fashion. A high permeability core is also suitable for this purpose. The wire that is going to be used will be in the range of #32 to #36 AWG. The core size will be approximately 0.4 to 0.6 inches (10 to 15 mm).

The B_{max} should be approximately one-half the saturation flux density (B_{sat}) at 100°C. A B_{max} of 1800 to 2500 G (0.18 to 0.25 T) is satisfactory. To determine the number of turns on the primary, use Equations 3.36a and 3.36b.

$$N_{pri} = \frac{V_{cc} \cdot 10^8}{4 \cdot f \cdot B_{max} \cdot A_c} \tag{3.36a}$$

where A_c is in cm^2, and B_{max} is in G.

$$N_{pri} = \frac{V_{cc}}{4 \cdot f \cdot B_{max} \cdot A_c} \tag{3.36b}$$

where A_c is in m^2, and B_{max} is in T.

Round upwards to the nearest integer all fractional turns in the results. Next multiply this result by the desired turns ratio to determine the number of secondary turns. The typical transformer will have a 1:1 turns ratio for a power MOSFET, and may be less that for a bipolar power transistor.

For dc input voltages that are more than 100VDC, a layer of Mylar tape should be placed between primary and the secondary(ies) and between any multiple secondaries. One should not completely trust the insulation breakdown voltage rating of the magnet wire since it could nick during the winding process.

3.5.9 Winding Techniques for Switchmode Transformers

The physical winding of switching power supply transformers is very important. It can make the difference between a switching power supply that is excellent and one that is noisy and poorly regulated. The design of switching power supply transformers is much more critical than that of 50/60Hz transformers.

There are three major factors in the winding of the transformer:

1. Must the power supply meet any safety regulatory specifications?
2. The windings should have tight electrical coupling.
3. The leakage inductance of all the windings should be as low as possible.

Some of these factors fight one another so one should use the best compromise practices.

Winding to meet safety regulations

If the input voltage of the switching power supply is greater than 40V at the peak, then the power supply comes under the regulation of one or more international safety regulatory agencies. Many of these agencies mirror each other's safety limits, but the designer should still review the requirements for the markets into which his or her company's products are being sold. The International Engineering Consortium (IEC) is the main standard-writing body, whose standards have been adopted by all of the European Community safety agencies. The remaining safety agencies, such as Underwriters Laboratories (UL) in the U.S., Canadian Standards Agency (CSA) in Canada, and VCCI in Japan, are working together to adopt a uniform set of safety standards based upon the IEC standard. This will allow one set of standards to be utilized all over the world. Until the "harmonized" standards are adopted, there will be differences among countries around the world.

There are also differing standards between the markets in each of these countries. For instance, the telecom market has vastly different safety requirements than the patient-contact medical market. So, it is important to determine the target market early in the product design process. These market differences will also be included in the IEC standards harmonizing efforts.

The typical cores used in "off-line" or 90 to 265VAC input switching power supplies are the E-E cores and their derivatives. These are bobbin-based cores and lend themselves for ease of manufacturing. The requirements of the safety agencies are quite clear in the area of transformer construction. The *creepage*, or the distance across a surface between the input and output windings of a transformer, can be no less than 4mm. To accomplish this, the transformer maker places a 2mm tape build-up on both ends of the winding area in the bobbin and places the winding between the border tape. The border tape creates a 4mm total creepage distance between isolated windings. The typical IEC compliant transformer can be seen in Figure 3–24.

An insulating sleeve must also be placed around the wires as they exit from the bobbin because the wire passes within the 4mm space required by the regulations. The distance between the input and output pins must also be 4mm

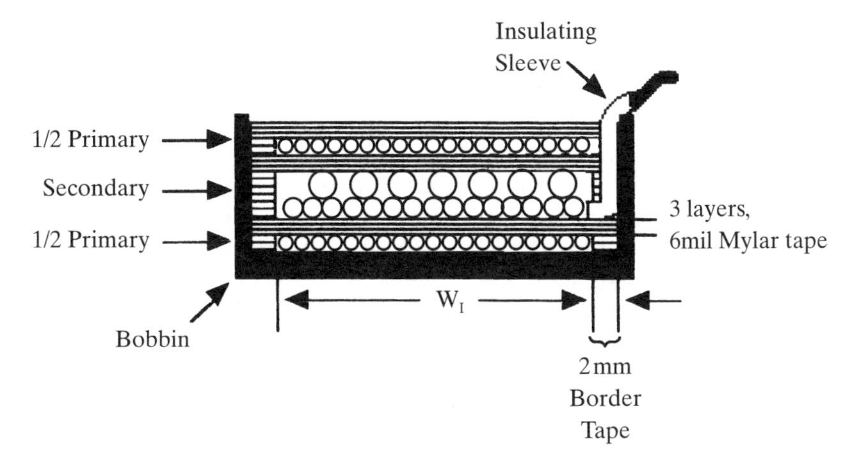

Figure 3–24 IEC-compliant, interleaved, off-line transformer.

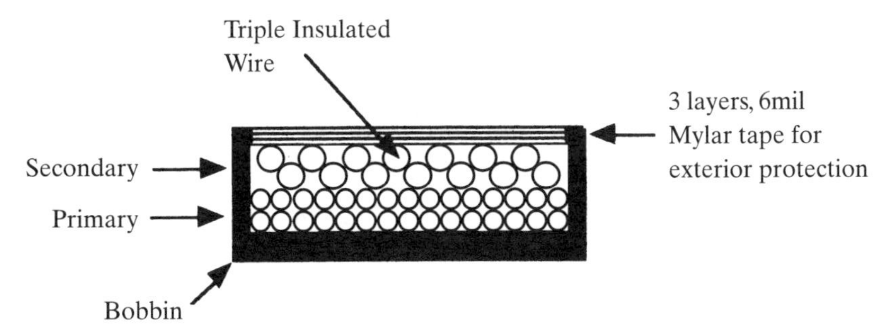

Figure 3–25 Off-line transformer using triple-insulated wire.

or more creepage distance apart from each other. This can also be accomplished by having "fins" molded between the pins on the bobbin and other similar structures.

The creepage distance between the opposing phases of the input (between the + and − dc and between the H1 and H2 (or hot and neutral) must be 3.2mm minimum.

The conductivity of a surface varies with the long-term pollution exposure in the operating atmosphere and its average humidity. The previously mentioned creepage spacings change for different applications, such as industrial, telecom, etc. The designer must refer to the appropriate specification.

The addition of tape, insulating sleeving, and pin spacing makes the final transformer much larger and higher in cost. This is because taping is a manual operation and takes a significant amount of time to perform.

Another method to meet the safety specifications is the use of triple-insulated wire for the secondary windings. Its use can reduce the transformer's size and leakage inductance compared to the border-tape method. *Triple-insulated wire* has three distinct, approved layers of insulation and it may be wound directly in contact with the primary windings. Whether Mylar tape is needed in or around the transformer depends upon the insulation system used in its construction. This type of transformer can be seen in Figure 3–25.

Winding for low leakage inductance

There are choices and techniques that can minimize the amount of leakage inductance a winding will exhibit. *Leakage inductance* is a measurable amount of inductance that does not couple to the core or to another winding. It behaves like a separate inductor in series with the winding lead. It is a parasitic element that causes spikes on the drain or collector of power switches and on the anodes of output diodes. That is because its flux cannot be loaded by the reflected impedances within the core.

An equation that estimates the amount of leakage inductance one might expect from a selected core and calculated winding is shown in Equation 3.37.

$$L_{\text{leak}} = \frac{K_1 (L_{\text{mt}}) n_{\text{x}}^2}{100 W_1} \left(T_{\text{ins}} + \frac{b_{\text{w}}}{3} \right) \tag{3.37}$$

where K_1 is equal to 3 for a simple primary and secondary winding, and is equal to 0.85 if the secondary winding is interleaved between two layers of primary windings.

L_{mt} is the mean length of a turn around the bobbin for the whole winding (in).

n_{x} is the number of turns contained in the winding being analyzed.

W_1 is the length of the winding from end-to-end (in).

T_{ins} is the thickness of the wire insulation (in).

b_{w} is the build (or thickness from the bobbin center-leg) of all the windings of the completed transformer (in).

The equation gives the major factors that affect the amount of leakage inductance the winding will exhibit. A major factor that is under the control of the transformer designer is the selection of a core that has a long center-leg. The longer the winding, the less leakage inductance will result. Keeping the number of turns of a winding to its minimum also will help greatly, since its influence is squared. Additionally, the coupling of the primary winding to the secondaries has a large affect on the primary leakage inductance. This is seen when the secondary is sandwiched, or interleaved, between two halves of the primary winding.

Another annoying transformer parasitic element is the *inter-turn capacitance*. This resembles small capacitors distributed between turns throughout the winding. Inter-turn capacitance is a problem in transformers with very high voltages across the primary winding. It is particularly a problem in off-line and high input voltage switching power supplies. It is caused by two adjacent turns on the same winding that reside at very different voltages. Equation 3.38 describes the energy that is stored between only two turns within a winding. Of course this energy is multiplied many times by all of the turns that are next to one another, but the equation is informative as to the causes of the capacitance. This energy is released as spikes during the transition times.

$$E_{(\text{stored})} = \frac{0.0194 V^2}{\ln\left(\frac{2s}{d}\right)} \tag{3.38}$$

where s is the space between the windings (m).

d is the diameter of the wire (m).

The distributed capacitances can store a great amount of energy when layers of turns are wound back and forth on top of one another. At the ends, their

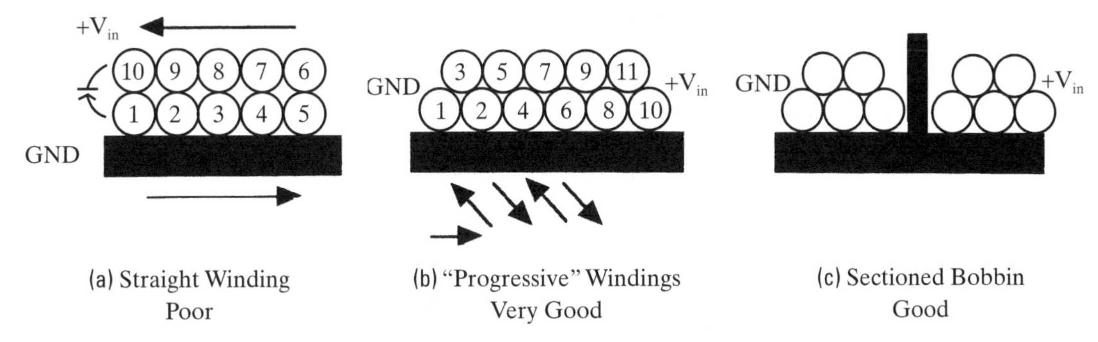

Figure 3–26 Winding technique for minimizing inter-turn capacitance.

voltages are very different and may even be close to the insulation breakdown. This will yield poor results. In Figure 3–26 three techniques are shown.

Progressive winding is when one to five turns wind across the first layer, then the wire is stepped back over that first layer and a second layer is placed on top. The most voltage that any turn sees is the number of turns in that progressive segment. A sectioned bobbin can be used to separate equal winding segments of the primary winding so that the greatest voltage difference within a section would be the input voltage divided by the number of sections.

The last technique is called *Z-winding*. After one layer is completed, the wire is run atop the bottom layer to the beginning side of the winding and then wound like the first layer. Its performance falls between the zig-zag method first mentioned (worst), and the sectioned bobbin and progressive techniques (best).

Low capacitance winding techniques also greatly reduces the dielectric stress of the wire insulation. This reduces the possibility of a arc jumping between two adjacent windings due to insulation breakdown.

Winding the transformer for tight coupling

Tight primary-to-secondary and secondary-to-secondary coupling is the ultimate goal of the transformer designer. Poor coupling causes the power signals to be delayed prior to reaching the output rectifier(s). The forward recovery period of the output rectifier(s) is also added to this delay. This makes the windings essentially unloaded during the switching transitions, which creates very large spikes on the windings due to the stored magnetization energy in the core. Add the energy stored in the leakage inductance and the inter-turn capacitance of the winding, and one has a problem.

Output cross regulation is affected by the amount of coupling between the secondary windings. *Cross-regulation* is how the variation of the load on one output affects the other outputs. It can be seen as the amount of "stiffness" in all of the outputs in a multiple output power supply in response to a change in load on any one output. Poor cross-regulation especially affects secondaries with wide differences in the turns ratios, that is, high voltage outputs mixed with low voltage outputs. There are two areas in which the designer can improve the output cross-regulation: the intrinsic cross-regulation that is designed into the transformer through its construction techniques, and the electrical cross-regulation found in Section 3.9 (voltage feedback). Maximizing the performance of both methods is usually needed for a reasonable power supply performance.

The first technique that enhances coupling between windings is *twisted pair winding*. This occurs when two or more wires are twisted together and then wound onto the bobbin at the same time. For 24 though 28 gauge wires about 3 twists per inch (or 1 twist per cm) is about right. Any tighter and the insulation could be damaged. This technique offers the best coupling by guaranteeing that the wires always will be placed adjacent to one another. Even windings of different lengths (turns) are helped by this technique. Having only a portion of the length twisted together, the coupling is improved.

Another winding technique is *filar winding*. This occurs when two or more wires are wound at the same time, but are not twisted. They rest next to the other wire(s) during most of the winding, but not all of it.

One wire product on the market is called Multiwire, with which two insulated wires are bonded together, which accomplishes the same purpose. It is also more easily handled by the transformer maker.

Of course filar or twisted pair winding techniques cannot be used between the primary and secondary when the peak primary voltage is more than 40 V. The safety agencies require that 3 layers of 1 mil thick Mylar tape (total thickness .006 inches with adhesive (0.167 mm)) be placed between the primary and secondary windings for input voltages less than 260 VAC. This ruins the coupling between the two windings. One method to improve the primary-to-secondary coupling is to *interleave* the windings (see Figure 3–24). This technique's cost in labor is higher than simply having the secondary layered atop of the primary. Hence, it is recommended that the interleave technique be used when the primary-to-secondary turns ratio exceeds 15–20 : 1. This would include switching power supplies that operate at 240 VAC or above and have a +5 VDC output or below. Figure 3–27 shows the effects of interleaving on a 480 VAC input, off-line flyback.

One can easily notice the difference in the spike energy between the two oscillographs. Typically, this energy is dissipated as heat within a clamp and/or snubber across the primary winding.

Using transformer winding techniques that are effective, even though they may be more costly to the transformer, can improve the overall performance of the power supply. This may save money for the whole power supply in the long run.

3.6 The Design of the Output Stages

The output stage rectifies and filters the high frequency ac switching waveform created by the power switch(es). The nontransformer-isolated topologies (buck, boost, buck-boost, and sepic) rectify and filter the power switch-generated ac signal directly. The transformer-isolated topologies have a transformer between the output stage and the power switches. The design of the output stage perhaps has the greatest effect on the efficiency of the power supply than any other stage, since the majority of the losses within the supply are seen in the output stage.

There are two basic types of output stage: those used in forward-mode converters and those used in boost-mode converters. The difference is the presence of the output filter inductor between the rectifier and the output filter capacitor in the forward-mode output stage. Figure 3–28 shows the common output stages.

For nontransformer-isolated switching power supply topologies, the output rectifier is connected directly to the power switch and the output stage operates in the half-wave mode. In transformer-isolated topologies, the output stage can

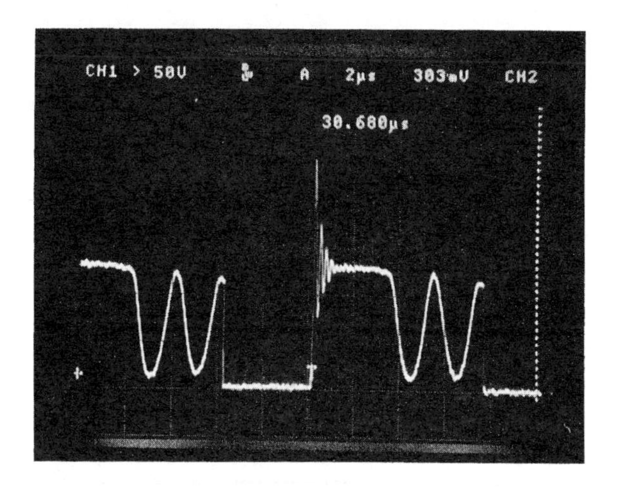

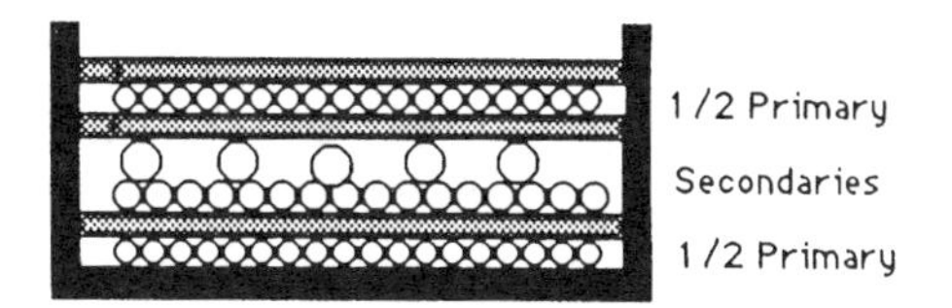

Transformer Winding Arrangement.
(Interleaved)

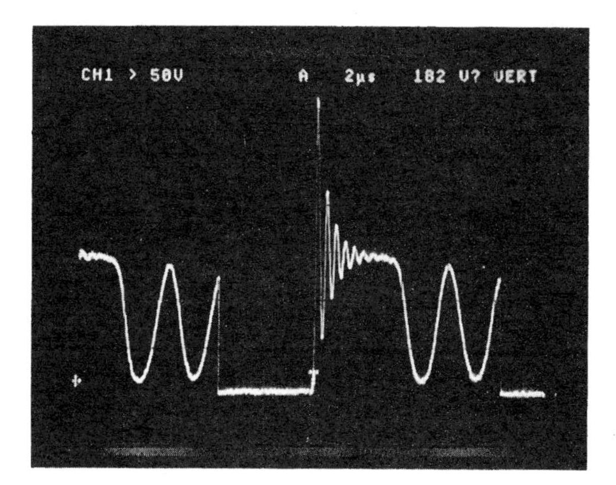

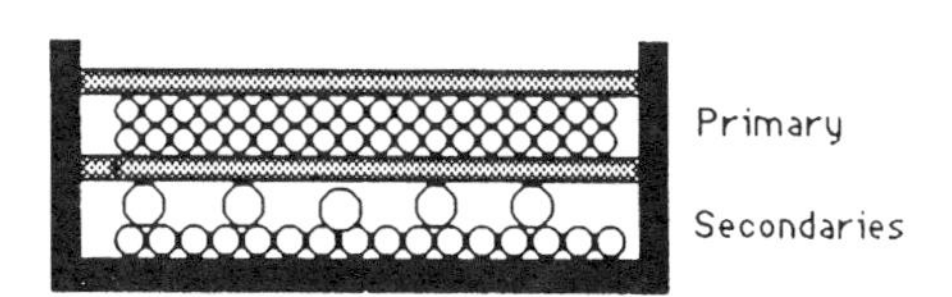

Transformer Winding Arrangement.
(Non-interleaved)

Figure 3–27 The effects of interleaving on the electrical waveforms of an off-line flyback converter. (Note spike amplitude and overall degree of "ringing.")

operate either in the half-wave mode (one transistor forward) (see Figure 3–28a) or in the full-wave mode (push-pull, half- and full-bridges). The secondary windings in the full-wave topologies can have a center-tap (Figure 3–28b) or be noncenter-tapped (Figure 3–28c) in which a full-wave rectifier bridge is used. The boost-mode topologies can only have a half-wave output (Figure 3–28d).

First, the designer should choose the type of rectification technology that is most appropriate for the application. The choice is whether to use *passive rectification* in which semiconductor rectifiers are used or *synchronous recification* in which power MOSFETs are placed in parallel with a smaller passive rectifier. Synchronous rectifiers are typically used in battery operated portable products where the added efficiency, usually an added two to eight percent, is important to extend the operating life of the battery or in applications where heat is important. In today's switching power supplies, passive rectifiers can dissipate 40 to 60 percent of the total losses within the power supply. Synchronous rectifiers affect only the conduction loss, which can be reduced by as much as 90 percent.

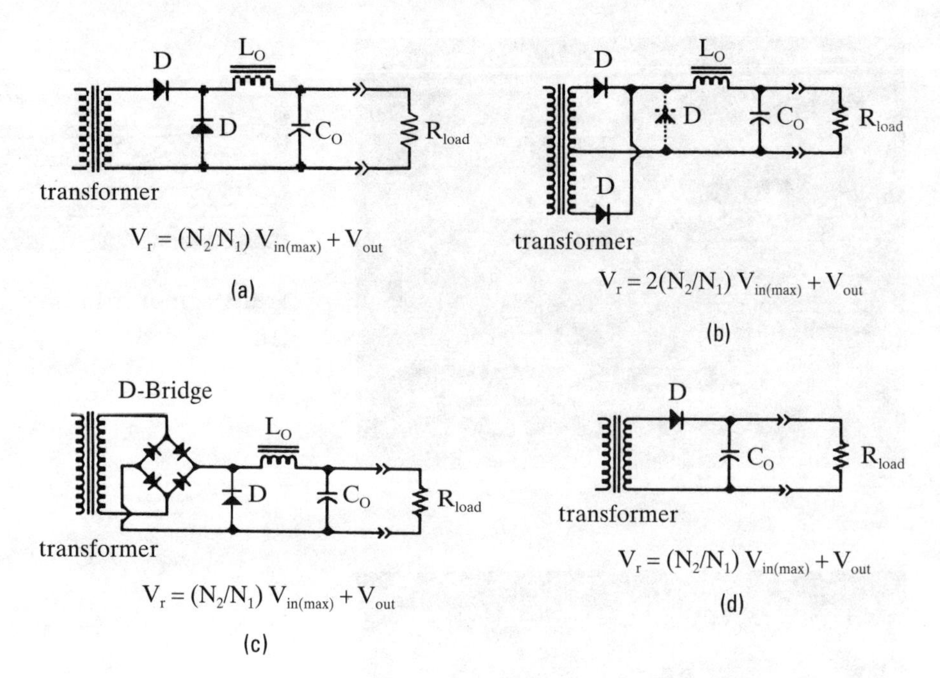

Figure 3–28 Output stages for forward and boost-mode converters: (a) half-wave forward-mode; (b) center-tapped forward-mode; (c) full-wave bridge forward-mode; (d) boost-mode.

Their use, though, is restricted only to continuous-mode, forward-mode outputs where the current flow can be guaranteed only in one direction. The designer should carefully consider the increase in efficiency verses the added cost of the drive circuitry for the end application.

3.6.1 The Passive Output Stage

The passive output stage is the traditional passive, semiconductor rectifier-based design. It is used for many applications that are not battery operated, and where switching power supply efficiencies of between 72 and 84 percent are acceptable. This would include many off-line applications where the heat generated by the supply is easily handled within the power supply.

The choice of rectifier technology has a large affect on the overall efficiency of power supply. There are common choices for rectifiers that tend to be dependent upon the output and input voltages. These two factors determine the maximum reverse voltage seen by the rectifier, which then dictates which type of rectifier to use. The input voltage is seen by the output rectifier multiplied by the transformer turns ratio and the type of secondary winding used. In outputs that are full-wave rectified, the rectifiers only see the input voltage multiplied by the secondary-to-primary turns ratio. Center-tapped secondaries present twice the voltage seen by the full-wave rectified output. Equation 3.39 gives the relationship for determining the minimum reverse blocking voltage required for a particular applications.

$$V_r > k(n_2/n_1)V_{in(max)} \quad \text{(forward)} \tag{3.39a}$$

$$V_r > (n_2/n_1)V_{in}(_{max} + V_{out}\,\text{(flyback)}) \tag{3.39b}$$

where k is 1 for full-wave secondary, and k is 2 for a center-tapped secondary.

The ultrafast diode has a 0.8 to 1.1 V forward voltage drop and exhibits a 35

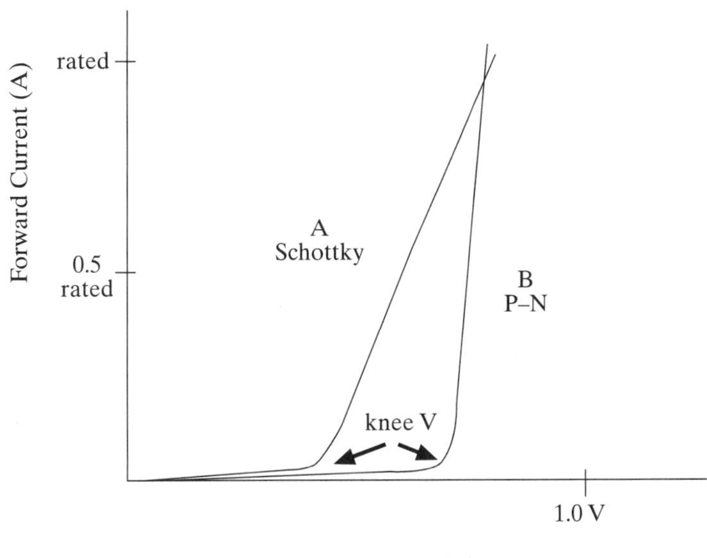

Figure 3–29 Forward conduction voltage characteristic of the Schottky versus the ultrafast diode.

to 85 nS reverse recovery time. All P-N diodes have a significant reverse recovery time, where the ultrafast-type diode exhibits the shortest time. The *reverse recovery time* of a P-N diode is caused by the charge stored within the P-N junction at the instant that the reverse voltage is applied to the diode. These minority carriers must then reverse direction and require a finite period of time to reach the junction boundary. This is seen in the circuit as a current momentarily flowing in the reverse direction, when the voltage has already reached its maximum reverse bias. This equates to a very large instantaneous power loss (refer to Section 4.1). Ultrafast diodes are used in outpuis stages when the reverse voltage seen by the diode is too great for a schottky diode.

Schottky rectifiers have a 0.3 to 0.6 V forward voltage drop and exhibit less than a 10 nS reverse recovery characteristic. Schottky diodes are generally more desirable, but have practical maximum reverse blocking voltages of about 40 to 50 V. This typically limits schottky diodes to those outputs of 15 VDC or less. Some schottky diodes have reverse blocking voltages as high as 200 V but exhibit a high junction capacitance characteristic that resembles the performance of a P-N reverse recovery characteristic.

Another difference between the ultrafast P-N and the schottky diode is the nature of the conduction characteristic of the diodes. The schottky diode has a higher conduction resistance which makes its forward voltage drop increase with increased forward current, as seen in Figure 3–29, curve A. This is caused by the resistive nature of the bulk resistance of the drift region of the silicon. The P-N-type diodes have a "flatter" forward conduction voltage characteristic more indicative of the forward drop of the actual P-N junction, as seen in Figure 3–29, curve B.

The use of a full-wave bridge causes an added rectifier loss in series with the output current, which may lower overall efficiency. If the difference in reverse voltage is great enough to use Schottky diodes instead of one P-N diode, usually the advantage falls to the Schottky diodes. The advantages would be a slightly smaller secondary winding and better reverse recovery performance.

3.6.2 Active Output Stages (Synchronous Rectifiers)

For those applications where high efficiency is important, synchronous rectification may be used on the higher current (power) outputs. Synchronous rectifier circuits are much more complicated than the passive 2-leaded rectifier circuits. These are power MOSFETs, which are utilized in the reverse conduction direction where the anti-parallel intrinsic diode conducts. The MOSFET is turned on whenever the rectifier is required to conduct, thus reducing the forward voltage drop to less than 0.1 V. Synchronous rectifiers can be used only when the diode current flows in the forward direction, that is in continuous-mode forward converters.

Figure 3–30 shows the common ways synchronous rectifiers are employed within switching power supplies.

A small Schottky rectifier with a current rating of about 20 to 30 percent of the MOSFET current rating (I_D) is placed in parallel with the MOSFET's intrinsic P-N diode. The parallel schottky diode is used to prevent the MOSFET's intrinsic P-N diode from conducting. If it were allowed to conduct, it would exhibit both a higher forward voltage drop and its reverse recovery characteristic. Both can degrade its efficiency of the supply by one to two percent.

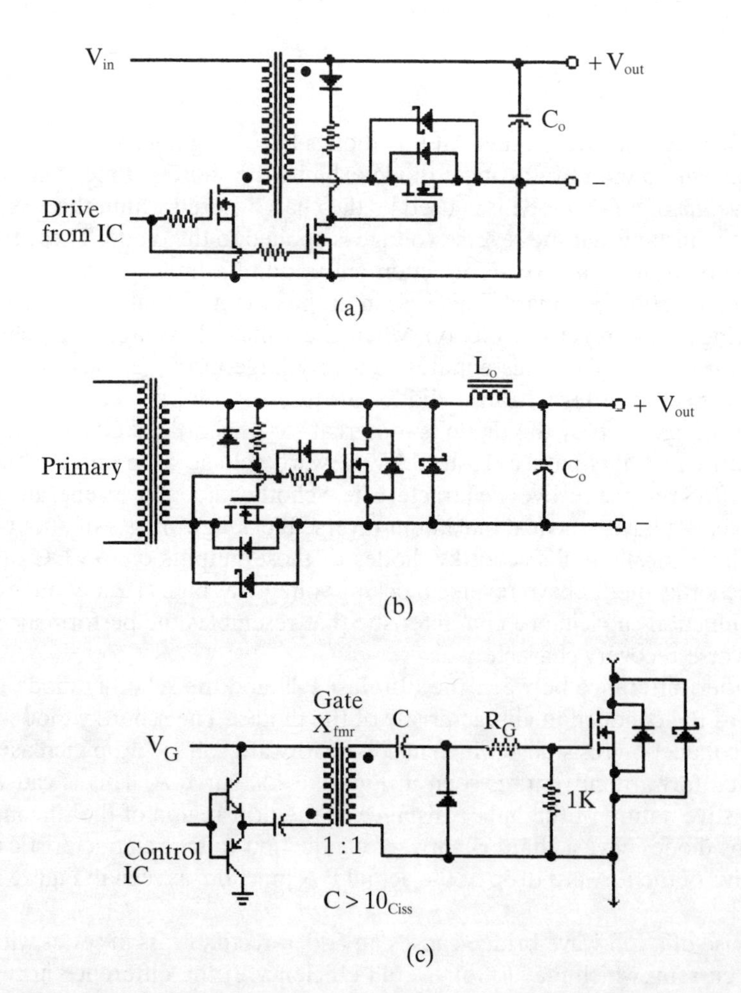

Figure 3–30 Common synchronous rectifier circuits: (a) nonisolated; (b) self-driven; (c) transformer-coupled.

The gates of the MOSFET must be precisely driven. If the gates turn off too slowly, then a *punch-through current* can occur between power switch MOSFET and the synchronous rectifier MOSFET. This is where both the MOSFETs are momentarily conducting at the same time. This unrestricted current would add to the losses and quickly lead to a failure. Figure 3–31 shows the critical periods in the operation of the synchronous rectifier.

The hidden loss for synchronous rectifiers is the charge needed to drive the gates of the MOSFETs. Refer to Section 3.7.2 for the information on the gate charge calculations. Essentially, one must drive a gate whose capacitance is between 800 and 2800 pF from OFF to ON and vise versa each cycle of operation. This loss increases linearly with switching frequency and gate capacitances (C_{iss} and C_{rss}).

3.6.3 The Output Filter

The output filter converts the rectified rectangular ac waveform into the dc output. Forward-mode converters have a two-pole *L-C* filter which produces the dc average of the rectified rectangular waveform. Boost-mode converters have a single-pole, capacitive input filter which produces a dc voltage which is the peak voltage of the rectified waveform. Both are reactive impedance filters and exhibit very little loss.

Designing the output filter choke (L_o) in a forward-mode converter is done first. This simple procedure can be seen in Section 3.5.5. A key design factor is to design the inductor to operate in the continuous current mode. The typical value of peak inductor current is 150 percent of the rated output current. The typical valley (minimum) current is about 50 percent of the rated output current.

A system-related consideration is the cross-regulation of the outputs. *Cross-regulation* is the degree of tracking of the output voltages when one or more of the loads change on their outputs. A manifestation of poor cross-regulation exhibits itself when a sensed output is loaded and the unsensed outputs greatly rise in voltage. To improve the cross-regulation in forward-mode converters, the technique of mutually coupled output filter chokes can be used. This places two output filter inductors from complementary output voltages (i.e., +/− 5 V, etc.) on the same core. This vastly improves the cross-regulation of those two outputs and exhibits much lower output ripple voltage (refer to Section 3.5.6).

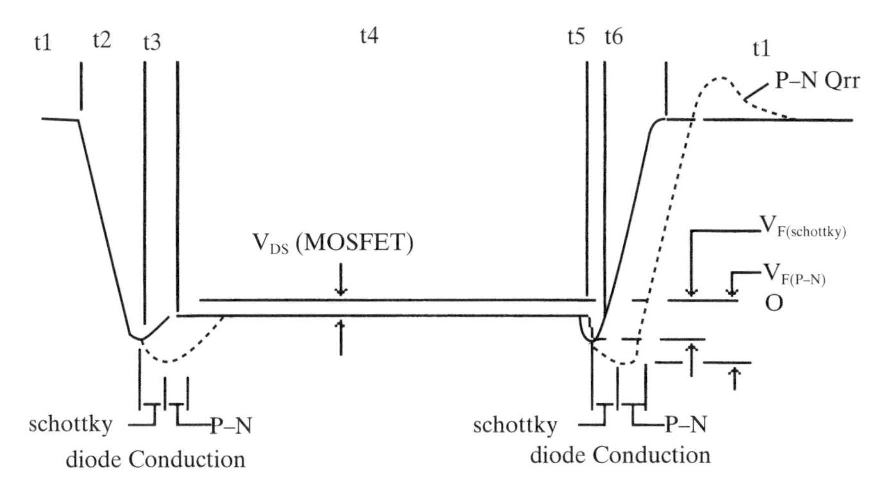

Figure 3–31 Critical time periods in the operation of a synchronous rectifier.

The calculation of the output filter capacitor (C_o) is the same for both the forward-mode and boost-mode output stages. Its value is simply determined by deciding how much peak-to-peak output ripple voltage one desires on the output voltage. *Output ripple voltage* is a small triangular ac waveform that rides atop the dc output voltage. For forward-mode converters, one can expect a typical output ripple voltage of $30\,mV_{p\text{-}p}$. In boost-mode outputs a value of 150 mVp-p is typical. If ripple-sensitive circuits are on a particular output, the designer should consider the addition of a dc filter stage following the output filter capacitor (refer to Section 3.5.7). To calculate the value of the output filter capacitor one executes Equation 3.40.

$$C_{out(min)} = \frac{I_{out(max)} \cdot (1-\partial_{(min)})}{f \cdot V_{ripple(pk-pk)}} \tag{3.40}$$

where I_{out} is the rated output current for that output (amps).
$\partial(\text{min})$ is the smallest estimated duty cycle at high input line and at light load (an estimate of 0.3 is good).
$V_{ripple(pk\text{-}pk)}$ is the desired peak-to-peak output ripple voltage (volts).

Especially within flyback-mode converters, the choice of the proper capacitor is extremely important. This is because the flyback-mode converters have no inductive impedance between the rectifier and themselves. This allows very high instantaneous values of current to enter and exit the capacitor. This high ac current flows through the ESR (*equivalent series resistance*) and ESL (*equivalent series inductance*) of the capacitor. The ESR causes the capacitor to get hot, shortens its life, and adds additional ripple voltage to the theoretical value used in Equation 3.40. ESL adds sharp step functions and spikes to the theoretical ripple voltage waveform. Tantalum capacitors are better than aluminum electrolytic capacitors in regards to superior ESR and ESL properties. In boost-mode converters, the approximate peak-to-peak ac ripple currents entering the filter capacitor is given by using Equation 3.41. Converting this peak-to-peak value to RMS, given by capacitor manufacturers, can be a difficult exercise in mathematics. As an approximation use about 33 percent of the peak-to-peak value as the RMS.

$$I_{ripple(pk-pk)} \approx \frac{2I_{out(av)}}{\partial_{min}} \tag{3.41}$$

Capacitor manufacturers have only begun to specify the use of their capacitors in high frequency switching power supplies. One must use care when reviewing capacitors for use in one's power supply. The ESR should be specified at a frequency greater than 1 kHz.

To properly design the capacitance for the output stage, one should place enough capacitors in parallel so that each capacitor operates at about 70 to 80 percent of its maximum ripple current rating. The sum of the capacitors should equal the final calculated value, but each capacitor should have the value of C_{tot}/n, where n is the number of capacitors in parallel.

Ultimately, the designer must try-out the final design and check the output ripple voltage and the temperature rise of each capacitor.

One last factor is the physical layout of the output stage when more than one output filter capacitor is used. The capacitors should be located radially symmetric from the output rectifier, and the printed circuit traces for the rectified voltage and the grounds should be of similar trace-widths and lengths. Any dissimilarity of these traces causes more series resistance and inductance to the remote capacitor. This makes the closest capacitor experience more of the ripple current and therefore get hotter. PCB traces also add to the ESR and ESL values to the capacitors. Refer to Section 3.14 for a more in-depth discussion on PC design.

A high frequency capacitor could also be placed in parallel with the larger capacitors. This is because the aluminum electrolytic and tantalum capacitors cannot absorb the very high frequency current components being presented to them. A .01 or .1 μF ceramic capacitor is well suited for this purpose.

3.7 Designing the Power Switch and Driver Section

The main purpose behind the power switch section is to convert the dc input voltage to a pulsewidth modulated ac voltage. The following stages can use a transformer to step-up or step-down the ac waveform, and finally the output stage converts the ac into the dc output(s). To accomplish the dc-to-ac conversion, the power switch operates only in the saturated and cutoff states. This makes the losses as low as possible.

There are two major types of power switches used today: the bipolar power transistor (BJT) and the power MOSFET. The IGBT (integrated gate bipolar transistor) is used in the higher power industrial applications, such as >>1 kW power supplies and electronic motor drives. The IBGT has a slower turn-off than does the MOSFET, so it is typically used for switching frequencies of less than 20 kHz.

3.7.1 The Bipolar Power Transistor Drive Circuits

The bipolar power transistor is a current driven device. To guarantee a "switch-like" operation, it must operate close or within its saturated state. For this to occur, the "on" base current must satisfy (also refer to Figure 3–29).

$$I_B \geq \frac{I_{C(max)}}{h_{FE(min)}} \tag{3.42}$$

where I_B is "on" base drive current.
$I_{C(max)}$ is the maximum anticipated collector current.
h_{FE} is the minimum specified dc gain of the transistor.

There are two types of base drive schemes. *Fixed base drive*, shown in Figure 3–33, drives the transistor into saturation over the entire conduction period. Since the collector current is almost always less than the maximum expected value, the transistor is nearly always being overdriven. Driving the transistor deep into saturation results in a slow turn-off time. The parameter storage time (t_S) is the time delay between the "off" signal to the base and when the collector begins to turn off. During this time the collector-to-emitter voltage continues to maintain a saturated voltage level. Although it is not a loss, it does require a shortening of the maximum duty cycle the power supply can use. The drive circuit should provide fast transitions in the base current (on and off) and draw the base voltage slightly negative.

The design philosophy behind the fixed-base drive circuit is to draw current from a relatively low voltage source (3 to 5 V) which is usually provided by an auxiliary winding on the power transformer. The resistor directly in series with the base (R2 as shown in Figure 3–33) should be on the order of 100 ohms. Its

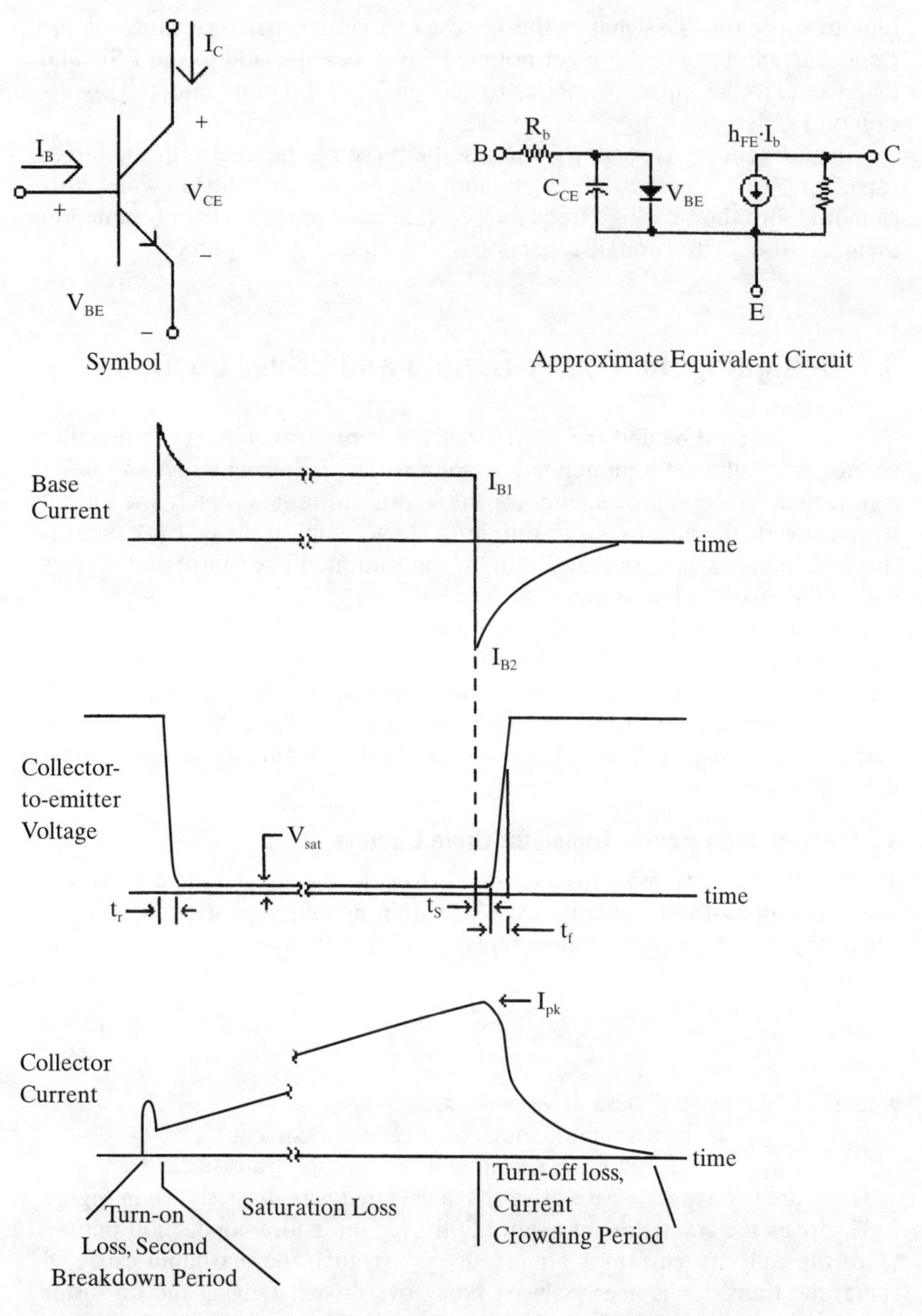

Figure 3–32 Waveforms for a bipolar power transistor within a PWM switching power supply.

purpose is to limit the dc current entering the base during turn on and turn off. A small capacitor on the order of 100 pF should be placed in parallel with this series resistor (R2). This is called a *base speed-up capacitor*, which provides a rapid positive and negative surge of current during the turn-on and turn-off transitions of the transistor. This reduces switching times and reduces the threat of second breakdown and current crowding. The resistor on the collector of the base driver transistor (R1 as shown in Figure 3–33) further controls the on-state

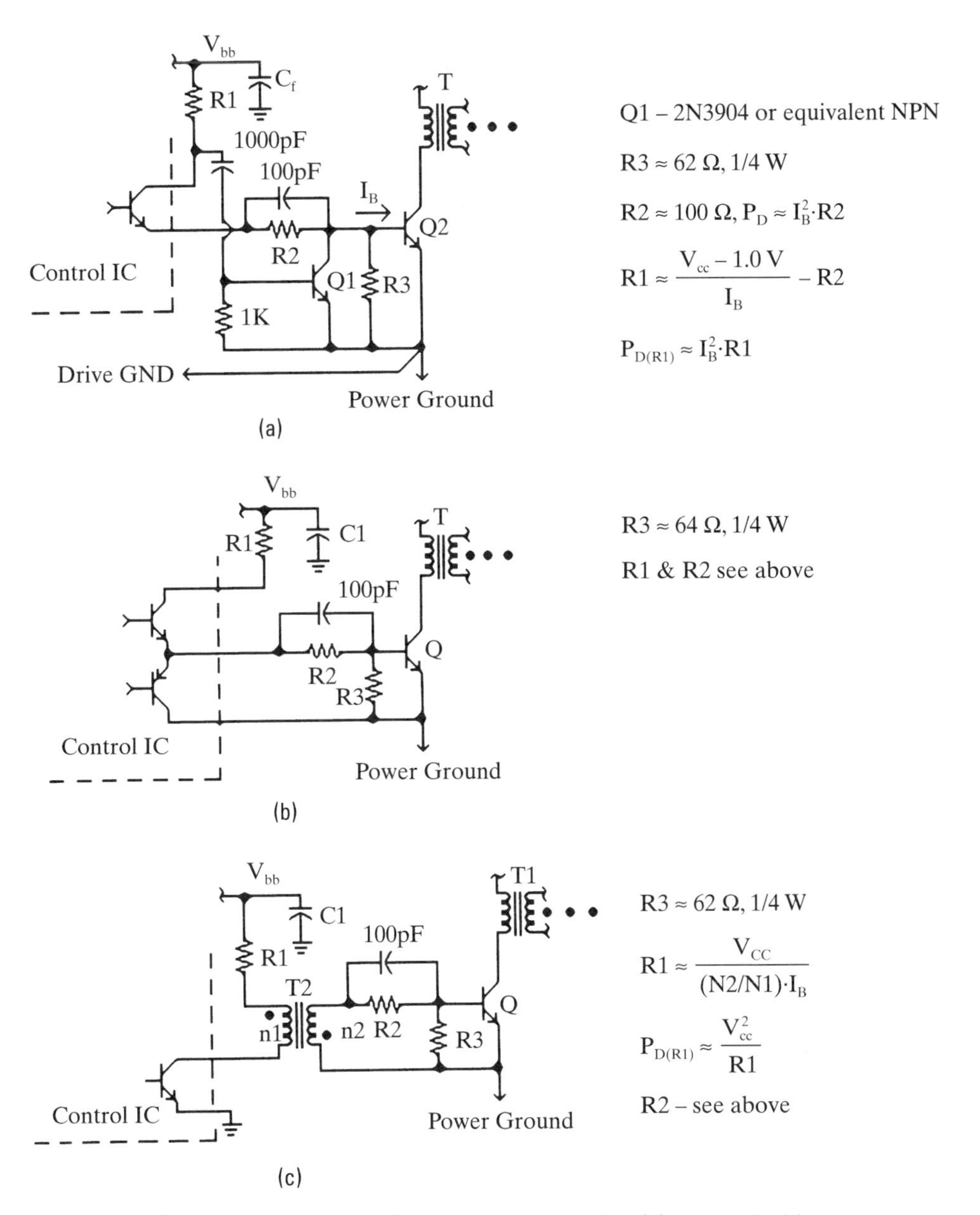

Figure 3–33 Fixed base drive circuits: (a) quasi-totem-pole drive; (b) totem-pole drive; (c) transformer-coupled drive.

base drive current. The base voltage should be checked with an oscilloscope and should go slightly negative during the turn-off transition, but should not exceed the base-emitter avalanche voltage rating (<5 V).

The second scheme, shown in Figure 3–34, is called *proportional base drive* and always drives the transistor at or just below the transistor's saturation state. The collector-emitter voltage is higher than with fixed base drive, but the transistor can now switch in about 100 to 200 nS. This is five to ten times faster than with fixed base drive. In practice, though, the fixed base drive scheme is used in the majority of low- to medium-power, low-cost applications. Proportional base drive is used for the higher-power applications.

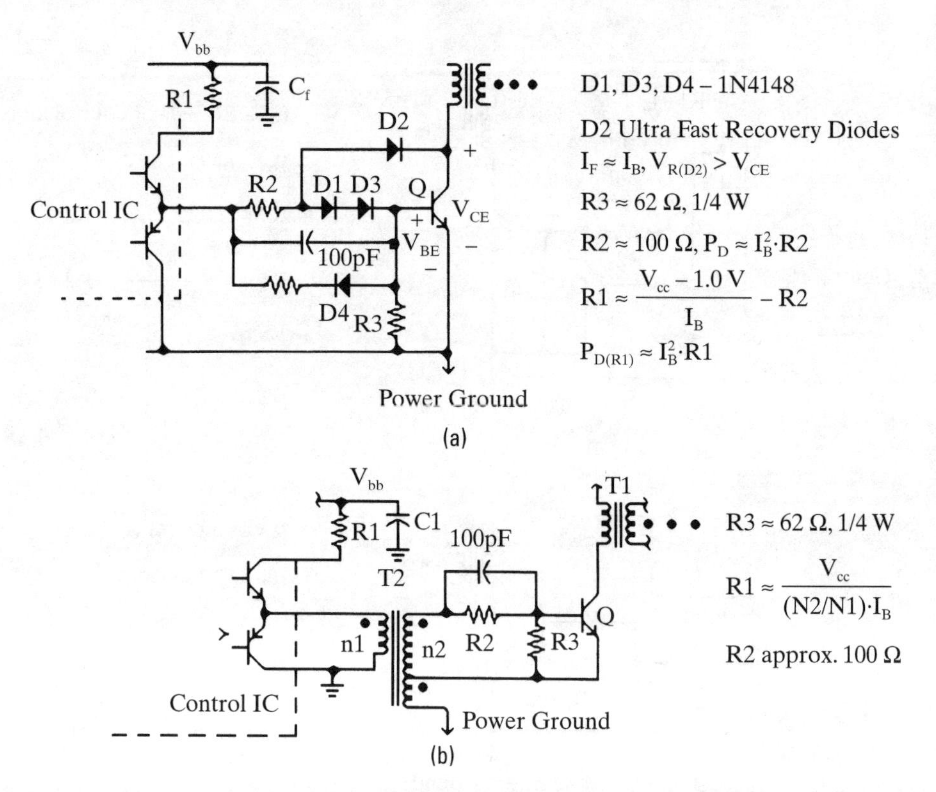

Figure 3–34 Proportional base drive circuits: (a) the Baker clamp; (b) transformer-coupled proportional base drive.

The last consideration is from which voltage to draw the base current. Since the base-emitter junction resembles a forward-biased diode, the maximum V_{BE} is 0.7 to 1.0 V. Ideally, a voltage source of 2.5 to 4.0 V is sufficient. If the base drive voltage source is too high, there will be a significant loss associated with driving the base.

In the initial breadboard, the voltage and current waveforms associated with the power transistor must be carefully scrutinized and it must be verified that they do not exceed the SOA limits. This is also the time to modify any values that enhance its switching characteristics, since it represents about 40 percent of the supply's entire losses. The drive schemes shown in Figures 3–33 and 3–34 are the common approaches to driving the bipolar transistor and will give the designer a very good staring point.

3.7.2 The Power MOSFET Power Switch

The power MOSFET is the most common choice as a power switch. Its cost and saturation loss are comparable to the bipolar transistor in most applications and it switches five to ten times faster. It is also easier to use in a design.

The MOSFET is a voltage-controlled current source. To drive a MOSFET into saturation, enough voltage must be applied between the gate-source terminals in order to pass more than the maximum expected drain current. The relationship between gate-to-source voltage and drain current is called *transconductance* or g_m. Power MOSFETs fall typically into two categories: standard MOSFETs, which must have about 8 to10v V_{gs} to guarantee full-rated drain

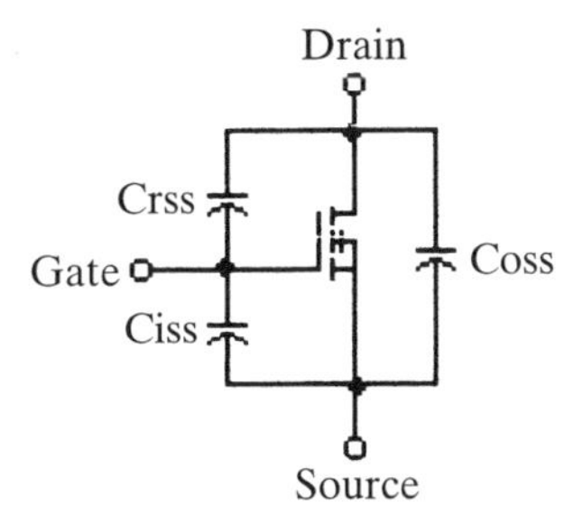

Figure 3–35 The symbol of a power MOSFET with the parasitic capacitances.

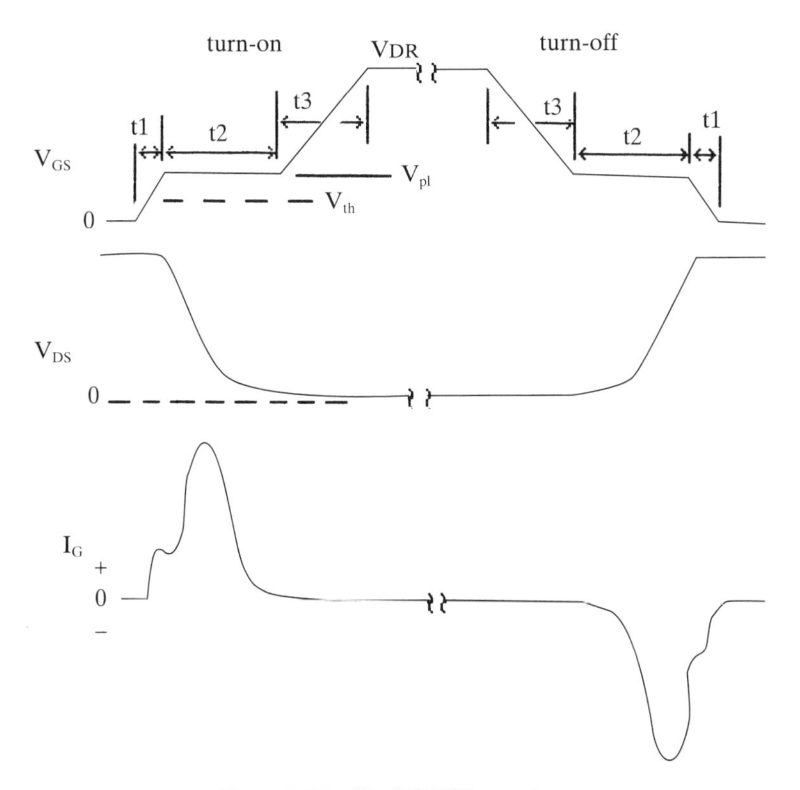

Figure 3–36 The MOSFET waveforms.

current, and logic-level MOSFETs, in which only 4.0 to 4.5 V is needed. Logic-level MOSFETs typically have low drain-to-source voltage ratings (<60 V).

MOSFETs can switch very quickly, typically from 40 to 80 nS is common. To drive the MOSFET this fast, one must consider the intrinsic parasitic capacitors which exist in all power MOSFETs (see Figure 3–35).

These capacitances are specified in each power MOSFET datasheet and are very important. C_{oss}, or drain-to-source capacitance, is considered in the drain loads, but does not directly enter into the drive design. The C_{iss} and C_{rss} have direct and calculable effects upon the switching performance of the MOSFET. Figure 3–36 shows the gate and drain waveforms of a typical N-channel MOSFET switching cycle.

The plateau in the gate drive voltage signal is caused by the opposing transition in the drain-to-source signal being coupled into the gate node through the

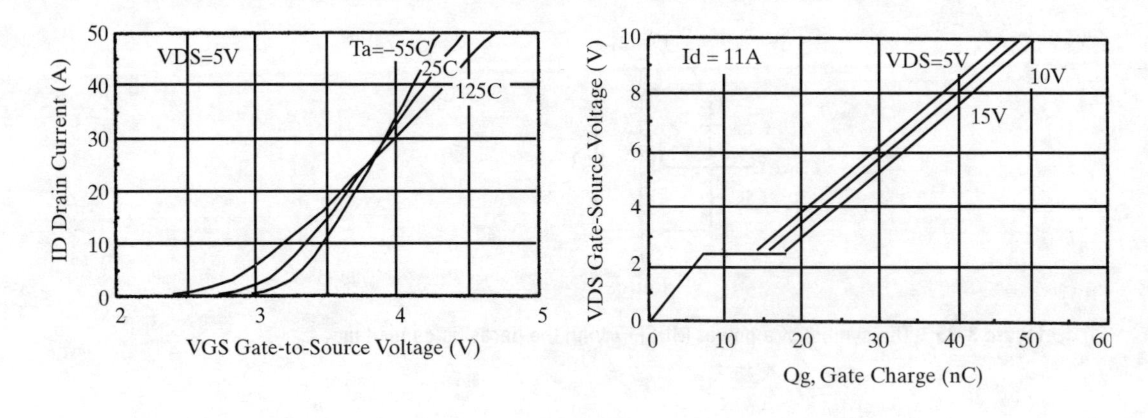

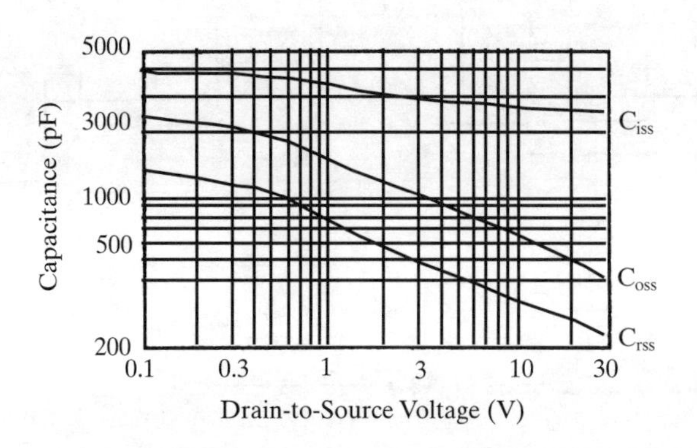

Figure 3–37 Representative datasheet curves for a MOSFET.

miller capacitor (C_{rss}). During this period, a large impulse is seen in the gate drive current. This plateau occurs at a voltage slightly higher than the rated gate threshold voltage and is $V_{TH} + I_D/g_m$. The value of the plateau voltage can also be determined by using the transfer function graph provided in every MOSFET data sheet (see Figure 3–37). For an inaccurate estimation, the threshold voltage can be used. The MOSFET datasheets provide the graphs seen in Figure 3–37.

These capacitances cause delays in the switching performance of the MOSFET. The drive circuits should be able to drive capacitive loads. First, the amount of charge required to transition the gate through each voltage level must be determined. Figure 3–37 yields this information when one subtracts the differences in gate drive voltage operating points from one another. The switching delay can be calculated from Equations 3.43 through 3.50.

Turn-on delays:

$$\begin{array}{lcc} & \textit{CMOS} & \textit{Bipolar} \\ \text{Turn-off Delay } t(1) \approx & \dfrac{Q_{(t1)}}{I_{OH}} & \dfrac{Q_{(t1)}\mathrm{R}_{\mathrm{eff(OH)}}}{V_{OH}} \end{array} \tag{3.43}$$

$$\text{Risetime } t(2) - \quad \frac{Q_{(t2)}}{I_{OH}} \quad \frac{Q_{(t2)}\mathrm{R}_{\mathrm{eff(OH)}}}{V_{OH} - V_{p1}} \tag{3.44}$$

$$t(3) - \frac{Q_{(t3)}}{I_{OH}} \qquad \frac{Q_{(t3)}\mathrm{R}_{eff(OH)}}{V_{OH} - V_{p1}} \tag{3.45}$$

$$\mathrm{R}_{eff(OL)} - \frac{V_{OL}}{I_{OL}} \tag{3.46}$$

Turn-off delays:

COMS *Bipolar*

$$\text{Turn-off Delay t(3)} - \frac{Q_{(t3)}}{I_{OL}} \qquad \frac{Q_{(t2)}\mathrm{R}_{eff(OL)}}{V_{OH} - V_{p1}} \tag{3.47}$$

$$\text{Falltime } t(2) - \frac{Q_{(t2)}}{I_{OL}} \qquad \frac{Q_{(t2)}\mathrm{R}_{eff(OL)}}{V_{OL} - V_{p1}} \tag{3.48}$$

$$t(1) - \frac{Q_{(t1)}}{I_{OL}} \qquad \frac{Q_{(t1)}\mathrm{R}_{eff(OL)}}{V_{OL} - V_{p1}} \tag{3.49}$$

$$\mathrm{R}_{eff(OH)} - \frac{V_{DR} - V_{OH}}{I_{OL}} \tag{3.50}$$

Bipolar-based drivers are more able to source the current surges needed by the MOSFET gate than CMOS-based drivers, which act more like a current-limited source and sink. Speed is typically controlled by placing a resistor in series between the driver and the gate. In switching power supplies, when rapid switching is required, a resistor of over 27 ohms is not advisable, since its switching speed would drop and its switching loss would rise significantly. Larger series gate resistors would create short periods of oscillation when the drain voltage is transitioning.

3.7.3 The IGBT as a Power Switch

The IGBT is a silicon hybrid composed of a power MOSFET on the gate terminal and an "unlatchable" SCR between the collector and emitter terminals. Its internal schematic can be seen in Figure 3–39.

The advantages of IGBTs over MOSFET devices are the savings in silicon die area and its bipolar collector current characteristics. It also has two disadvantages: higher saturation voltage due to two series P-N junctions, and the fact that IGBTs can have a long turn-off "tail," which adds to the switching loss. The "tail" loss has limited the switching frequencies to less than 20 kHz. This has made it ideal for industrial electronic motor drives where the switching frequency is just beyond the human audio range.

IGBTs have been the target of much research by semiconductor companies and the tail time has been shortened significantly. Originally, the tail time was about 5 uS; now the tail time is about 100 nS and is improving. The saturation voltage has also been improving from approximately 4 V to less than 2 V. Although this is a problem for low-voltage dc-dc converters, it has a great deal of appeal for off-line and industrial, high-power converters. As a personal judgment, I would consider IGBTs for those converters with input voltages greater than 220 VAC and 1 KW.

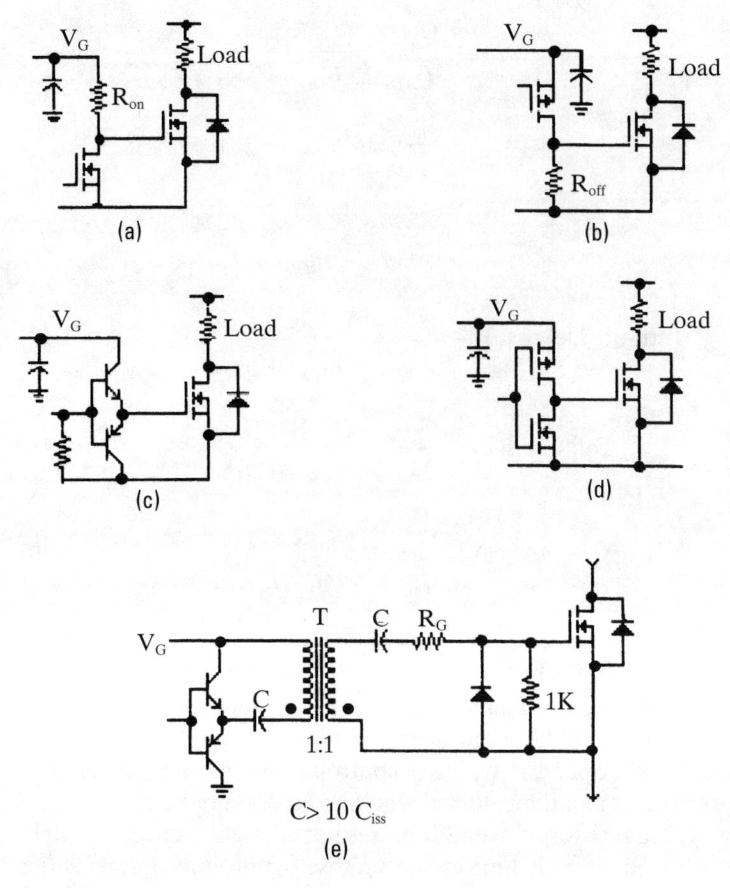

Figure 3–38 MOSFET drive circuits: (a) passive turn-on; (b) passive turn-off; (c) bipolar totem-pole; (d) MOS totem-pole; (e) transformer-isolated drive.

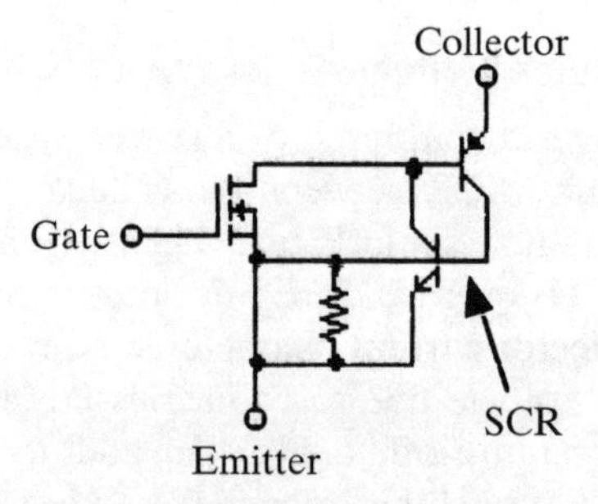

Figure 3–39 The IGBT.

Driving the IGBT is identical to the MOSFET. It has similar gate driving characteristics and a MOSFET driver IC works very well with an IGBT.

3.8 Selecting the Controller IC

Today, one has many choices in the area of controller-integrated circuits. One must decide early in the design process how much functionality and integration is desired in the IC, such as whether the power switch should be inside the IC, the high voltage start-up circuit, soft start circuit, synchronization circuits, etc. This is driven by the application and the extra functions that are desired.

Extreme caution must be practiced reviewing candidate ICs. The first page of the data sheet does not always completely describe the IC. Subtle factors involving the method of control, how the output drivers are to be used, and how the protection circuit operates must be further investigated. This requires careful review of the internal IC and applications schematics. An example where confusion creeps in is the newer generations of buck controller ICs for less than 3.3 V outputs. These ICs are designed to operate from a higher supply voltage than their input power line and drive the MOSFETs at a very low voltage level. The traditional buck controller operates from its input voltage and drives the MOSFET from the input voltage. One cannot differentiate these two ICs apart from the first page of the datasheet and it is even more difficult to tell them apart if they are presented as a result of a search on a supplier's website. This could cost you valuable time.

Today, more than ever, the new control ICs are targeted narrow application niches. This typically includes removing as many pins as possible from the package to minimize external circuits. This severely limits the flexibility of an IC. This makes the selection process even more important.

3.8.1 Short Overview of Switching Power Supply Control

This short overview may be redundant from the earlier part of this chapter, but it is appropriate. The primary purpose of the controller is to maintain a constant output voltage for a large range of load currents. A negative feedback loop is used for this purpose. All power supply controllers, both linear and switching, sense the output voltage. The rated output voltage is reduced to the level of a voltage reference somewhere within the control IC. This *feedback voltage* is placed into the inverting input of a high-gain op amp called the *voltage error amplifier*. The reference voltage is placed into the non-inverting input of the same op amp. The output of the amplifier represents a greatly amplified version of the difference between the reference and the output voltage. This output voltage is called the *error voltage*. This error voltage is then used to control the amount of energy the power supply wants to pass to the load. It can have a positive value, indicating that the output is too low and the power supply needs to supply more energy to the output. Conversely, a negative value indicates that the output is too high and the thruput energy should be reduced.

Current is typically sensed so that the power supply does not exceed its power ratings. There are two methods of measuring current: average output current and instantaneous current. Average current circuits operate in a very similar manner to the voltage feedback loop described previously. Current is typically measured as a voltage across a resistor placed in series with the current to be measured. This voltage is amplified or used in its very small value. This voltage is then placed into the inverting pin of an operational amplifier and a reference voltage representing the desired maximum value of the output. When the current gets too high, the *current error voltage* swings from positive to negative, indicating that the output current has exceeded the desired maximum amount. This signal can be used to override the voltage error signal and reduce the energy thruput of the power supply.

Instantaneous current sensing is used to protect the power semiconductors. A current sensing resistor is placed in the power switch current path and its voltage represents the instantaneous current flowing through the power device. This voltage is then taken to a very fast analog comparator where if a predetermined voltage is ever exceeded, the power device is instantly turned off. This offers very nice protection of the power device.

3.8.2 Selecting the Optimum Control Method

The selection of the control IC is very important. If the wrong choice is made, it could result in supply instability and waste valuable time. The designer should understand the subtle difference between the various forms of control. In general, forward-mode topologies usually have voltage-mode controllers, and boost-mode topologies usually have current-mode control. This is not a rock-solid rule since every method of control can be used for every topology, with mixed results.

Voltage-mode control

This method of control can be seen in Figure 3–40. The important trait of voltage-mode control is that the error voltage is placed into a PWM comparator and compared against a clock generated sawtooth waveform. As the error voltage rises and falls the pulsewidth of the output signal increases and decreases. *To identify a voltage-mode control IC, look for a capacitor R-C oscillator and see if the capacitor sawtooth voltage goes to a comparator with the error voltage.*

There are two versions of overcurrent protection on voltage-mode control ICs. The older version is *average current foldback*. Here, the output current is

Table 3–6 PWM Control Methods

Control Method	Optimum Topology	Issues
Voltage-mode with average OC foldback	Forward-mode	OC foldback slow, could cause power switch failures
Voltage-mode with pulse-to-pulse OC limiting	Forward-mode	Very good OC protection; usually hi-side current sensing
Hysteretic current-mode	Forward- and boost-mode	Heavily patented; few control ICs
Current-mode, turn-on with clock	Boost	Very good OC protection; many ICs; typically GND-driven SW

Quasi-Resonant and Resonant-transition Control Methods

Control Method	Optimum Topology	Issues
Constant off-time	Zero voltage switching QR	Variable frequency High frequency limit needed
Constant on-time	Zero current switching QR	Variable frequency Low frequency limit needed
Phase modulated	PWM forward-mode full bridge	Fixed frequency

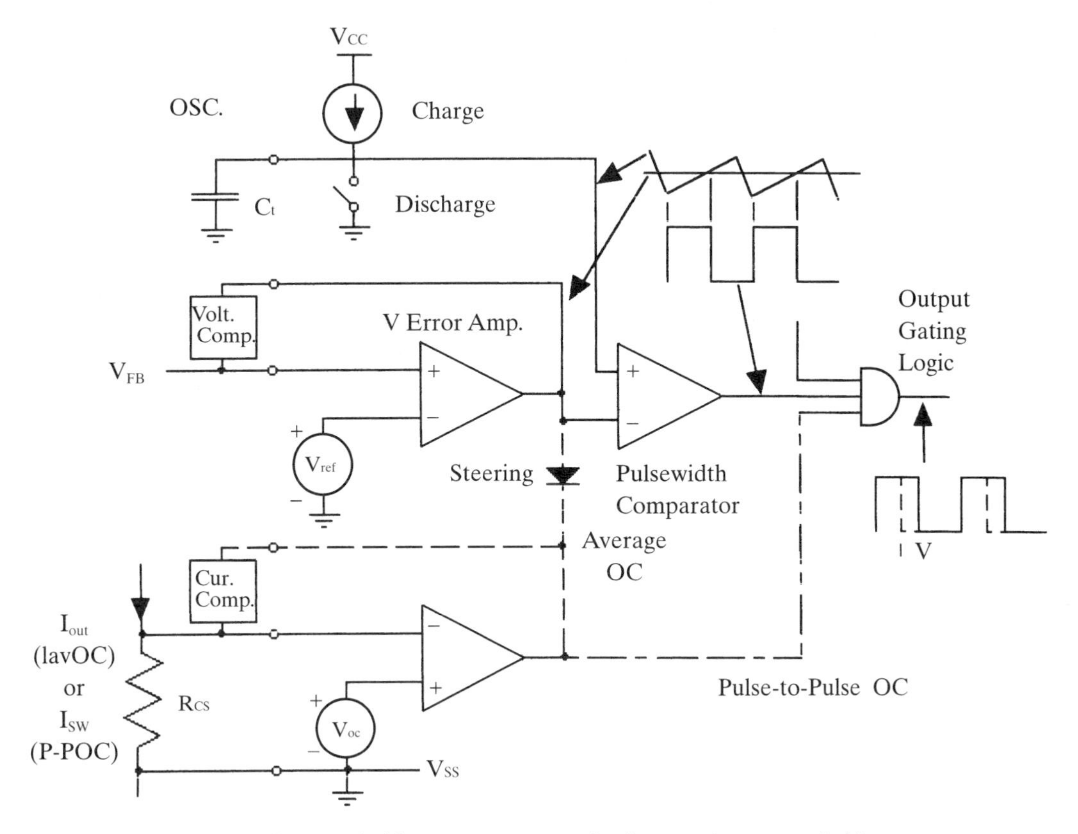

Figure 3–40 Voltage-mode control with average output and pulse-to-pulse current limiting.

sensed with a resistor in series with the load. The current signal may be amplified and inputted to a compensated current error amplifier. The current amplifier then senses when the output current approaches its preset limit and will override the voltage error amplifier and limit the current, if the current attempts to grow larger. Average current foldback has one inherent flaw when used by itself as an overcurrent protection, its response is too slow to prevent damage to the power switch if a sudden short circuit is applied to its output and cannot sense when the magnetic elements are entering the saturated state. These can create exponential currents in a matter of several microseconds and destroy the power switch.

The second method of overcurrent protection is *pulse-to-pulse overcurrent protection*. This method guarantees a maximum safe power switch current. A current sensing element (resistor or current transformer) is placed in series with the power switch(es). It views the instantaneous current flowing through the power switch and will instantly cutoff the power switch if a preset instantaneous current limit is exceeded. This circuit must be very fast and will protect the power switch from all forms of instantaneous overcurrents including core saturation. It is not a form of current-mode control since this protection limit is fixed and not influenced by external parameters.

One last form of voltage-mode control is very rudimentary. It could be called *hysteretic voltage-mode*. In this form of control, a fixed frequency oscillator is gated "ON" only when the output voltage has fallen to below a limit dictated by the voltage feedback loop. It is sometimes called "hiccup-mode" because the power switch occasionally bursts on and then returns to a constant off state.

There are a few control ICs and integrated switching power supply ICs that have this mode of control. It produces a fixed amount of ripple on the output voltage which varies in frequency in proportion to the load current.

Current-mode control

Current-mode control is best used in topologies where the linear slopes within the current waveforms are higher. This would be the boost-mode topologies such as boost, buck-boost, and flyback.

Current-mode methods control the peak (and sometimes the minimum) current excursion points flowing though the power switch. This equates to the excursions of the flux density within the magnetic core. Essentially the magnetic behavior of the core is being regulated. The most common form of current-mode is "turn-on with clock." This is when a fixed frequency oscillator sets a flip-flop and a high speed current comparator resets the flip-flop. The "1" state of the flip-flop is when the power switch is conducting.

The threshold for the current comparator is set by the output of the voltage error amplifier. If the voltage error amplifier indicates that the output voltage is too low, then the current threshold is raised to allow more energy to reach the load. The converse is true too.

Current mode control has an inherent overcurrent protection. The high-speed current comparator provides pulse-to-pulse current limiting. This form of protection is a *constant power form of overload protection* (see Section 3.11). This form of protection folds back the current and voltage to maintain a constant power into the load. This may not be optimum for all products, especially where the typical failures slowly increase the failure current. Another form of overload protection can also be placed in the circuit.

Another form current-mode control is called *hysteretic current-mode control*. Here both the peak and the valley currents are controlled. This is obviously better for continuous-mode forward for boost converters. It is somewhat complicated to set-up, but it does offer very fast response times. It is not a very common method of control and its frequency varies.

Other modes of control

Some IC makers today take many liberties in creating new modes of control or switching control methods at some operating points to enhance the overall efficiency of the target power supply. This practice can be confusing and may not operate in all applications other than the application it was design. For example, some buck controller ICs will allow the inductor to enter the discontinuous-mode, by dropping the frequency of operation. The feedback loop stability may change, so they use some obscure control method to compensate for the anticipated instability.

Voltage Hysteretic Control. This is what people used to call "hiccup-mode." A simple comparator is used to view the output voltage. If the voltage falls below a certain limit, the PWM loop turns on for a period of time until this limit is surpassed (plus some hysteresis voltage). It guarantees that the output voltage ripple is equal to or greater than the amount of hysteresis voltage in the control circuit.

Variable Frequency Control. At light loads, fixed frequency control methods loose efficiency because of the fixed switching losses. Some controllers will

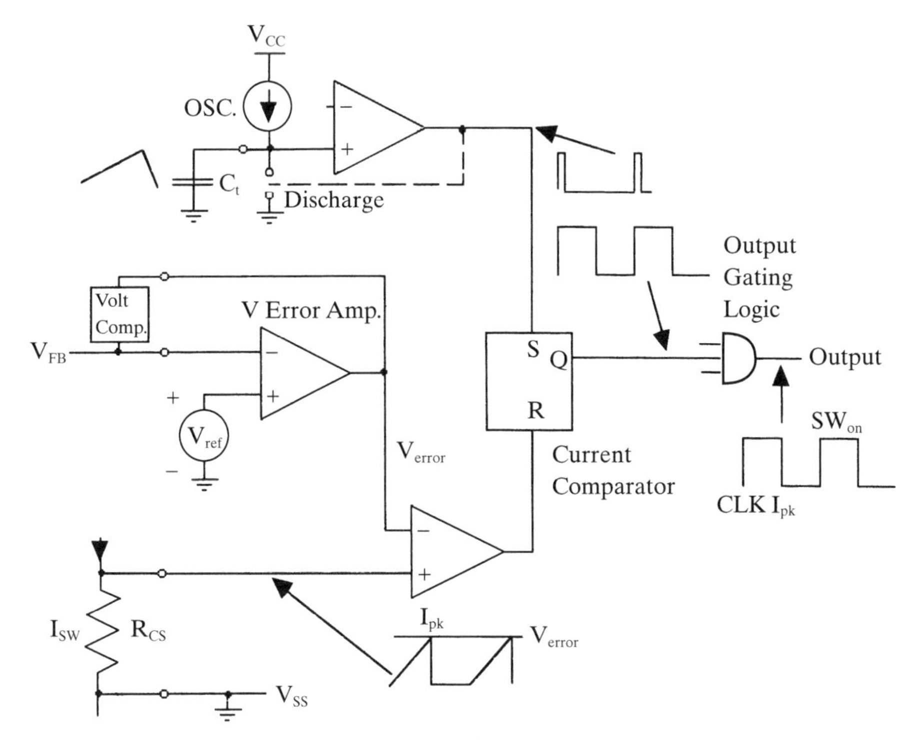

Figure 3–41 Turn-on with clock current-mode control.

switch to a variable frequency clock at light loads, but retain the same method of control.

Summary. Be very careful in selecting a control IC for your application!

3.9 Designing the Voltage Feedback Circuit

The only function of the voltage feedback loop is to hold the output voltage(s) at a constant value. Complications arise in areas such as transient load response, accuracy of the output(s), multiple outputs, and isolated outputs. All of these individually can be nightmares for the designer, but if the design approaches are understood then each factor can easily be satisfactorily addressed.

The heart of the voltage feedback loop is a high-gain operational amplifier which is called an *error amplifier*, which is nothing more than a high-gain amplifier that amplifies the difference between two voltages and creates an error voltage. In power supplies, one of these voltages is a reference voltage and the other represents the level of the output voltage. The output voltage is typically divided down to the voltage of the voltage reference prior to being presented to the error amplifier. This creates a "zero error" point for the error amplifier. If the output deviates from this "ideal" value, the error voltage output of the amplifier changes significantly. This error voltage then is used by the power supply to provide a correction to the pulsewidth to bring the output voltage back to its ideal value.

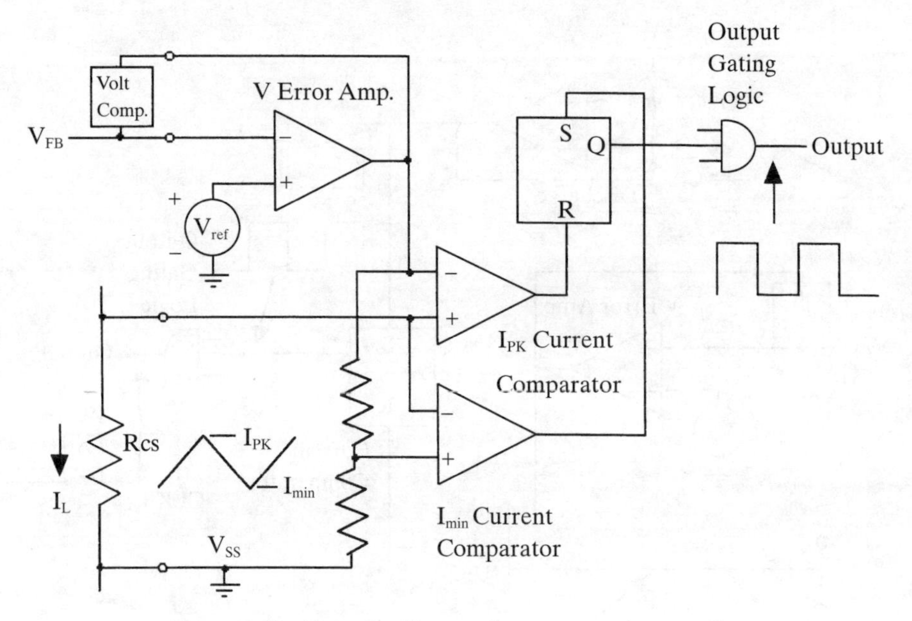

Figure 3–42 Example of hysteretic current-mode controller.

The major design issues surrounding the error amplifier are that it should have a high gain at dc which promotes good output load regulation and have a good high frequency response which promotes good transient load response. *Output load regulation* is how closely the output voltage is maintained with a change in the load of the sensed output. *Transient response* is how quickly the output voltage returns to its rated value after a step response to the load has occurred. These issues come under the area of feedback loop compensation, which is covered in detail in Appendix B.

An example of an elementary voltage feedback application is the nonisolated, single-output switching power supply. If we neglect the error amplifier compensation, then the design is quite simple. Let us examine a situation where a 5 V output is regulated and a 2.5 V reference is provided within the control IC. This can be seen in Figure 3–43.

To begin the process, one decides how much sense current is to be drawn through the output voltage resistor divider. For the sake of reasonable values to be calculated for the error amplifier compensation values, resistance values in the range of 1.5 to 15 K should be used in the upper leg of the resistor divider. Let us use a sense current of 1 mA as the resistor divider sense current. This makes the lower resistor in the divider (R_1)

$$R_1 = 2.5\,\text{V}/.001\,\text{A} = 2.5\,\text{K}$$

The accuracy of the output voltage is directly affected by the tolerance of the resistors used in the voltage divider and the accuracy of the voltage reference. All of the tolerances add to determine the resulting accuracy. That is, if two one percent resistors are used within the divider, and a two percent reference was also used, a four percent accuracy can then be expected in the final output voltage. Some additional error is introduced by the input offset voltage of the amplifier. Its error contribution is the value offset voltage divided by the divide ratio of the resistor divider. So, if an amplifier has a 10 mV maximum offset

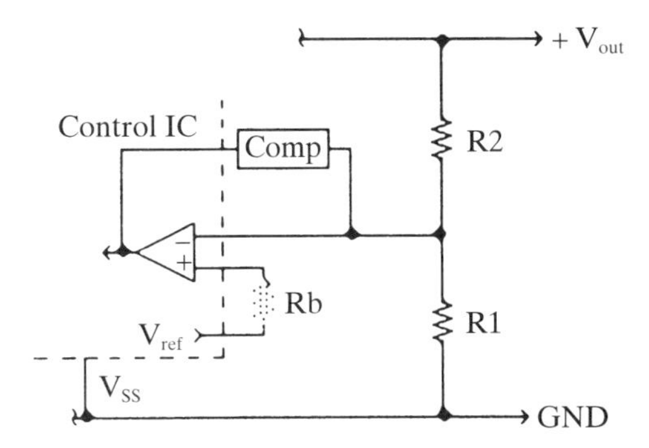

Figure 3–43 A nonisolated voltage feedback circuit.

voltage in this design example, one can expect a output voltage error of 20 mV that will drift with temperature.

Continuing the design example, let us use the closest 1 percent resistor value –2.49 K. This makes the actual sense current

$$I_S = 2.5\,\mathrm{V}/2.49\,\mathrm{K} = 1.004\,\mathrm{mA}$$

The upper resistor in the resistor divider (R_2) is then

$$R_2 = (5.0\,\mathrm{V} - 2.5\,\mathrm{V})/1.004\,\mathrm{mA} = 2.49\,\mathrm{K\,ohms}$$

That completes its set-up. Later, the compensation around the amplifier needs to be done to set the dc gain and the bandwidth performance.

If the power supply has multiple outputs, then cross-regulation of the outputs is a concern. Usually only one or several outputs can be sensed by the voltage error amplifier. The unsensed outputs can then only be regulated by the intrinsic cross-regulation abilities of the transformer and/or output filters. This can be quite poor, meaning that a change in the sensed output's load causes the unsensed outputs to change significantly. Conversely, if an unsensed output's load is changed, it is not adequately sensed through the coupling in the transformer to the sensed output, to have it exhibit good regulation.

To greatly improve the cross-regulation of the outputs, one can sense more than one output voltage; this is called *multiple output sensing*. It usually is not practical to sense all of the outputs and it really is not needed. An example of the improvement of cross-regulation can be seen in a typical multiple output flyback converter with a +5 V, a +12 V and a –12 V output. When the +5 V output goes from half-rated load to full load, the +12 V goes to +13.5 V, and the –12 V goes to –14.5 V.

This indicates the poor intrinsic cross-regulation capabilities of the transformer, which can be slightly improved by using filar winding techniques described in Section 3.5.9. If the +5 V and the +12 V outputs were both sensed, and the +5 V output was then loaded as described previously, then the +12 goes to +12.25 V, and the –12 V goes to –12.75 V.

Multiple output sensing is done by using two resistors in the top of the voltage sensing resistor divider. The top-end of each resistor goes to a different positive output voltage as seen in Figure 3–44.

The center node of the resistor divider becomes a current summing node where a portion of the total sense current comes from each of the sensed output voltages. The higher power output, and usually the output that requires a tighter output regulation, require the majority of sense current. The lighter loaded

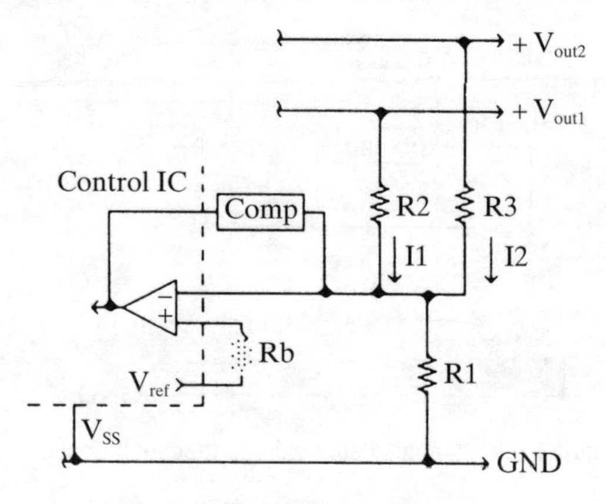

Figure 3–44 Multiple sense outputs.

output requires the balance of the sense current. The percentage of sense current from each indicates how well regulated each output is.

Let us again examine the power supply with +5 V, +12 V, and −12 V outputs. Since +/−12 V loads usually provide power to op amplifiers that are relatively immune to variations on their Vcc and VEE supply lines, their voltage regulation can be looser. Using the same situation as the first example in this section, R_1 is 2.49 K and a sense current of 1.004 mA.

The first step is to assign a current split. The less sense current drawn from a particular output, the poorer its output regulation. Let us assign a current split of 70 percent for the +5 V and 30 percent for the +12 V. R_2 becomes

$$R_2 = (5.0\,\text{V} - 2.5\,\text{V})/(0.7)(1.004\,\text{mA}) = 3{,}557\,\text{ohms}$$
$$R_2 = 3.57\,\text{K (closest value)}$$

The resistor to the +12 V (R_3) is

$$R_3 = (12\,\text{V} - 2.5\,\text{V})/(0.3)(1.004\,\text{mA}) = 31.5\,\text{K ohms}$$

The closest value is 31.6 K.

Improvements in all combinations of loading are experienced when the outputs are multiple sensed.

The last arrangement of voltage feedback is the *isolated feedback*. This is used when the input voltage is considered lethal to the operator of the equipment (>42.5 VDC). The two accepted methods of electrical isolation are *optical* (optoisolator) or *magnetic* (transformer). This section will talk about the more common method of isolation, when an optoisolator is used to isolate the lethal portions of the circuit from the operator portion. The optoisolator's C_{trr} (current transfer ratio (or I_{out}/I_{in})) drifts with temperature, can degrade slightly with age, and typically has wide tolerance from part to part. C_{trr} is the current gain of an optoisolator measured in a percentage. To compensate for these variations in the optoisolator and to eliminate the need for a potentiometer, the error amplifier should be placed on the secondary side (or input) of the optoisolator. The error amplifier will sense the deviation in the output caused by the optoisolator drift and adjust the current accordingly. The schematic of the typical isolated feedback circuit is shown in Figure 3–45.

The usual choice for a secondary error amplifier is the TL431 which has a temperature compensated voltage reference, and an amplifier within a three-leaded package. It does need a minimum of 1.0 mA continuous current flowing

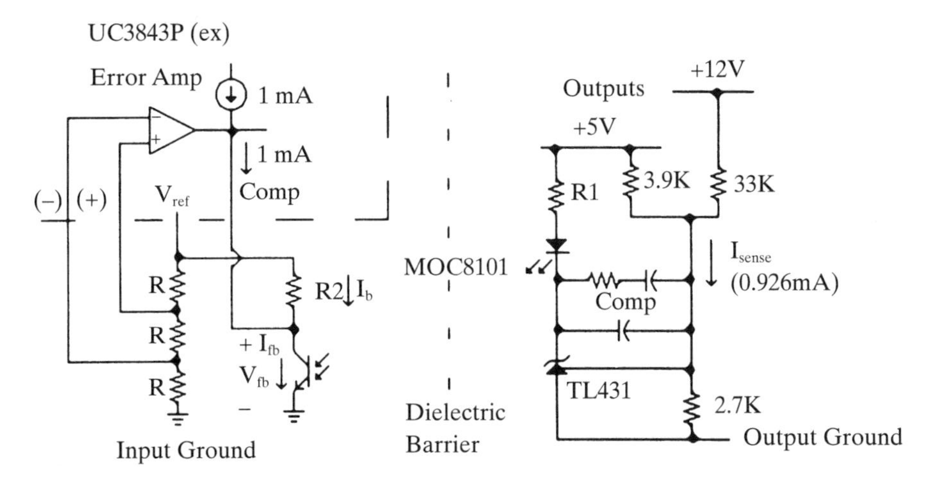

Figure 3–45 Example of an optoisolated voltage feedback circuit.

into its output pin for proper operation, the output signal will then add to that bias current.

In this example, the error amp on the control IC (a UC3843AP) is disabled by wiring the inputs of the error amp so that the output is guaranteed to be high. The values of R are not important (say 10 K each). The compensation pin has a 1 mA current source internally and has a "high" voltage of +4.5 V for the full output condition.

The network that sets the output pulsewidth on the compensation pin is a current summing network. R_1 guarantees that the operating current from the TL431 coupled through the optoisolator, does not load the 1 mA internal pull-up current source in the control IC and that +4.5 V is achieved on this pin when full output pulsewidth is desired. This worst-case minimum current at full output is

$$I_{fb(min)} = I_{cc(max)} \cdot C_{trr(max)}$$
$$I_{fb(min)} = (1.2\,\text{mA})(130\%) = 1.56\,\text{mA}$$

This makes R_1 equal

$$R_1 = (0.5\,\text{V})/(1.56\,\text{mA} - 1.0\,\text{mA})$$
$$R_1 = 892\text{ ohms} - \text{make } 820\,\text{ohms (safety margin)}$$

The optoisolator must provide more current for the compensation pin to reach its minimum output point of +0.3 V. To do this the current transmitted from the optoisolator must be

$$I_{fb(max)} = (4.5\,\text{V} - 0.3\,\text{V})/(820\,\text{ohms}) = 5.12\,\text{mA}$$

Resistor R_2 can now be determined by adding the other maximum voltage drops of the LED of the optoisolator and the terminal voltage of the TL431:

$$R_2 = [5\,\text{V} - (1.4\,\text{V} + 2.5\,\text{V})]/(5.12\,\text{mA})$$
$$R_2 = 214\text{ ohms} - \text{make } 200\,\text{ohms (safety margin)}$$

The resistors used in the sensing of the output voltage are the same as used in the previous example using cross-sensing. Only the error amplifier compensation remains to be done to complete this section (Appendix B). I must caution the designer in this design: Tolerances and temperature drift play a very large part in the isolated feedback design and they must be accommodated in the part value calculations. Optoisolators, for instance the C_{trr}, can vary over a range of 300 percent, which may necessitate a potentiometer to be added to the circuit. Some optoisolators are sorted by their manufacturers to a narrower C_{trr} range, but it is rare and not liked by them. The voltage reference should also be a temperature-compensated variety, such as provided in the TL431.

The issue of output accuracy from unit to unit usually requires that the reference used be trimmed to two percent variation or less, and the resistors within the resistor voltage divider be one percent. The accuracy of the outputs then are these tolerances added together plus any inaccuracies within the transformer turns.

There can be many variations on the voltage feedback design and only the most straightforward approaches have been shown as these are the most common types of approaches.

3.10 Start-up and IC Bias Circuit Designs

The start-up and bias supply circuit provides the operating voltage for the control IC and power switch drive sections. This circuit is sometimes called the *bootstrap start-up circuit*. Since all of the power drawn and delivered by this circuit is considered a loss, it is important to keep its overall function as efficient as possible.

The bootstrap start-up circuit gains in importance for higher input voltages. For input voltages greater than 20 VDC where the controller IC and the power switch cannot directly use the input voltage for their own power, a start-up/bias circuit must be used. Its function is basically that of a shunt or series linear regulator in order to provide a relatively stable voltage for the controller and power switch drive circuits.

For starting a power supply from a totally unpowered state, such as experienced when input power is first applied to the power supply, current must be drawn from the input power line. The voltage rating of the start-up circuit must be greater than the highest anticipated input voltage including any voltage surges that might get passed the input EMI filter section of the power supply. For this circuit, a little thought must be given to its desired functionality. There are several common functions that the start-up circuit can perform and its function should be appropriate for the operational requirements of the overall system.

1. Full functionality of the control/power switch circuit during a short-circuit on the power supply's output with immediate return to operation upon the removal of the short.
2. The power supply enters a *hiccup* restarting mode of operation during short-circuits and restarts upon the removal of the short-circuit.
3. Enter a complete shut-down state thus turning off the system during a short-circuit. The input power must be turned off and then back on in order to restart the power supply.

The first two modes of operation are the common approaches to the start-up circuit's functionality and it is recommended where the system has removable sections such as telephone systems or card-cage systems, or where routine service is expected and service people may inadvertently create a short-circuit within the load. The shut-down method of functionality is appropriate for important instruments where erroneous operation can be detrimental to the instrument or its operator.

In products where a little added loss is not important, a simple Zener shunt regulator is typically the approach as seen in Figure 3–46. Here the start-up current is constantly being drawn from the input power line, even during steady-state operation. If the start-up current is less than the current required to operate the IC and driver (~0.5 mA), then the power supply has a hiccup mode of recovery. If the start-up current is enough to power the circuit (~10 to 15 mA), then the power supply remains in a overcurrent foldback state during short-circuits and immediately recovers upon removal of the short-circuit. The difference is the amount of power lost within the start-up circuit during operation. The amount of hysteresis voltage in any LVI (low voltage inhibit) circuit within the control IC affects the hiccup restart of the supply. An IC supply bypass capacitor of 10 μF or more is needed to store enough energy to get the supply started before its voltage drops and resets the LVI. In general, the larger the hysteresis voltage, the more certain the supply will start on the first attempt.

For off-line switchers, where continuous current drawn from the input line would represent a significant loss, cutting off the start-up current during steady-state operation is recommended. After the entire power supply has stabilized in steady-state operation, the IC and drive circuits can draw all of their power from the auxiliary winding on the transformer. Here the conversion efficiency is about 75 percent as compared to the 5 to 10 percent. This method is shown in Figure 3–47 is such a circuit. The circuit is a high-voltage current-limited, linear regulator, and cutting off its start-up current is accomplished by reverse-biasing the diode on the emitter and the base-emitter junction during steady-state operation. The small signal transistor must have a $V_{CEO(SUS)}$ greater that the maximum input voltage. The collector resistor sustains nearly all the loss. Only a small bias current continues to flow into the base of the transistor and Zener diode.

Once again, the designer can choose whcther to make the supply operate in the hiccup mode restarting or to have the control and driver section continue to operate during a short-circuit condition. Choosing the value of the collector resistor for 0.5 or 15 mA current selects the respective operation.

A variation on this method is called the *shut-down on overcurrent* circuit shown in Figure 3–48. Here the start-up circuit is a discrete, high-voltage, one-shot circuit that is active only during the start-up sequence, but turns completely off thereafter. If an overcurrent foldback condition is experienced, the IC and drive circuits no longer have a voltage supply from which to draw current. This

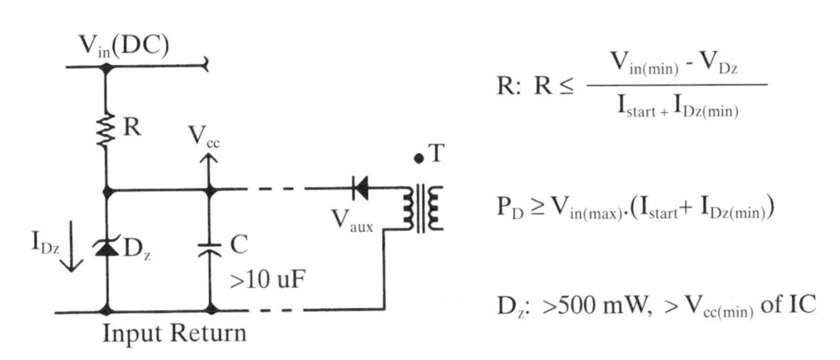

Figure 3–46 The Zener diode controller supply regulator.

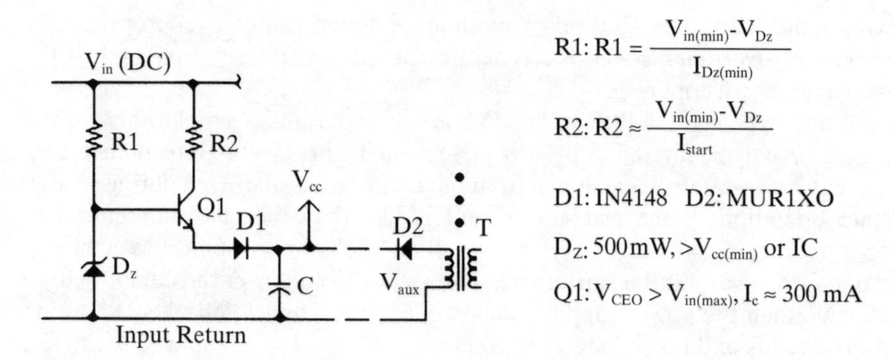

Figure 3–47 The high-voltage linear regulator bootstrap start-up circuit (used only at start-up and foldback periods).

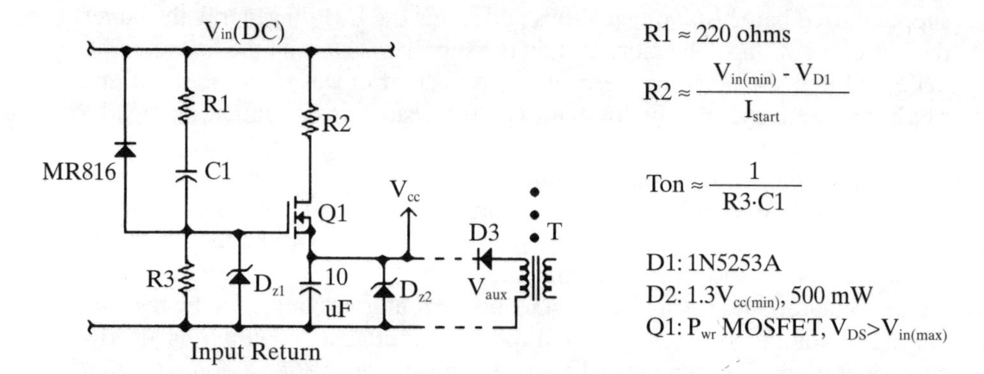

Figure 3–48 The overcurrent shut-down circuit.

turns off the entire power supply until the input voltage is turned off and then turned back on.

The designs previously described have worked particularly well for me, but there are many varying implementations along the same principles of operation. If a different design approach is being contemplated, keep one thing in mind—the start-up period of the switching power supply is the most traumatic period in its operational life. That is, more failures occur during the start-up period than any other period of operation. The sequencing of the various bias supplies is very critical to the supply. The power switch drivers should always be fully powered before the control IC begins the switching process. If it is not, the power switches will not operate in a saturated manner and may therefore fail due to overdissipation.

Another factor is the rated breakdown voltage of resistors. For a one-quarter watt resistor, the rated breakdown voltage is 250 VDC. The one-half watt resistor is 350 VDC. To design safely, two resistors in series should be used on all branches that are connected to the input line in off-line converters.

3.11 Output Protection Schemes

Protection schemes for the outputs of a power supply should be developed carefully to complement the use and functionality of the end product. Although these circuits are used only in the event of an unusual system failure, one does not want additional damage or additional user aggravation to be derived from the activation of a protection function. One should also consider the nature of the loads, their likely failure modes and causes, and the limits beyond which further damage is caused.

Protection of the load and of the power supply from failures in the load should be an important consideration within all switching power supply designs. It is important to know what failures are likely to occur within the switching power supply and the load. An exercise that is frequently required in military designs is what is called a FMEA (failure modes and effects analysis), where each component is hypothetically assumed to fail open-circuited and then short-circuited. With such failures, how does each failure affect the other sections of the circuit? This anticipation of failures can make a power supply design robust. It is the responsibility of the power supply designer to provide protection to the load circuitry from anomalies encountered from the input line and failures within the supply and the load circuitry. Often, protection schemes can be cascaded to provide redundant protection in the event of a failure in a protection circuit. A fuse or a circuit-breaker usually provides such a back-up function.

When considering the method(s) of protection used within a power supply and any power busing system, it is important to consider the end user of the product and the product's function. The repair philosophy of the product should be carefully reviewed. If the product will have service people routinely "with their fingers in the circuits," then a circuit-breaker, a self-restarting bootstrap start-up circuit, and a simple overcurrent foldback circuit is the desired combination of circuits. If the product's function is noncritical and its operation can be easily done without, then a fuse, an overcurrent shut-down type of start-up circuit, and overcurrent foldback and/or crowbar circuit may be desired. Some protection methods may necessitate the equipment being brought into a repair shop for checking prior to being reused. The types of protection philosophies fall into three categories:

1. Repair after failure (fuses, fusistors, etc.).
2. Recovery after failure (circuit-breakers, overcurrent foldback, overvoltage override circuits, etc.).
3. Shut-down after failure, but recoverable when the fault is removed (overcurrent shut-down type of start-up circuits, etc.).

Choosing the most appropriate combination of protection schemes can make a large difference in the customer's perceptions of the product and hence in its reputation and sales.

There are basically three types of overcurrent protection. These are shown in Figure 3–49. Current-mode control or any primary-side peak current controller offers a constant output power limiting when a short-circuit is "soft," but does eventually foldback both the voltage and current into hard shorts. When an overcurrent failure presents a lower and lower resistance to the supply, the output voltage decreases, but the output current can continue to increase. This may produce burning of printed circuit board traces and components. This reducing of the output voltage can be useful in conjunction with a shut-down on failure type of start-up circuit. Constant current limiting (see Figure 3–50) is produced when the voltage across a current sensing resistor is amplified and compared to a reference voltage. When the trip point is exceeded, the output current is held constant for any greater load. Output current must be sensed for this type of overcurrent protection. Overcurrent foldback limiting only occurs when a small amount of the output voltage is used as a threshold reference for

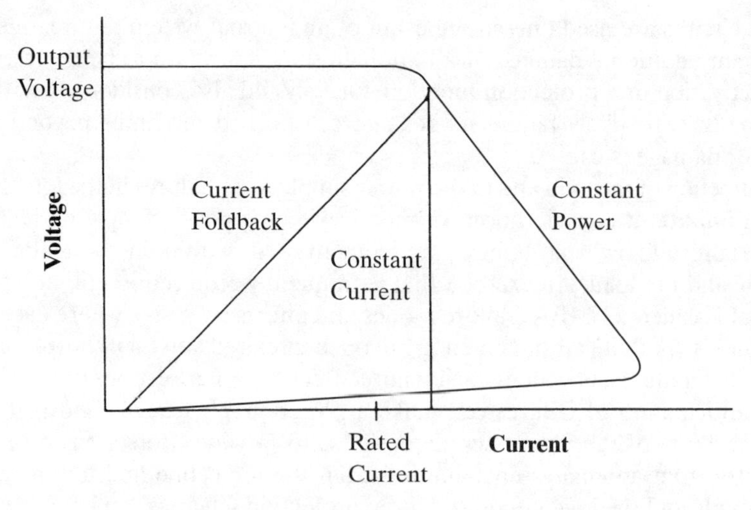

Figure 3–49 Types of overcurrent protection.

the trip point of the overcurrent knee. When the current through the current sense resistor is large enough to exceed the trip point, the voltage is folded back thus also folding back the output current in proportion to the output voltage. This protects the load circuitry from burning (see Figure 3–51).

Overvoltage protection on the output lines can take two forms, the brute force approaches, and the overvoltage override approaches. The *brute force* approaches are the overvoltage clamps and crowbars, as seen in Figure 3–52. These approaches assume that the power supply has failed and cannot limit the current into the load. The most problematic topology is the buck topology, where a shorted series pass unit directly connect the input to the output. The buck converter must definitely have a crowbar circuit (if I_{out} is greater than 1 A) or an overvoltage Zener clamp (if I_{out} is less than 1 A) on its output. For all the other topologies, especially the transformer-isolated topologies, this is not a possibility and therefore a crowbar is a nuisance. Overvoltage Zener diodes on the outputs can serve as useful insurance in the case of a voltage feedback loop failure or poor output cross-regulation. An input fuse or circuit breaker is required with these forms of overvoltage protection.

The overvoltage override methods assume that the power supply is still operating and the voltage feedback has become open-circuited, or that one of the outputs has become light loaded and its voltage rises above the maximum specification. These methods have a separate comparator or transistor and resistor dividers wired to each output. The comparator or transistor would then override the error amplifier. These are shown in Figure 3–53.

Choosing the form(s) of protection is always is a matter of economics and board space. Be creative with your approaches to protection, but check them out carefully under all operating conditions.

3.12 Designing the Input Rectifier/Filter Section

The input rectifier/filter circuit plays a largely unappreciated role within the switching power supply. The typical input rectifier/filter circuit is composed of three to five major subsections: an EMI filter, possibly a start-up surge current

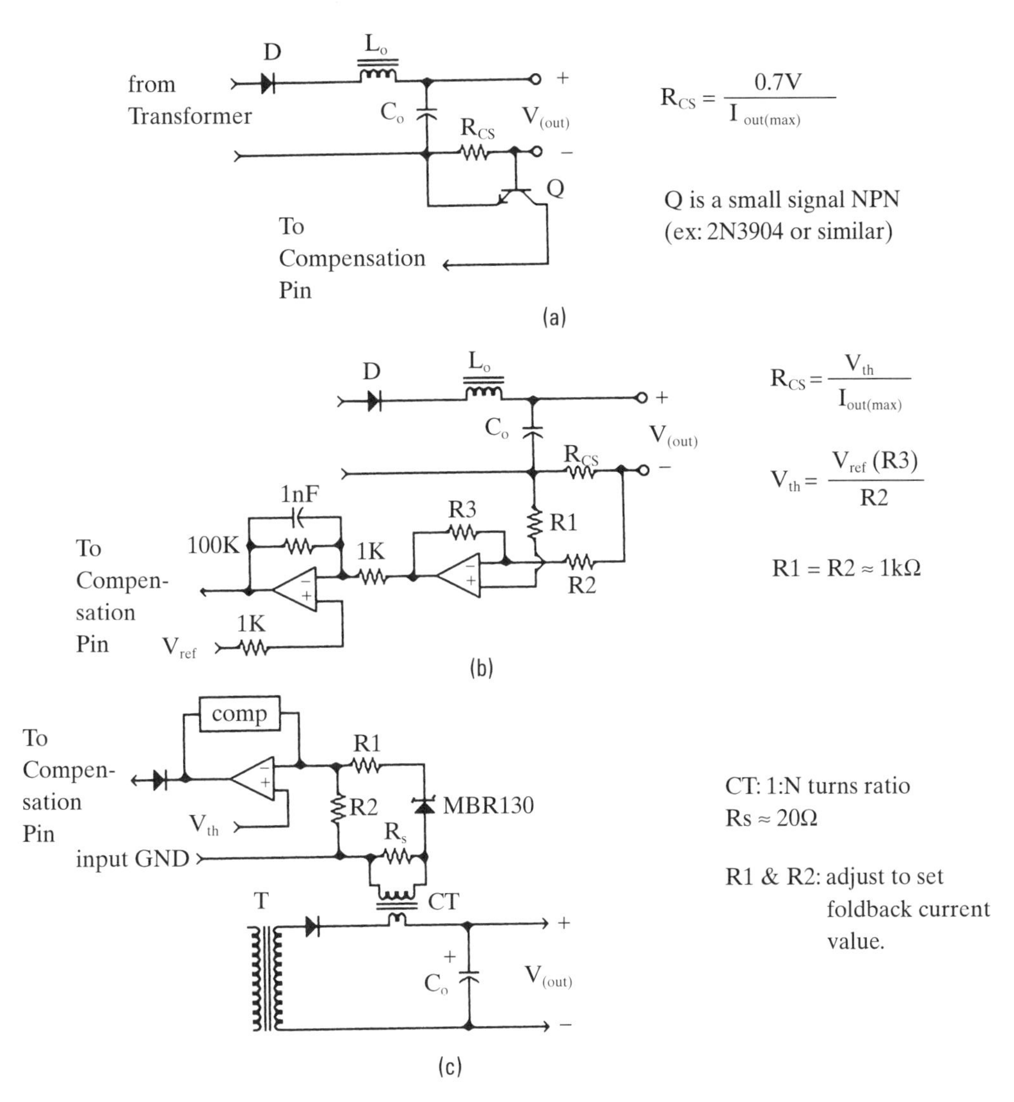

Figure 3–50 Methods of constant current limiting: (a) discrete overcurrent limiting (constant-current limiting); (b) precision resistive current-sensing overcurrent protection (constant-current limiting); (c) use of a current transformer to sense ac current.

limiter, a surge voltage suppressor, a rectification stage (for off-line applications), and the input filter capacitor. Many new ac off-line power supplies are required to have power factor correction (PFC). For these applications, design the input stage described here, then refer to Appendix C. The designs of the typical input rectifier/filter circuits are shown in Figure 3–54 for both the dc and ac applications.

For off-line switching supplies, the first task is to select the input rectifiers. These are standard recovery diodes, such as the 1N400X rectifiers. The major parameters under consideration are average forward current (I_O), surge current (I_{FSM}), the dc blocking voltage (V_R), and the anticipated power dissipation (P_D). The surge current occurs at power-up and can be greater than 10 times the average RMS input current. It is caused by the fully discharged input filter capacitor taking the classic impulse of current when a step change in voltage

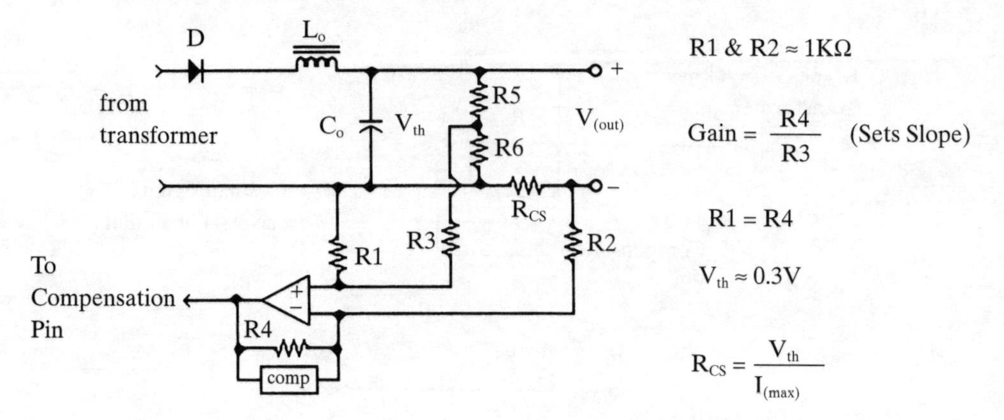

Figure 3–51 Overcurrent foldback.

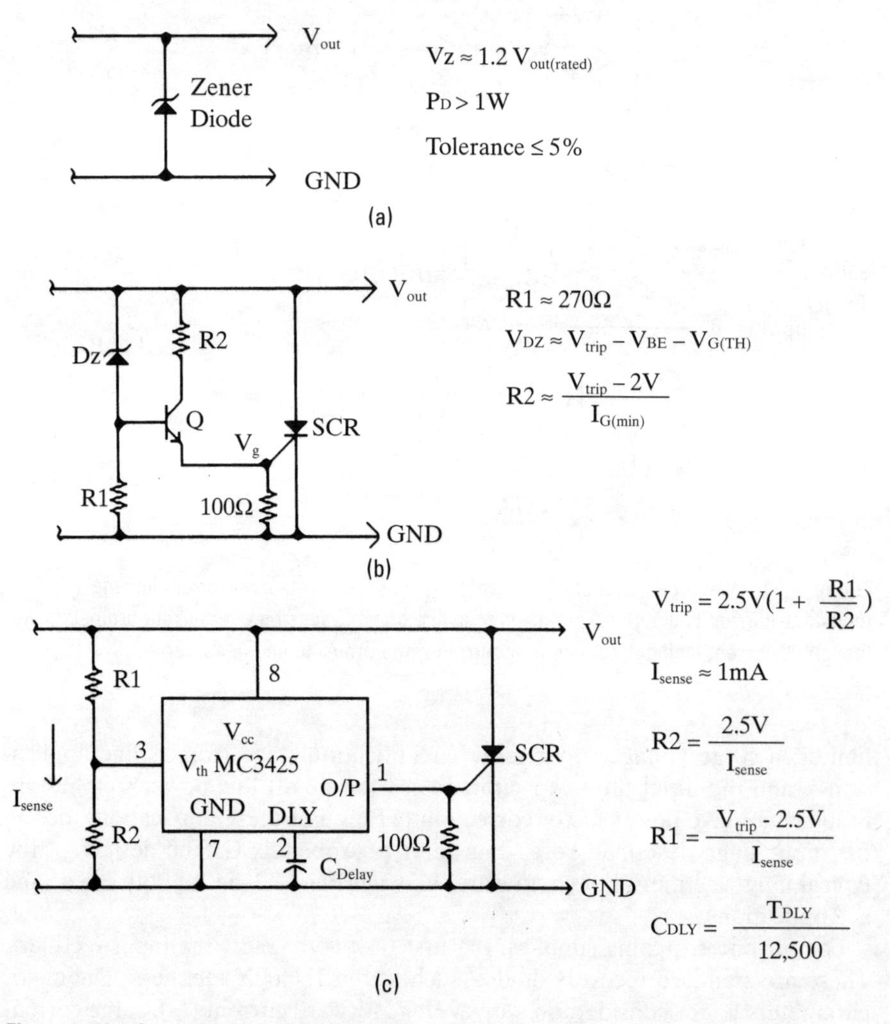

Figure 3–52 Overvoltage protection schemes: (a) Zener diode clamp; (b) overvoltage crowbar; (c) precision overvoltage crowbar.

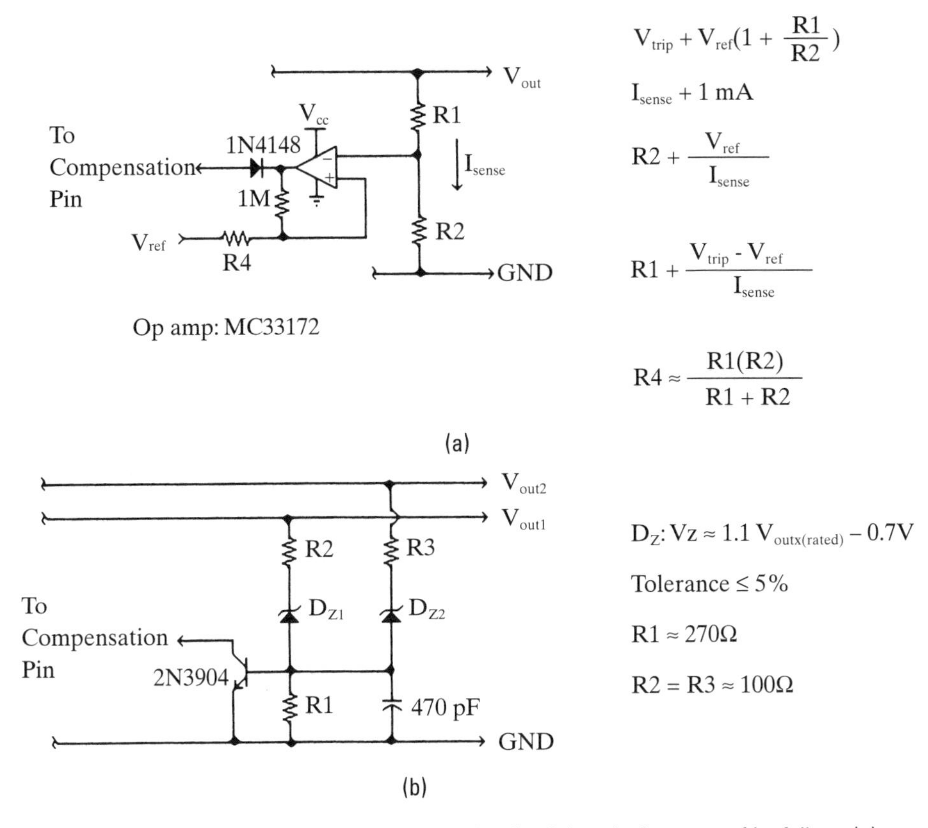

Figure 3–53 Methods of overriding the voltage feedback loop in the event of its failure: (a) overvoltage override circuit (when voltage loop goes open-circuited); (b) multiple-output overvoltage override circuit.

occurs on its terminals. A thermistor is usually added just following the EMI filter to protect the ac rectifiers. The "cold" value of the thermistor is typically between 6 and 12 ohms. After start-up, the thermistor heats up and should assume a "hot" resistance value of approximately 0.5 to 1.0 ohm.

The average current through the input rectifiers is a thermal consideration. Referring to Figure C–1 in Appendix C, you can see the actual waveform flowing through the nonpower factor-corrected ac rectifiers. The peak current can be as much as five times the average dc current flowing through the diode. This causes greater heating of the rectifier. To compensate for this, a higher current diode is selected to minimize the peak current forward voltage drop and to rid the die of the heat. In summary, the minimum diode ratings should be

$$V_R \geq 1.414 \cdot V_{in(p-p)(max)} \tag{3.51}$$

$$I_F \geq 1.5 \cdot I_{in(DC)(max)} \tag{3.52}$$

$$I_{FSM} \geq 5 \cdot I_F \tag{3.53}$$

The typical diodes for this application are

If the current is <1 A: 1N400X
<1.5 A: 1N539X
<3 A: 1N540X
<6 A: MR75X

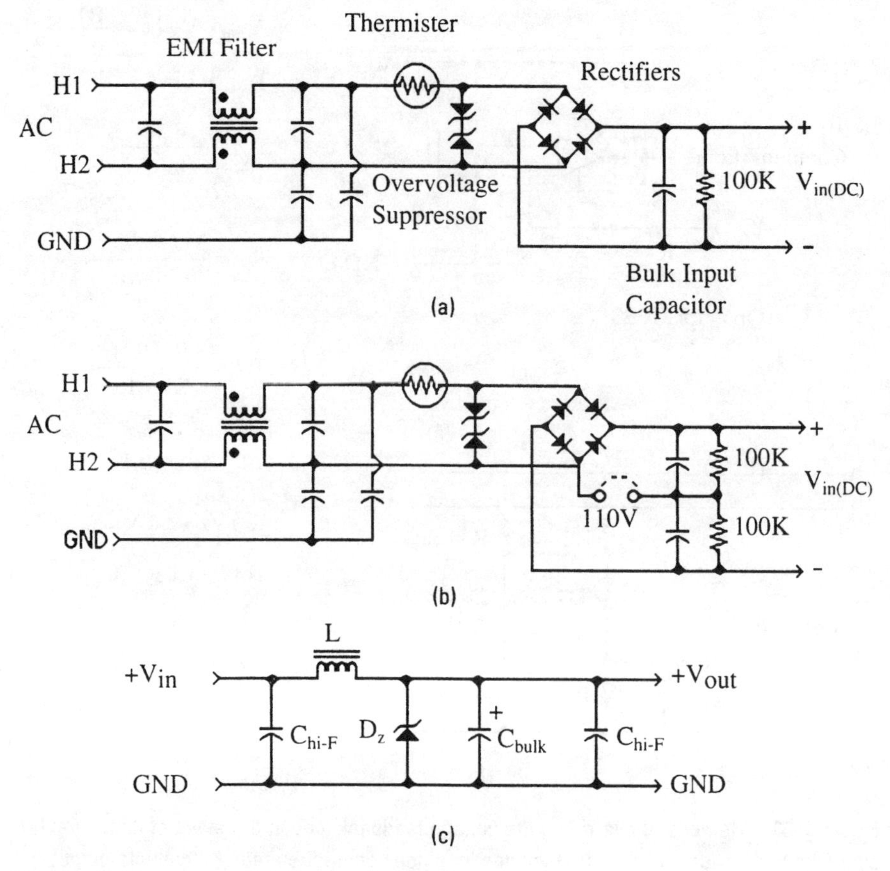

Figure 3–54 Typical ac and dc input filter circuits: (a) ac input filter circuit for a single or universal input power supply (common-mode EMI filter shown); (b) a voltage doubling ac input circuit for 110 V and 220 V ac inputs; (c) single dc bus input filter.

The next operation is to calculate the value of the input filter capacitor. The designer needs to decide how much ripple voltage the supply can tolerate on the dc input line. The less ripple voltage desired, the larger the capacitor, and the larger the surge current during the power-up period. There are three major considerations when selecting a capacitor for this function: the value to yield the desired ripple voltage, the voltage rating, and the ripple current rating of the capacitor.

The typical design goals for the ripple voltage are about five to eight percent of the peak rectified input ac voltage for ac off-line converters and about 0.1 to 0.5 volts peak-to-peak in dc-dc converters. The value of the bulk input capacitor can be found by

$$C_{in} = \frac{0.3P_{in(av)}}{f_{in}\left(V_{in(min)}\right)\left(V_{ripple(p-p)}\right)^{2}} \tag{3.54}$$

where: f_{in} is the minimum rated ac line frequency for off-line supplies.
V_{in} is the minimum peak rectified value of the ac line.
V_{ripple} is the peak-to-peak droop in voltage desired across the input capacitor.

The voltage rating of the capacitor should be

(off-line) $$V_W > 1.8V_{in(RMS)} \tag{3.55}$$

(dc-to-dc) $$V_W > 1.5V_{in(max)} \tag{3.56}$$

In ac off-line converters, aluminum electrolytic capacitors are used as the input filter capacitor. They have proven to be more rugged than any other type of capacitor in the hazardous environment existing on the ac line. Choosing the final capacitor depends mostly upon the expected operating temperature range, quality level, and case dimensions.

The input capacitors in dc-dc converters are much more critical. DC-DC converters typically draw much higher ripple currents at the switching frequency of the supply. These currents cause internal heating of the input filter capacitors and shorten their operating life, if they are not selected properly. The input filter capacitor needs to have a low ESR and a high ripple current rating. The entire current waveform seen on the power switch within the power supply is flowing into and out of the input filter capacitor. The input line, because of the existence of series impedances in the wiring, cannot provide the high frequency current pulses demanded by the power switch inside the switching power supply. The input capacitor serves a very important role of receiving charge at a low frequency from the input line and supplying charge to the power switch at high frequency. Therefore, the current experienced by the power switch is completely seen by the input filter capacitor.

The designer must convert the current waveform observed at the power switch into a worst-case RMS value. Converting a triangular or trapezoidal current waveform into an RMS value depends upon the waveforms peak amplitude and its duty-cycle. One can estimate the RMS value of a waveform by breaking a waveform into simpler waveforms whose RMS value is known. For example, a trapezoidal waveform can be broken into a rectangular wave whose RMS highest value is 50 percent of the peak value (50 percent duty cycle) and a triangular waveform whose highest value is about 33 percent of the peak value. These individual estimated values are then added to estimate the worst case total RMS value.

It is unusual to be able to find one capacitor to handle the entire ripple current of the supply. Typically one should consider paralleling two or more capacitors (n) of 1/n the capacitance of the calculated capacitance. This will cut the ripple current into each capacitor by the number of paralleled capacitors. Each capacitor can then operate below its maximum ripple current rating. It is critical that the printed circuit board be laid out with symmetrical traced to each capacitor so that they truly share the current. A ceramic capacitor (~0.1 μF) should also be placed in parallel with the input capacitor(s) to accommodate the high frequency components of the ripple current.

Ahead of this section is the EMI filter. The design of the single-line dc filter inductor is found in Section 3.5.7. This inductor has a relatively high value of dc current flowing through it and wants to isolate the high frequency switching noise from the input power bus.

In ac off-line applications the common-mode choke is typically used and its design procedure can be found in Appendix E. The filters resemble "pi" filters, but are actually bi-directional *L-C* filters. The important function is to filter the noise generated by the switcher before it exits via the input power lines.

The capacitors used for this function are high-voltage film or ceramic capacitors that exhibit very good high frequency characteristics. These capacitors range in value from 0.005 to 0.1 μF. Care must be taken as to the working voltage rating of the capacitors. Off-line converters must pass regulatory testing which places an extra voltage stress upon them. The test is called the dielectric withstanding voltage test or "HIPOT." Any capacitor placed between the input lines to the earth ground lead (green wire) must be able to withstand this test voltage. These voltages are: UL, 1700 VRMS (2500 VDC); for VDE, IEC, and CSA, 2500 VRMS (3750 VDC). To pass the regulatory agencies in the European commonwealth, special capacitors intended just for this application must be used. These capacitor families are tested and certified for these ac EMI filter applications.

The surge suppression section should be physically placed behind the EMI filter inductor and before the ac rectifiers (off-line) and before the input filter capacitor (dc-inputs). All surge suppressors need the series impedance of the EMI inductor to keep them from exceeding their instantaneous energy ratings. The EMI inductor drastically reduces the peak transient voltage and spreads it out over time, thus reducing the instantaneous effects of the transient upon the transient suppressor. This improves the survivability of the suppressor. Be aware that different technologies of surge suppressors have different series inherent resistance characteristics. The metal-oxide varistor (MOV) has a relatively high resistance when it is conducting. Semiconductor transorbs exhibit a lower resistance. The series resistance affects how much additional voltage will appear across its terminals during the surge period. For example, a 180 V MOV may rise to 230 volts at the peak of the transient. This should be considered when selecting both the input capacitor and the surge suppressor. MOVs, although inexpensive, do degrade after suppressing a few high-energy transients and exhibit a higher leakage current. The "trip" voltage of the surge suppressor should be higher than the maximum high input operating voltage specification of the supply so that it does not conduct during normal operation. For example, for the 110 VRMS line, a 180 to 200 V trip voltage is typically used.

A comment on RFI and this circuit: The EMI filter stage should be placed as close to the entrance of the power feed wire into the enclosure as possible. The conducted EMI from outside can be reradiated into the case if the wire length is too large before the filter. Conversely any RFI may be picked up by the long length of power cord inside the case and radiated outside where the regulatory testers will feel they've earned their pay by "nabbing" another errant product. Refer to Appendix E for noise control methods.

3.13 Additional Functions Normally Associated with Power Supplies

Additional functions are added to the basic switching power supply so that the supply's operation better complements the function of the overall product. The overviewed functions are those routinely found in some switching power supplies, but certainly are not all of the possible functions. These functions may include: synchronization, undervoltage alerts, input undervoltage inhibit, and partial supply shut-down.

3.13.1 Synchronization of the Power Supply to an External Source

Synchronization allows the switching power supply to be phase-locked with another circuit in the external product. This may be necessary for circuits which include CRTs (cathode ray tubes), analog-to-digital and digital-to-analog converters, etc. The noise generated by the switching power supply could interfere with these circuits. For example, lines may be displayed on a CRT at the instant when the power switch and retifiers turn on or off. Noise spikes picked up in the sensitive input signals of a comparator can cause it to compare at the wrong instant within an A-to-D converter. To solve this, it may be desirable to lock the clock within the supply to the timing of these sensitive circuits to try to place any noise spikes in a harmless point in time.

Some power supply control ICs have synchronization inputs for this purpose. For those ICs which have an oscillator, but not a synchronization pin, the circuit in Figure 3–55 can be used. The frequency on the IC must be set lower than the synchronization signal. The sync signal causes the oscillator to prematurely time-out.

3.13.2 Input, Low Voltage Inhibit

Many times it is desirable to shut the product off during *brownout conditions*. These are conditions where the input voltage falls below the minimum specification for normal operation. This condition can cause a voltage-mode controlled power supply to enter a type of "latch-up" mode where the supply would quickly jump to its maximum duty cycle and would no longer control the output regulation. This could be destructive to the supply and the loads when the input voltage returns to normal levels. Also, at the lower input voltages, the higher peak currents flowing through the power switches can cause them to fail due to overdissipation. To avoid this, a simple voltage comparator sensing the input line is needed as in Figure 3–56.

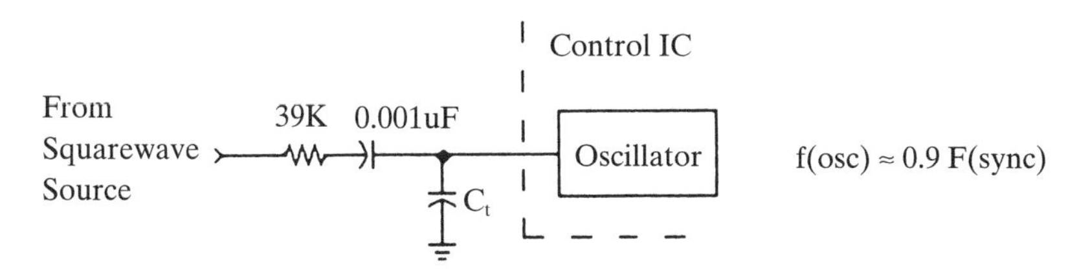

Figure 3–55 Synchronizing a switching power supply (no sync pin on the IC).

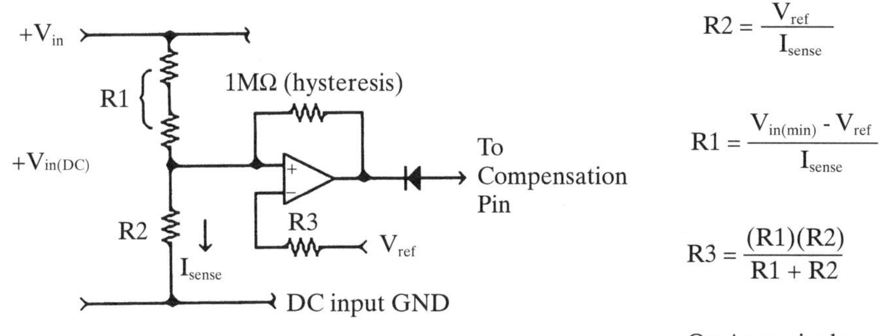

Figure 3–56 Input undervoltage shutdown circuit.

3.13.3 Impending Loss of Power Signal

For products that include microprocessors, floppy or hard disk drives, or any other device where a sudden loss of power during its operation could cause damage to the product or injury to the user, it is recommended that a power fail signal be issued by the power supply. For a simple microprocessor system, a simple low voltage sensor on the +5 volt line may be sufficient. If the product includes any electromechanical devices which may require a finite time to institute an orderly shut down, a longer period of time may be required. This means sensing the power supply's input voltage and making the input filter capacitor larger so that the supply can maintain a regulated output longer. Caution must be practiced since large input filter capacitors will significantly increase the turn-on surge current through any input rectifiers, which can cause rectifier failures. This method may yield as much as 8 to 15 mS of additional time to shut the device down. These circuits can be seen in Figure 3–57.

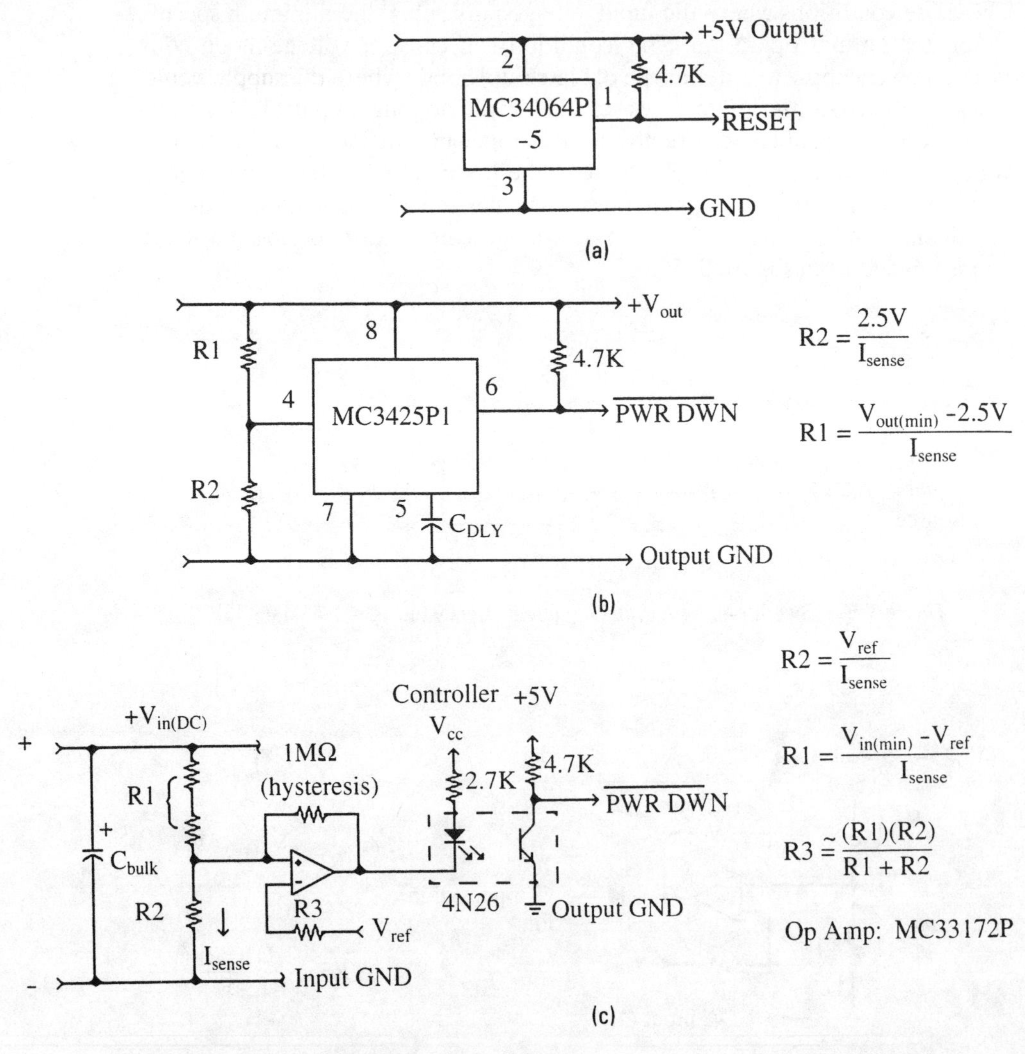

Figure 3–57 Power fail indicator/alert circuits: (a) a 5 V under-voltage indicator; (b) an under-voltage indicator for any voltage; (c) power-down signal derived from the input line (longest warming period).

3.13.4 Output Voltage Shut-down

Sometimes it is desirable to turn off a section of the product but leave other portions operating. This is typically the case in laptop computers for power savings or consumer appliance applications, such as CRT monitors, televisions, and stereo equipment.

Many consumer appliances must place themselves in the standby mode when the user has walked away from the product. This type of operation is called "green certified." This certification dictates that the product must use less than 5 watts of power while in the standby mode. The switching power supply alone would have quiescent operating current greater than 5 watts, so a very low power switching power supply, with much lower quiescent currents, can be used to power only that portion of the product needed to reactivate the product upon demand.

In the case of power factor corrected ac power supplies, the main converter can be turned off and the circuit to remain active could use the auxiliary voltage on the power factor transformer. The power factor circuit would run at a much reduced pulsewidth and/or frequency.

The portable equipment market usually needs to turn off any temporarily unused circuits to extend its battery life. In this case, a simple series MOSFET switch can be used. The RDS(ON) should be as low as possible to minimize the power dissipation within the MOSFET switch. These approaches are shown in Figure 3–58.

3.14 Laying Out the Printed Circuit Board

The final stage in the design of any switching power supply is the physical design of the printed circuit board (PCB). If it is designed improperly, the PCB could contribute to the supply's instability, and radiate excessive electromagnetic interference (EMI). The role of the designer is to insure a good PCB design by understanding the physical operation of the circuit.

Switching power supplies contain signals that are rich in high frequencies and any PCB trace can act as an antenna. A trace's length and width affect its resistance and inductance, which in turn affects its frequency response. Even traces containing dc signals can pick-up RF signals from neighboring traces and cause circuit problems or even reradiate the interfering signal again. All traces that carry ac current must be made as short and wide as possible. That means that any power-handling components that connect to a trace and to other power traces must be located adjacent to one another. The length of a trace is directly proportional to the amount of inductance and resistance the trace will exhibit. Its width is inversely proportional to the trace's inductance and resistance. The length dictates the wavelengths to which the trace will respond. The longer its length, the lower the frequency the trace can receive and transmit and the more RF energy it is susceptible to.

3.14.1 The Major Current Loops

There are four current loops inside of every switching power supply. Each of these loops must be kept separate from one another. They are listed in order of their importance to a good PCB layout.

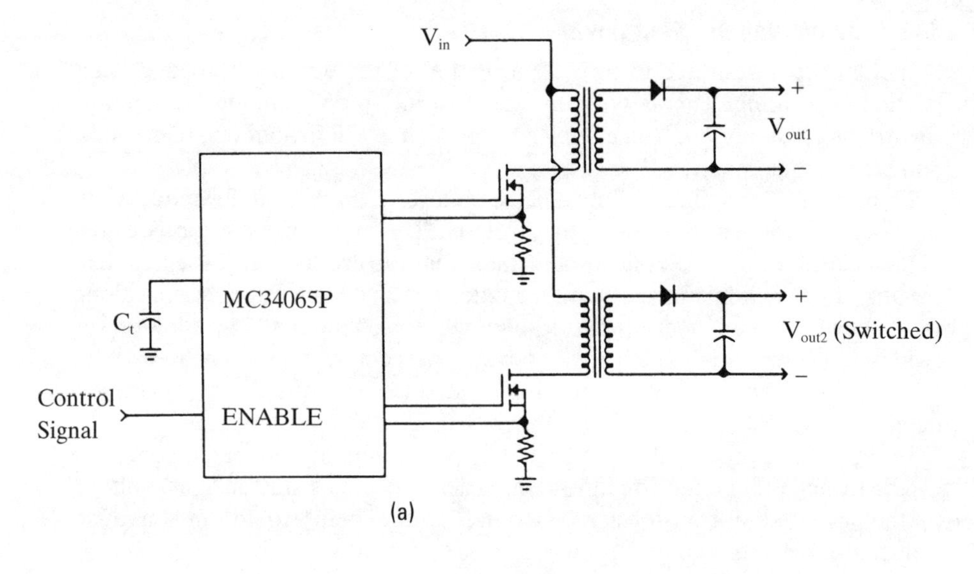

(a)

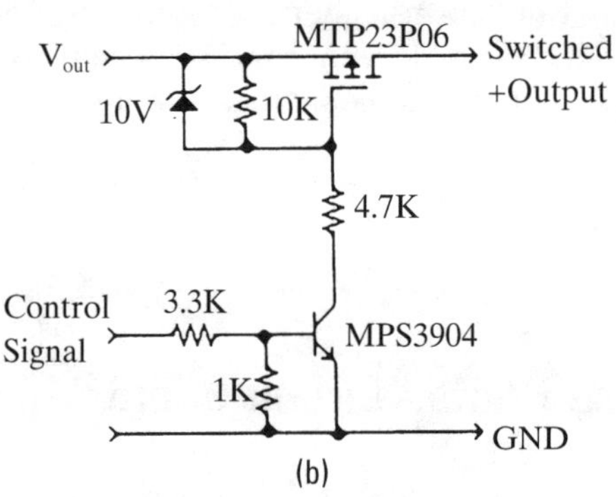

(b)

Figure 3–58 Methods of switching power buses: (a) dual switching power supply with enable; (b) a dc power switch.

1. The power switch ac current loop.
2. The output rectifier ac current loop.
3. The input source current loop.
4. The output load current loop.

These loops can be seen in Figures 3–59 a, b, and c for the three major switching power supply topology types.

The input source and output load current loops typically do not present problems. The current within these loops is largely composed of dc flow with some small ac current summed onto the waveform. These two loops typically also have special filters to discourage the escape of ac noise into the surrounding environment. The terminals of the input and output filter capacitors should be the only places where the input and output current loops, respectively, connect to the power supply. The input loop charges the input capacitor with a near-dc current but is unable to supply the high frequency current pulses required by the switching power supply. The filter capacitor's main role is to perform this wide band, energy reservoir function. The output filter capacitor similarly stores the high frequency energy from the output rectifier and allows the output load loop to remove the energy in a dc fashion. Therefore, the terminals of the input

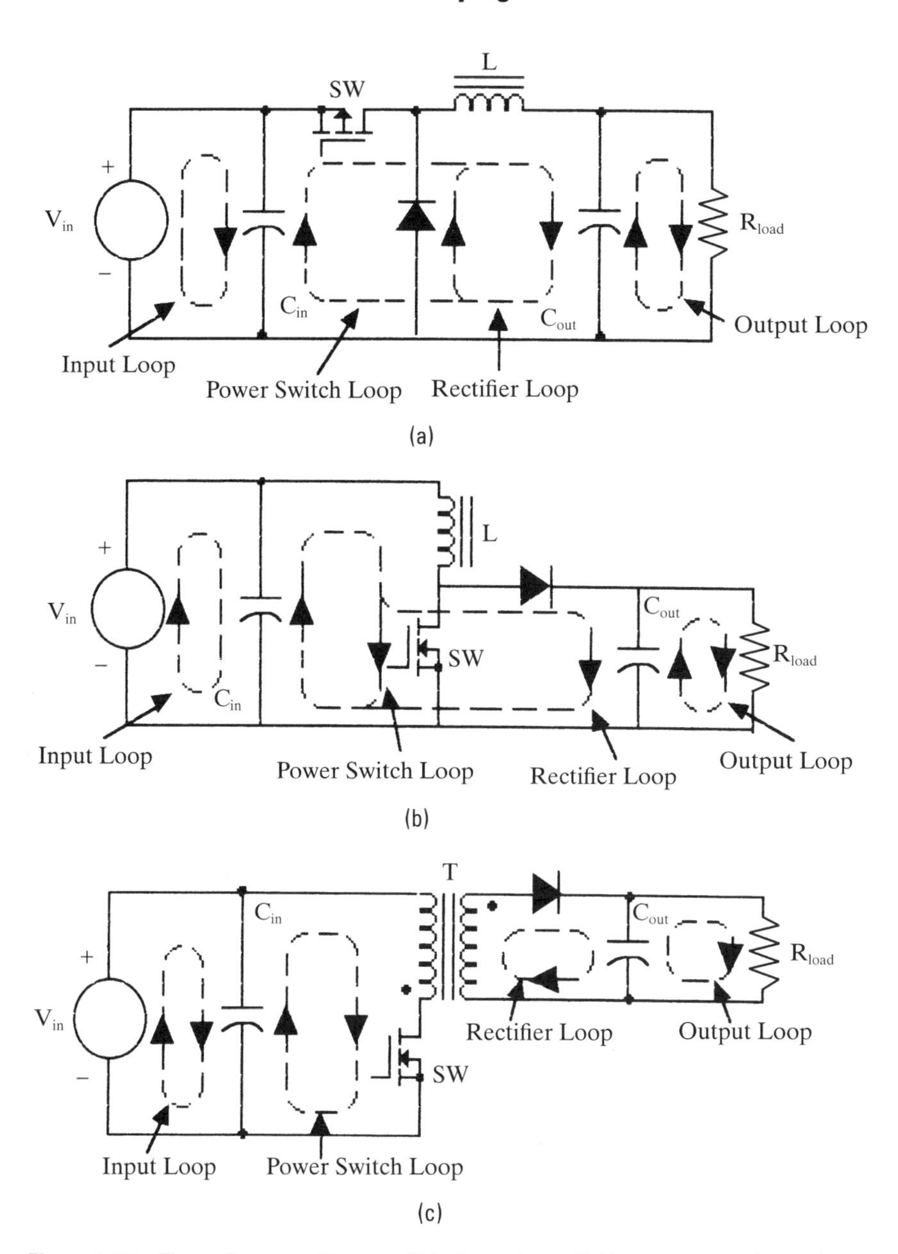

Figure 3–59 The major current loops within the major switching power supply topology types: (a) the nonisolated buck converter; (b) the nonisolated boost converter; (c) the transformer-isolated converter.

and output filter capacitors are very important. If the connections between the input or output loops and the power switch or rectifier loops are not routed directly to the capacitor's terminals, then ac energy will go around the input or output filter capacitors and escape into the environment via the input and output current loops.

The power switch and rectifier ac-current loops contain very high trapezoidal current waveforms typical in PWM switching power supplies. These waveforms are rich in harmonics which extend far above the basic switching frequency. These ac currents can have peak amplitudes two to five times that of the

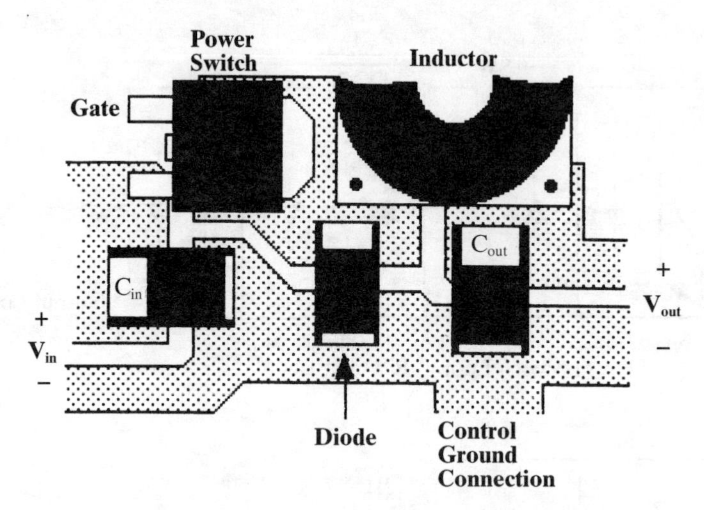

Figure 3–60 A good layout for a buck converter.

continuous input or output dc currents. The transition times are typically about 50 nS. These two loops, therefore, have the greatest ability to create electromagnetic interference (EMI).

These ac current loops should be routed before any other traces in the power supply. The three major components that make up each loop: the filter capacitor, the power switch or rectifier, and the inductor or transformer must be located adjacent to one another. The components must also be oriented such that the current path between them is as short as possible. A good example of a layout of the power section of a buck (or step-down) converter can be seen in Figure 3–60.

The traces within these loops also have the greatest affect on the converter's measured efficiency. If a significant voltage drop is experienced on any of these traces, the converter will appear less efficient, because it is operating at a lower voltage (hence higher current). But, if DVMs or meters are used to calculate its efficiency, the input voltage readings will be higher than the converter is actually experiencing, thus creating a misleadingly larger VxI product.

3.14.2 The Grounds Inside the Switching Power Supply

The grounds represent the bottom branch of the current loops discussed earlier. Grounds, though, serve a very important function as the common point of reference for the circuitry. Therefore, grounds must be carefully placed in the layout. Intermingling these grounds will cause problems with the stability of the power supply.

There is an additional ground to consider and that is the *control ground*. This is the ground connected to the control integrated circuit and all of its associated passive components. This ground is extremely sensitive and should be placed after the other ac current loops are placed. There are very specific points where the control ground connects to the other grounds. The connection, in general, is at the common-end of any component across which the control IC wants to sense a small voltage. These points would include the common end of the current sense resistor in a current-mode switching converter, and the bottom end of the output voltage resistor divider. The purpose is to create a low-noise Kelvin connection between the sensing components and the sensitive inputs to

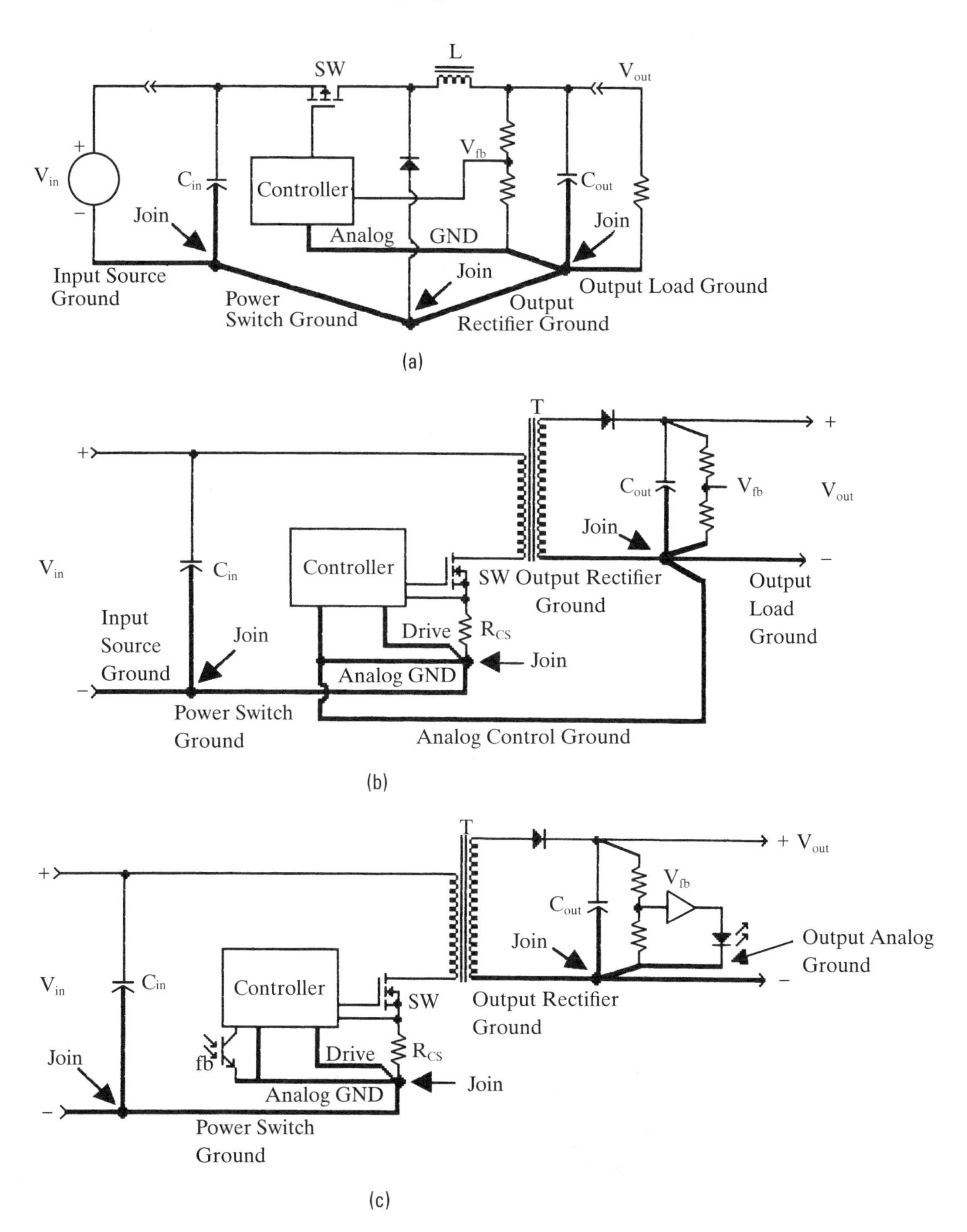

Figure 3–61 The ground arrangements for the major converter topologies: (a) the nonisolated dc/dc converter; (b) the nonisolated, transformer-coupled converter; (c) the isolated, transformer-coupled converter.

the voltage error or current amplifiers. If the control ground is connected at any other points, noise generated within those loops will be summed onto control signals, therefore adversely affecting the control IC's proper functioning.

The grounds for the three major switching power supply topology types can be seen in Figure 3–61.

Each of the high current grounds must be short and have wide traces. As a general rule, the common terminal of the filter capacitors should be the only points where the other grounds couple into the high current AC grounds. The only exception may be the control ground.

3.14.3 The AC Voltage Node

There is one node within each switching power supply that has the highest ac voltage compared to the others. This node is the ac node found at the drain (or collector) of the power switch. In nonisolated dc/dc converters, this node is also connected to the inductor and catch (or output) rectifier. In transformer-isolated topologies, there are as many ac nodes as there are windings on the transformer. Electrically, they still represent a common node, only reflected through the transformer. Special attention must be paid to each ac node separately.

This node(s) presents a different problem. Its AC voltage can be easily capacitively coupled into any adjacent traces on different metal layers, as well as radiate EMI. Unfortunately, it is generally the trace that must also act as a heatsink for both the power switch and the rectifiers, especially in surface mount power supplies. Electrically, the trace wants to be as small as possible, but thermally, it wants to be large. There is one good compromise in the surface mount designs, and that is to make the top PCB island identical to the bottom PCB island and connect them with numerous vias (or thru-hole connections). This can be seen in Figure 3–62.

This technique greatly reduces the capacitive coupling to other traces while more than doubling the heatsink volume and surface area. In thru-hole applications, one must keep the other signals and grounds, away from the high voltage trace and any of its heatsinks. In off-line converters, the earth ground can pick-up energy from this node through the heatsink (separated from the drain tab with an insulator) and allow it to exit the product via the ac plug.

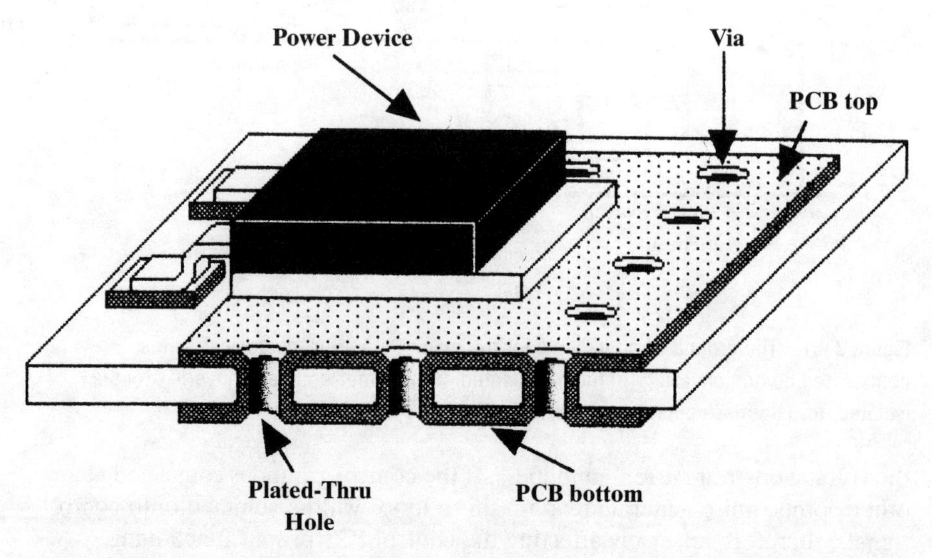

Figure 3–62 A good method for enhancing the heatsinking capabilities of a PCB island and minimizing capacitive coupling to other PCB traces.

3.14.4 Paralleling Filter Capacitors

Frequently, capacitors are placed in parallel to reduce the combined equivalent series resistance (ESR) of the filter capacitor function. This practice also allows each capacitor to "share" a portion of the overall ripple current such that each capacitor can operate within its ripple current specification. Equal "sharing" of the ripple current can only occur when the trace impedance between each of the capacitors and the source of the ripple current is identical to one another. This means that the traces between the rectifier or the power switch must be of equal length and width for each capacitor.

The temptation is to place capacitors in a row and wire them sequentially as seen in Figure 3–63a. This arrangement causes the capacitors closest to the power switch or rectifier to experience much more ripple current than the other capacitors further away. This shortens the life of the closer capacitors. Figure 3–63b shows a better way to arrange the traces for paralleled capacitors.

The designer should attempt to keep the capacitors "radially symmetric" from the ripple current source for both sides of the loop.

3.14.5 The Best Method of Creating a PCB for a Switching Power Supply

The best method of creating a layout for a switching power supply is analogous to its electrical design. The best design flow is as follows.

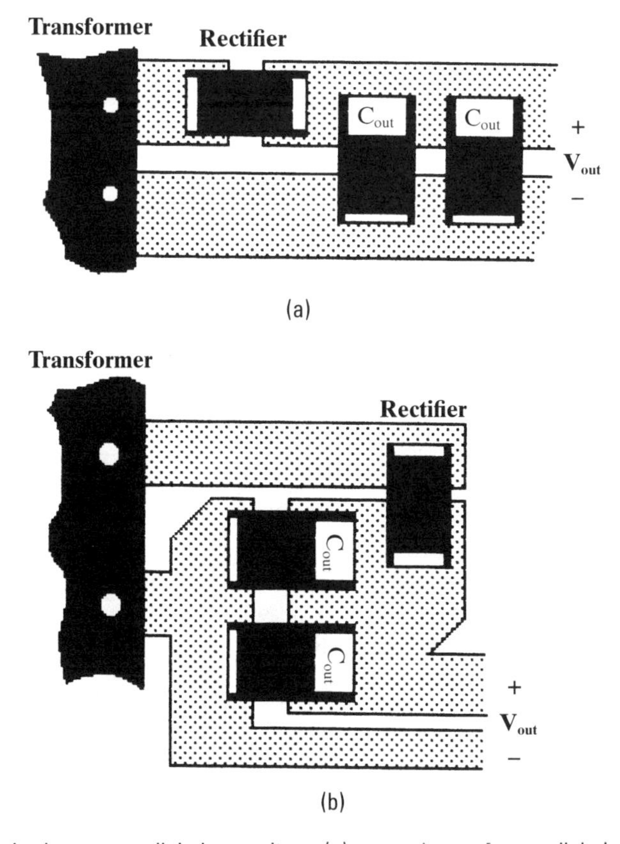

Figure 3–63 Laying out paralleled capacitors: (a) a poor layout for paralleled capacitors; (b) a better layout for paralleled capacitors.

1. Place the transformer or inductor.
2. Layout the power switch current loop.
3. Layout the output rectifier current loop(s).
4. Connect the control circuitry to the ac power circuits.
5. Layout the input source loop and input filter.
6. Layout the output load loop(s) and output filters.

As a good rule, the designer should "fill" the PCB with metal. That is, there should not be large areas of bare fiberglass. Grounds and power traces should be widened to fill these blank areas. This is a good practice for two reasons: first, the converter radiates heat better, and secondly, large copper areas capture and dissipate RF energy better by promoting eddy currents to flow.

Of course, the power supply is usually left to the last moment, so its allocated area is usually too small and its location is poor. This all means that the flow outlined above usually cannot be strictly followed, so each designer must do the best he or she can, by understanding the important electrical factors.

3.15 PWM Design Examples

Today's switching power supply market has two approaches to the design of switching power supplies: the *brut-force* design from the ground-up style, and the copy out of the data sheet style. Although there are many good controller ICs being made in the market today, many of which have a "pre-packaged" design procedure conveyed in the data sheet, the purpose of this book is to give a deeper, more fundamental understanding of the topic. Therefore, the following design examples contain controller ICs that require design expertise to complete the power system. These design examples contain basic truths about the design of switching power supplies that transcend the "power supplies made easy" methods of some of the IC suppliers. The design of the magnetic components, feedback loop compensation, and input and output filtering still hold true.

The following examples, though tedious, should provide a good insight into the design process of a PWM switching power supply.

3.15.1 A Board-level 10 Watt Step-down Buck Converter

Application

This switching power supply can be used for a board-level regulator where a linear regulator would create too much heat for the board to dissipate. A loosely regulated pre-regulator produces a distribution voltage of between +10 to +18 V. The output voltage of the board-level regulator is +3.3 V.

This design example purposely avoids the use of a highly integrated buck controller IC. It would be instructional to show the selection and design process involved in a switching power supply. Refer to Figure 3–65

Design specification

Input voltage range:	+10 V – +14 VDC
Output voltage:	+5 VDC
Maximum output current:	2 A
Output ripple voltage:	+30 mV p-p
Output regulation:	+/– 1%

"Black box" predesign estimates

Output power: $(+5.0\,V)(2\,A) = 10.0\,W$ (max)
Input power: P_{out}/Est. Effic. $= 10.0\,W/0.80 = 12.5\,W$
Power switch loss: $(12.5 - 10\,W)(0.4) = 1.0\,W$
Catch diode loss: $(12.5 - 10\,W)(0.6) = 1.5\,W$

Input average currents

Low input: $(12.5\,W)/(10\,V) = 1.25\,A$
High input: $(12.5\,W)/(14\,V) = 0.9\,A$
Estimated peak current: $1.4\,(I_{out(rated)})$
$1.4(2.0\,A) = 2.8\,A$

The desired frequency of operation is 100 kHz.

Inductor design (refer to Section 3.5.5)
The worst case operating condition is at the high input voltage.

$$L_{min} \frac{(V_{in(max)} - V_{out})(1 - V_{out}/V_{in(max)})}{1.4 I_{out(min)} f_{sw}}$$

$$L_{min} \frac{(14\,V - 5\,V)(1 - 5\,V/14\,V)}{1.4(0.5\,A)(100\,kHz)} = 82.6\,\mu H$$

where: $V_{in(max)}$ is the maximum possible input voltage.
V_{out} is the output voltage.
$I_{out(min)}$ is the lightest expected load current.
f_{sw} is the operating frequency.

The inductor is to be a surface mount toroid on a J-leaded plastic mounting board. There are standard surface mount inductors available from many suppliers. I have chosen the part DO3340P-104 from Coilcraft.

Selection of the power switch and catch diode
Power switch. The power switch is going to be a P-channel power MOSFET. The maximum input voltage is 18 VDC. Therefore, a V_{DSS} rating of +30 VDC or higher will be satisfactory. The peak current is 2.8 A. It is also desired to keep the heat dissipation less than 1 watt so the estimated RDS(on) should be less than:

$$R_{DS(on\text{-}max)} = P_{D(est)} \Big/ (I_{pk(est)})^2$$
$$= (1\,W) \Big/ (2.8\,A)^2 < 127 \text{ milliohms (maximum)}$$

I will select the ubiquitous *FDS9435* P-channel MOSFET with an on-resistance of 45 milliohms packaged in an SO8 package.

Catch diode. The catch diode needs to be a schottky diode to minimize the conduction loss and the switching loss of the function. The diode that has a reasonable forward voltage drop at the 3 A peak current is the *MBRD330* with a 0.45 V drop at 3 A (at +25°C).

The output capacitor (refer to Section 3.6)
The value for the output capacitor is determined by the following equation.

$$C_{\text{out(min)}}\ \frac{I_{\text{out(max)}}\left(1-DC_{\text{(min)}}\right)}{f_{\text{sw}}\left(V_{\text{ripple(p-p)}}\right)}$$

$$C_{\text{out(min)}}\ \frac{(2\,\text{A})(1-5\,\text{V}/14\,\text{V})}{(100\,\text{kHz})(30\,\text{mV})}=429\,\mu\text{F}$$

The major concern of both output and input filter capacitors is the ripple current entering the capacitor. In this application, the ripple current is identical to the inductor ac current. The maximum limits of the inductor current is 2.8 A for I peak and about one-half the maximum output current or 1.0 A. So the ripple current is 1.8 A p-p or an estimated RMS value of 0.6 A (about one-third of the p-p value).

I will be using a surface mount tantalum capacitor because they typically exhibit about 50 percent of the ESR of electrolytic capacitors. I will also derate the rating of the candidate capacitors by 30 percent at +85°C ambient temperature.

The best candidate capacitors are from AVX, which have very low ESR and thus can handle very high levels of ripple current. These capacitors are exceptional and not typical. One piece of these could handle the output demands.

AVX:
TPSE477M010R0050 470 μF (20%), 10 V, 50 mohm, 1.625 Arms
TPSE477M010R0100 470 μF (20%), 10 V, 100 mohm, 1.149 Arms

Nichicon:
F751A477MD 470 μF (20%), 10 V, 120 mohms, 0.92 Arms

There are very few surface mount capacitors with the desired value, voltage rating, and low ESR all in one part. It would be more conservative to parallel two capacitors of no less than one-half the desired capacitance value. This allows many more second sourced capacitors to be used and lowers the ESR. Lets use two 330 μF, 10 V tantalum capacitors in parallel. The Candidates now become:

KEMET:
T510X337M010AS 330 μF (20%), 10 V, 35 mohms, $2.0A_{\text{rms}}$

Nichicon:
F751A337MD 330 μF (20%), 10 V, 150 mohms, $0.8A_{\text{rms}}$

The input filter capacitor

This capacitor experiences the same current waveform at the power switch, which is a trapezoid with an initial current of about 1 A rising to 2.8 A with very sharp edges. This capacitor has much more rigorous operating conditions than the output filter capacitor. I will estimate the RMS value of the trapezoidal current waveform as a piecewise superposition of two waveforms, a rectangular 1 A peak waveform and a triangular waveform with a 1.8 A peak. This yields an estimated RMS value of 1.1 A. The value of the capacitor is then calculated as:

$$C_{\text{in}}=\frac{P_{\text{in}}}{f_{\text{sw}}\left(V_{\text{ripple(p-p)}}\right)^2}=\frac{12.5\,\text{W}}{(100\,\text{kHz})(1.0)^2}$$

$$C_{\text{in}}=125\,\mu\text{F}$$

Capacitors at the higher voltages have smaller capacitance values. There will be two 68 μF capacitors in parallel. The candidates are:

AVX: (2 required per system)
TPS686M016R0150 68 µF (20%), 16 V, 150 mohms, 0.894 A_{rms}

AVX: (3 required per system)
TAJ476M016 47 µF (20%), 16 V, 900 mohms , 0.27 ohms (net)

Nichicon: (3 required per system)
F721C476MR 47 µF (20%), 16 V, 750 mohms, 0.19 ohms (net)

Selecting the controller IC
The desired features one might look for in a buck controller IC would be:

1. Ability to operate directly from the input voltage.
2. Pulse-to-pulse overcurrent limiting.
3. Totem-pole MOSFET drivers.

There are many buck controller ICs on the market, but the one that was chosen is the UC3873. The internal reference presented to the voltage error amplifier is 1.50 V+/– 2 percent.

Setting the frequency of operation (C3)
From the datasheet, one sets the frequency by the following equation.

$$C_t = 1/[(15\,\text{k})f_{\text{sw}} = 1/(15\,\text{k})(100\,\text{kHz})$$
$$C_t = 666\,\text{pF the closest value is } 680\,\text{pF}$$

Current sense resistor (R1)
The type of protection on this IC is a pulse-to-pulse current sense that cuts off the power switch instantly upon exceeding a 0.47 V threshold. I will allow a 25 percent margin between the expected peak current and the protection cutoff threshold (protection 1.25(2.8 A) = 3.5 A).

$$R1 = (0.47\,\text{V})/(3.5\,\text{A}) = 0.134\,\text{ohms}$$

The closest standard value of sub-one ohm resistors is 0.1 ohms.

Voltage sensing resistor divider (R3 and R4)
R4 (bottom resistor)

$$R4 = (1.5\,\text{V})/(1\,\text{mA}) = 1.49\,\text{K ohms } 1\%$$

This makes the actual sense current 1.006 mA.
R3 (top resistor):

$$R3 = (5.0\,\text{V} - 1.5\,\text{V})/(1.006\,\text{mA}) = 3.48\,\text{K ohms } 1\%$$

Compensating the voltage feedback loop (refer to Appendix B)
This is a voltage-mode, forward converter. To produce the best transient response time, a 2-pole, 2-zero method of compensation is going to be used.

Defining the control-to-output characteristic
The output filter pole is determined by the filter inductor and capacitor and is a –40 dB/decade rolloff. Its nominal corner frequency is

$$f_{\text{fp}} = \frac{1}{21\pi\sqrt{L_o C_o}}$$
$$f_{\text{fp}} = \frac{1}{21\pi\sqrt{(100\,\mu\text{H})(660\,\mu\text{F})}} = 619\,\text{Hz}$$

The zero caused by the output filter capacitor(s) is (ESR is two 120 mohms in parallel)

$$f_{zesr} = \frac{1}{2\pi R_{esr} C_o} = \frac{1}{21(60\,\text{m}\Omega)(660\,\mu\text{F})} = 4020\,\text{Hz}$$

The intrinsic absolute gain of the power circuit at dc is

$$A_{DC} \approx V_{in} / \Delta V_{error} = (14\,\text{V})/(3.0\,\text{V}) = 4.66$$
$$G_{DC} = 20\,\text{Log}(A_{DC}) = 13.4\,\text{dB}$$

Locating the compensating poles and zeros in the error amplifier

The gain cross-over frequency of the closed loop should not be any higher than 20 percent of the switching frequency (or 20 kHz). I have found that gain cross-over frequencies of 10 kHz to 15 kHz are quite satisfactory for the majority of applications. This yields a transient response time around 200 uS.

$$f_{xo} = 15\,\text{kHz}$$

First one assumes that the final closed loop compensation network will have a continuous –20 dB/decade slope. To achieve a 15 kHz cross-over frequency, the amplifier must add gain to the input signal and "push-up" the gain curve of the Bode plot.

$$G_{xo} = 20\,\text{Log}(f_{xo} / f_{fp}) - GDC = 20\,\text{Log}(15\,\text{kHz}/619\,\text{Hz}) - 13.4\,\text{dB}$$
$$G_{xo} = G2 = +14.3\,\text{dB}$$
$$A_{xo} = A2 = 5.2\,(\text{absolute equivalent})$$

This is the gain needed at the midband plateau (G2) to achieve the desired cross-over frequency.

The gain exhibited at the first set of compensating zeros is

$$G_1 = G_2 + 20\,\text{Log}(f_{ez2} / f_{ep1}) = +14.3\,\text{dB} + 20\,\text{Log}(310\,\text{Hz}/4020\,\text{Hz})$$
$$G_1 = -8\,\text{dB}$$
$$A_1 = -0.4\,(\text{absolute equivalent})$$

To compensate for the double filter pole, I will place two zeros at one-half the filterpole frequency:

$$f_{ez1} = f_{ez2} = 310\,\text{Hz}$$

The first compensating pole will be located at the capacitor's ESR frequency (4,020 Hz).

$$f_{ep1} = 4{,}020\,\text{Hz}$$

The second compensating pole is just used to maintain high frequency stability by depressing the gain above the cross-over frequency.

$$f_{ep2} = 1.5 f_{xo} = 22.5\,\text{kHz}$$

Now one can begin calculating the component values within the error amplifier.

$$C_7 = \frac{1}{2\pi(f_{xo})(A2)(R3)} = \frac{1}{21(15\,\text{kHz})(5.2)(3.48\,\text{k}\Omega)}$$
$$C_7 = 586\,\text{pF make } 560\,\text{pF}$$

$$R_2 = (A1)(R1) = (0.4)(3.48\,\text{k}\Omega) = 1.39\,\text{k}\Omega \text{ make } 15\,\text{k}\Omega$$

$$C_6 = \frac{1}{2\pi(f_{ez1})(R2)} = \frac{1}{2\pi(310\,\text{Hz})(1.5\,\text{k}\Omega)}$$

$$C_6 = 2.9\,\mu\text{F or } 2.2\,\mu\text{F}$$

$$R_5 = R2/A2 = (1.5\,\text{k}\Omega)/(0.4) = 3.75\,\text{k}\Omega \text{ make } 3.9\,\text{k}\Omega$$

$$C_{10} = \frac{1}{2\pi(f_{fez2})(R5)} = \frac{1}{2\pi(22.5\,\text{kHz})(3.9\,\text{k}\Omega)}$$

$$C_{10} = 1814\,\text{pF make } 1800\,\text{pF}$$

Refer to Figure 3–64.

3.15.2 Low-Cost, 28-Watt PWM Flyback Converter

Application

This power supply is going to provide power for a piece of process control instrument. The instrument receives its power from a +24 V bulk power supply which also provides transformer isolation from the input bus voltage to the unit. Please refer to Figure 3–66.

Specifications

V_{out}: +5 VDC at 2 A maximum current, 0.5 A minimum.
+12 VDC at 0.5 A
–12 VDC at 0.5 A
+24 VDC at 0.25 A

V_{in}: 18 to 36 VDC required operational range
+24 VDC nominal input line voltage

"Black box" pre-design considerations (refer to Section 3.4)

$$P_{out} = (5\,\text{V} \cdot 2\,\text{A}) + (12\,\text{V} \cdot 0.5\,\text{A}) + (12\,\text{V} \cdot 0.5\,\text{A}) + (24\,\text{V} \cdot 0.25\,\text{A})$$
$$= 28\,\text{Watts}$$

$$P_{in} = P_{out}/\text{effic(est)} = (28\,\text{W})/.75$$
$$= 37.3\,\text{Watts}$$

$$I_{in(high)} = P_{in}/V_{in(low)} = (37.3\,\text{W})/18\,\text{V}$$
$$= 2.07\,\text{Amps}$$

$$I_{in(op)} = P_{in}/V_{in(nom)} = (37.3\,\text{W})/24\,\text{V}$$
$$= 1.55\,\text{Amps}$$

This indicates that a #18 AWG wire or equivalent should be used on the primary winding of the transformer.

$$I_{pk} \approx 5.5 \cdot P_{out}/V_{in(min)} = 5.5(28\,\text{W})/18\,\text{V}$$
$$= 8.55\,\text{Amps}$$

Select the frequency of operation of the power supply to be 40 kHz (or $T_{on(max)}$ = 12.5 uS).

Designing the flyback transformer (refer to Section 3.5.5)

$$L_{pri} = V_{in(min)} \cdot T_{on}/I_{pk} = (18\,\text{V})(12.5\,\mu S)/8.55\,\text{A}$$
$$= 26.3\,\mu\text{H}$$

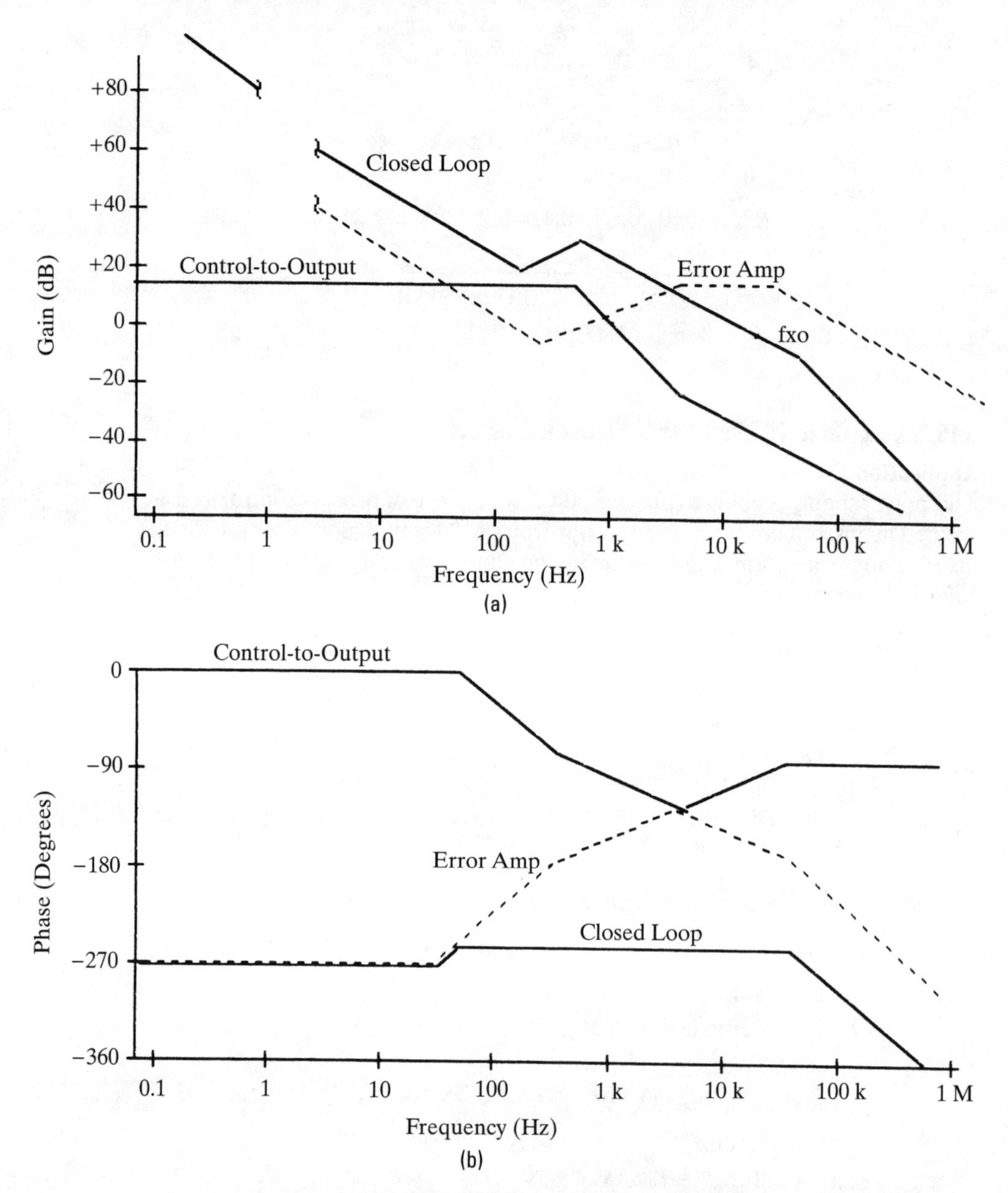

Figure 3–64 The gain and phase Bode plots for the design example 3.15.1: (a) the phase plot for the buck converter; (b) the phase plot for the buck converter.

Check the energy throughput of the core

$$P_{\text{out(est)}} = fL_{\text{pri}} \cdot (I_{pk})^2 / 2 = 40{,}000(26.3\,\mu H)(8.55)^2 / 2$$
$$= 38.45\,\text{Watts—OK}$$

I will use an MPP toroid core. The core sizing procedure is contained in Section 3.5.5. Estimate the core size required:

$$E_{\text{L}} = \text{“}LI^2\text{”} = L_{\text{pri}} \cdot (I_{\text{pk}})^2 = (.0263\,\text{mH})(8.55\,\text{A})^2$$
$$= 1.92$$

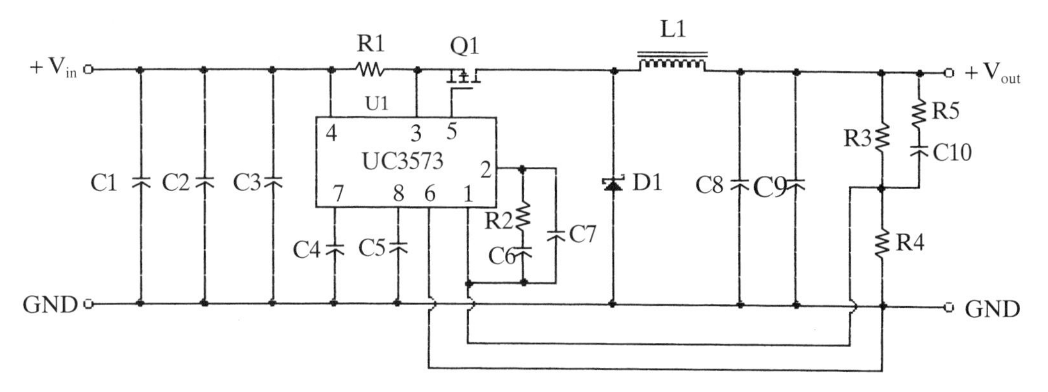

Figure 3–65 A 10-watt buck (step-down) converter.

Reference to Figure 3–21 yields the part number 55310-A2 with a permeability of 125. This core has an AL of 90 mH/1,000 T.

$$N_{pri} = 1{,}000\sqrt{\frac{L_{pri}}{A_L}}$$

The number of turns for the primary is:

$$N_{pri} = 1000(0.0263\,\text{mH}/90\,\text{mH})^{1/2}$$
$$= 17.09 \text{ turns} - \text{round down to 17 turns}$$

The number of secondary turns on the lowest voltage (+5 V) output (assuming a Schottky rectifier) is

$$N_{sec} = \frac{N_{pri}(V_o + V_D)(1 - \partial_{max})}{V_{in(min)} \cdot \partial_{max}}$$
$$N(+5\,\text{V}) = 17\,\text{T}(5.0 + 0.5\,\text{V})(50\%)/(18\,\text{V})(50\%)$$
$$= 5.19 \text{ turns round down to 5 turns}$$

The other windings (assuming ultrafast rectifiers) are

$$N_2 = \frac{(V_2 + V_{D2})N_1}{(V_1 + V_{D1})}$$
$$N_{(+12\,\text{V})} = (12.0 + 0.9\,\text{V})(5\text{T})/(5.0 + 0.5\,\text{V})$$
$$= 11.73 \text{ Turns round to 12 turns } (-12\,\text{V winding has the same})$$
$$N_{(+24V)} = (24.0 + 0.9\,\text{V})(5\text{T})/(5.0 + 0.5\,\text{V})$$
$$= 22.6 \text{ turns—round to 23 turns}$$

The amount of error in each output is

+/–12 V: +0.3 volts
+24 V: +0.4 volts acceptable

I chose an autotransformer style secondary which means that the higher voltage windings will include the lower voltage windings. The turns associated with this and the wire sizes are

+5 V: 5 turns, #17 AWG (use 3 strands of #22 AWG)
+12 V: 7 turns, #21 AWG

−12 V: 12 turns, #21 AWG
+24 V: 11 turns, #26 AWG
pri: 17 turns, #19 AWG (use 2 strands of #22 AWG)

Transformer winding technique

It is not economical to bifilar wind all the windings of the transformer together. Instead a selective bifilar winding technique will be used. The primary winding will be twisted together with the +24 V winding and the +12 V and −12 V windings will be twisted together prior to winding them on the core. The +24 V winding will act as the catch winding for the primary and will reduce the voltage spike during the turn-off of the power switch.

The +5 V winding is wound first, distributed equally around the circumference of the toroid. Next the primary/+24 V bundle is wound evenly around the core. Lastly, the +12 V and −12 V bundle is wound on the core. This winding can physically "squeeze" between the last winding.

For production considerations, the typical mounting approach is to place the finished toroid on a terminal board and pot the toroid assembly. This makes the assembly impervious to handling damage and allows the assembly to be easily placed into a PCB. Estimated parts cost approximately $2.50 U.S.

Selecting the power switch and rectifiers

Power Switch. There is a distinct advantage in using a power MOSFET in this application. It should have less drive and switching losses.

$$\begin{aligned} V_{DS(min)} &> (V_{out} + V_D)(N_{pri}/N_{sec}) + V_{in(max)} \\ &> (24.4\,\text{V} + 0.9\,\text{V})(17\,\text{T}/23\,\text{T}) + 36\,\text{V} \\ &> 54.7\,\text{V (neglecting any spikes caused by leakage } L) \\ &\quad \text{use } 100\,\text{V} \end{aligned}$$

I_D: For flyback-mode converters it is a good idea to select a power switch average current rating of about 1.5 times the maximum average input current of the supply. Another consideration is the loss. By overspecifying the current the $I2R_{DS(on)}$ loss (conduction loss) can be reduced with very little penalty on cost and input capacitance.

$$I_{D(min)} > 1.5(2.07\,\text{A}) = 3.11\,\text{Amps}$$

Use the MTP10N10M. I have selected the current sensing style of power MOSFET since I wish to implement a current-mode controller and this will reduce my sensing losses by three orders of magnitude.

Rectifiers.

$+5\ V_{output}$

$$\begin{aligned} V_{R(min)} &> V_{out} - [-V_{in(max)}(N_{sec}/N_{pri})] \\ &> +5\,\text{V} + 36(5/17) = 15.6\,\text{V} \end{aligned}$$

$$I_{F(min)} \approx I_{out(max)} = 2\,\text{Amps}$$

Use the 1N5824 (3 A).

+/−12 V output (see above procedure). Use the MUR110 (D5 and D7).
+24 V output. Use the MUR110 (D4).

The output-filter section (refer to Section 3.6)

The values for the output filter capacitors are to be determined by using Equation 3.36.

$$C_{out(min)} = \frac{I_{out(max)} \cdot (1-\partial_{(min)})}{f \cdot V_{ripple(pk-pk)}}$$

$C_{out(+5\,V)} = 480\,\mu F$ at 10 V—uses 2 ea. 220 μF at 10 V tantalum (C14 and 15) capacitors in parallel to reduce height and to reduce the ESR.

$C_{out(+/-12\,V)} = 122\,\mu F$ at 20 V—use 150 μF 35 V Tantalum (C12 & C16)

$C_{out(+24\,V)} = 60\,\mu F$ at 35 V – use 2 ea. 47 μF at 35 V (C11)

The switch-mode controller

In attempting to select a controller IC one should make a list of the important desired for the design. Also make a "nice but nonessential" list.

Essential	"Nice—but"
Low parts count	Undervoltage lockout
Current-mode control	Low I_{sense} threshold
MOSFET driver output (totem-pole)	50 percent duty cycle limiting
Single output driver	
Low cost	

After reviewing the list of popular controller ICs, the UC3845P appears to satisfy all the above requirements. (I also chose it because of its instructional nature.)

Referring to the data sheet in the Motorola "Linear and Interface Integrated Circuits" data book, the basic schematic implementation is given in the application figures. The designer need only determine the values for the timing resistor and capacitor, and the current-sense resistor. All of the other components are involved with the Vcc supply and the feedback compensation, which will be designed later. Looking at the "Timing Resistor *vs*. Oscillator Frequency" graph and wishing to operate the supply at a nominal 40 kHz, one determines a value of

$$C_t = C8 = 2000\,\text{pF}$$

$$R_t = R_4 = 22\,\text{K ohms}$$

R_{sense}:

$$R_s = VI_s(n/I_{pk})$$

$$= .6\,\text{V}(8.5\,\text{A}/1800) = 127\,\text{ohms make } 120\,\text{ohms}$$

This value will no doubt need to be adjusted during the breadboard stage.

The voltage feedback section (refer to Section 3.9)

In order to enhance the cross-regulation of the multiple outputs, it is desired to sense a portion of all the positive output voltages. To do this, one must investigate the technology of the circuits that must draw their power from the outputs. The hypothetical loads, in this case, are

+5 V — A microcontroller and 74HC logic with a +/– 10 percent tolerance on the V_{DD}.

+/–12 V — Has operational amplifier centered analog circuitry. These tend to be immune to fluctuations in their supply voltages.

+24 V — These are external process interfaces that have a +18 V low-line limit. The +5 V logic should be the tightest regulated.

First select a voltage divider sense current—a nominal 1 mA. Determine the lower resistor (R10 + R11) in the voltage divider:

$$R_{10} + R_{11} = V_{\text{ref}} / I_{\text{sense(est)}} = 2.5\,\text{V} / 1\,\text{mA}$$
$$= 2.5\,\text{K—round up to } 2.7\,\text{K ohms}$$

Let us add a 1 KΩ potentiometer to the divider to adjust the final output voltages during the final test stage of manufacturing. The "pot" will be wired as a rheostat with the wiper being connected with the upper lead. The weakness of potentiometers is that the wiper gets "noisy" on open circuits. Wiring the "pot" in this fashion causes the output voltages to fall if the wiper opens. Otherwise the supply would go to its maximum duty cycle and destroy the loads. Assume the "pot" is adjusted mid-way so R10 then becomes

$$R_{10} = 2.7\,\text{K}\Omega\text{—}500\,\Omega = 2.2\,\text{K}\Omega$$

The true sense current is

$$I_{\text{sense(act)}} = V_{\text{ref}} / (R_{10} + R_{11}) = 2.5\,\text{V} / 2.7\,\text{K} = 0.96\,\text{mA (use this value)}$$

Select the proportion of sense current to be drawn from each output: +5, 60 percent; +12, 20 percent; +24, 20 percent.

Determine the values of the upper divider resistors:

$$R_{\text{u-sense}} = (V_{\text{out}} - V_{\text{ref}}) / (I\% \cdot I_{\text{sense(act)}})$$

+5: $R_7 = (5.0 - 2.5\,\text{V})/(0.6(0.96\,\text{mA})) = 4340\,\Omega$ round up to 4.7 KΩ
+12 V $R_8 = (12.3 - 2.5\,\text{V})/(0.2(0.96\,\text{mA})) = 51\,\text{K}\Omega$
+24 V $R_9 = (24.4 - 2.5\,\text{V})/(0.2(0.96\,\text{mA})) = 114\,\text{K}\Omega$ make 110 KΩ

I will leave the feedback-loop compensation to the end.

The input filter section (refer to Section 3.12)
Cin.

$$C_{\text{in}} \cong \frac{2P_{\text{out}}}{f \cdot (V_{\text{ripple(p-p)}})}$$
$$C_{\text{in}} = 2(37.3\,\text{W}) / [(40{,}000\,\text{Hz})(1\,Vp - p)]$$
$$= 186\,\mu\text{F}$$

Place 2 each 100 µF, 50 V aluminum electrolytic capacitors and a 0.1 µF, 100 V ceramic in parallel.

Lin. Since we have a single input wire with a common ground, I will use an MPP toroid. Referring to the manufacturer's "Normal Magnetization Curves," I see that 20 Oe provides a dc bias at less than half the saturation flux density of the core. The recommended permeability is 125 μ. The estimated core size should be a magnetics P/N 55120-A2 using two strands of #20 AWG. The number of turns needed is

$$N = \frac{H \cdot 1_{\text{m}}}{.4 \cdot \pi \cdot I_{\max}}$$
$$N = (20\,\text{O})(4.11\,\text{cm}) / .4\pi(2.04\,\text{A})$$
$$= 32 \text{ Turns}$$

Start-up section (refer to Section 3.10)

Although, the input line is low enough to draw all the bias current to run the control IC and the MOSFET, doing so would waste approximately 1.2 W, or a 4.2 percent efficiency loss. A start-up circuit that provides current from the input only during start-up or overcurrent foldback is best. The IC and MOSFET can receive power from the +12 V output during normal operation. Refer to schematic in Figure 3–59:

D1: Use an 11 V, 500 mW Zener—1N5241
R1: R1 = (18 V − 11 V)/(0.4 mA) = 17.5 KΩ—make 18 KΩ
Q1: Use an MPSA05
R2: R2 = (18 V − 12 V)/(5.0 mA) = 1.2 KΩ
D2: Use a 1N4148
D3: Use an MBR030.

Feedback loop compensation (refer to Appendix B)

To exhibit the tightest output regulation and the fastest transient response time, I have decided to use the single pole-zero method of compensation. The control-to-output characteristic curves for a current-mode controlled flyback-mode converter are of a single pole nature so a single pole-zero method of compensation should be used. The +5 V output is the highest power output and is the highest sensed, so it is the primary output. The placement of the filter pole, ESR zero, and DC gain are

$$A_{DC} = \frac{(36\,\text{V} - 5\,\text{V})^2}{36\,\text{V} \cdot 2.5\,\text{V}} \cdot \frac{5\text{T}}{17\text{T}}$$
$$= 3.14$$
$$G_{DC} = 20\,\text{Log}(3.14) = 9.94\,\text{dB}$$
$$f_{fp(hi)} = \frac{1}{2\pi(5\,\text{V}/2\,\text{A})(440\,\mu\text{F})}$$
$$= 144\,\text{Hz} \quad (\text{at rated load } (2\,\text{A}))$$
$$f_{fp(low)} = \frac{1}{2\pi(5\,\text{V}/0.5\,\text{A})(440\,\mu\text{F})}$$
$$= 36.2\,\text{Hz} \quad (\text{at light load } (0.5\,\text{A}))$$

The control-to-output curves can be seen in Figure 3–67.

The gain cross-over frequency should be less than fsw/5 or

$$f_{xo} < 40\,\text{kHz}/5 = 8\,\text{kHz}$$

Determine the amount of gain needed to boost the closed loop function to 0 dB at the gain cross-over frequency (see Equation B.24).

$$G_{xo} = 20\log(f_{xo}/f_{fp(hi)}) - G_{DC}$$
$$= 20\log(8{,}000/144) - 9.94\,\text{dB}$$
$$= 24.95\,\text{dB} \quad (\text{needed for Bode Plot only})$$
$$A_{xo} = 52.4 \quad (\text{absolute gain—needed later})$$

Locate the compensating error amplifier zero at the location of the lowest manifestation of the filter pole or

$$f_{ez} = f_{fp} = 36.2\,\text{Hz}$$

Locate the compensating error amplifier pole at the lowest anticipated zero frequency caused by the ESR of the capacitor or

$$f_{ep} = f_{p(ESR)} = 20\,\text{kHz (approximated)}$$

Knowing the upper resistor in the +5 V voltage sensing divider (4.7 K),

$$C_7 = \frac{1}{2\pi A_{xo} \cdot R_1 \cdot f_{xo}}$$
$$= 1/2\pi(52.4)(4.7\,\text{K})(8\,\text{kHz})$$
$$= 80\,\text{pF make } 82\,\text{pF}$$
$$R_3 = A_{xo} R_7$$
$$= 52.4(4.7\,\text{K})$$
$$= 246\,\text{K—round to } 270\,\text{K}$$
$$C_6 = \frac{1}{2\pi f_{ez} R_2}$$
$$= 1/2\pi(36.2\,\text{Hz})(270\,\text{K})$$
$$= .016\,\mu\text{F make } .015\,\mu\text{F}$$

This completes the design of the feedback loop compensation elements, and the error amplifier curves and the overall plots are also included in Figure 3–66. This also completes the design of the major portions of the switching power supply. The schematic is shown in Figure 3–67.

Parts list

C1	0.1 µF, 100 V, ceramic
C2–3	100 µF, 50 V, aluminum electrolytic
C4	0.1 µF, 100 V, ceramic

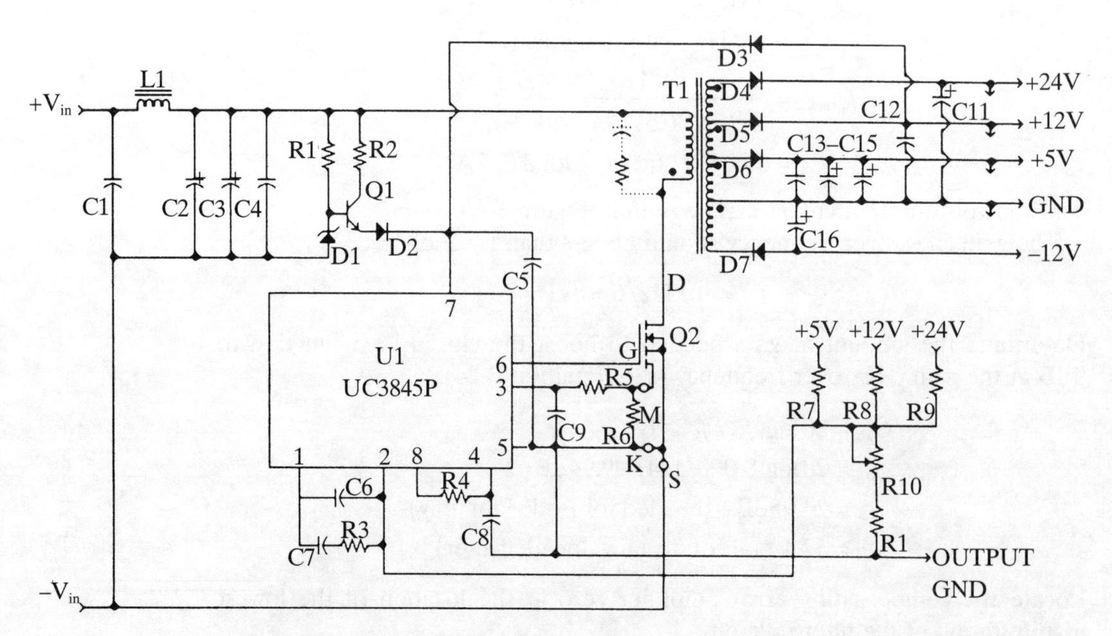

Figure 3–66 Schematic for design example 3.15.2. A 28 W current-mode, flyback dc-to-dc converter.

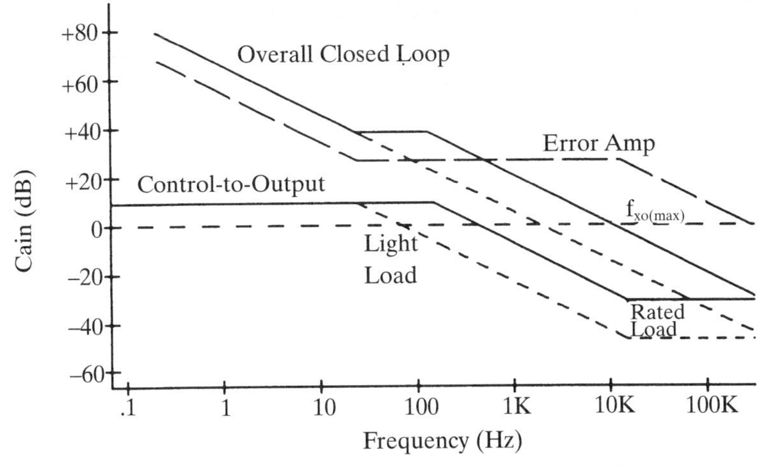

A. The Gain Plot for the Power Supply

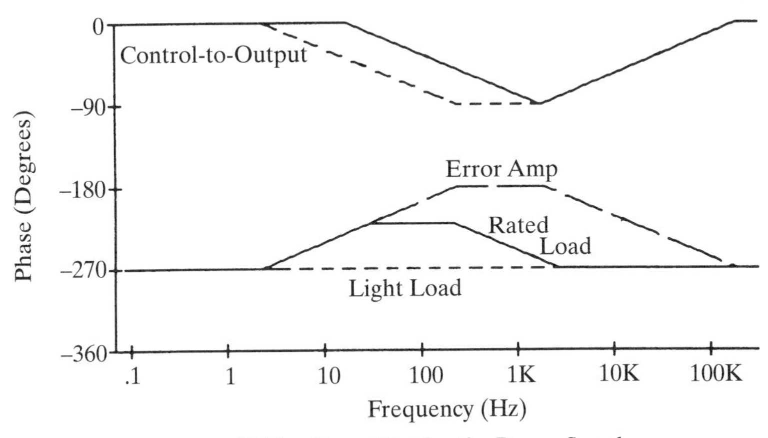

B. The Phase Plot for the Power Supply

Figure 3–67 The gain and phase Bode plots for design example 3.15.2 (compensation design).

C5	10 μF, 20 V, tantalum
C6	0.015 μF, ceramic
C7	82 pF, 50 V, ceramic
C8	0.022 μF, 50 V, ceramic
C9	470 pF, 50 V, ceramic
C10	47 μF, 35 V, tantalum
C11	100 μF, 20 V, tantalum
C12	0.1 μF, 50 V, ceramic
C13–14	220 μF, 10 V, tantalum
C15	100 μF, 20 V, tantalum
D1	11 V, 500 mW, Zener 1N5241
D2	1N4148
D3	MBR030
D4	MUR110
D5	MUR110
D6	1N5824
D7	MUR110

L1	Inductor—see text
Q1	MPSA05
Q2	Pwr MOSFET—MTP10N10M
R1	18 K, 1/4 W
R2	1.2 K, 1/2 W
R3	270 K, 1/4 W
R4	18 K, 1/4 W
R5	1 K, 1/4 W
R6	120 ohms, 1/4 W
R7	4.7 K, 1/4 W
R8	51 K, 1/4 W
R9	110 K, 1/4 W
R10	1 K, Variable Resistance
R11	2.2 K, 1/4 W
T1	Transformer—custom
U1	IC. UC3845AN

3.15.3 65 Watt, Universal AC Input, Multiple-output Flyback Converter

This switching power supply can be used for electronic products that plug into ac power sources ranging from 85 VAC to 240 VAC. This particular switching power supply can be "scaled" to provide from 25 watts to 150 watts of output power. This supply can be used for products such as small office PBXs, etc. Please refer to Figure 3–72.

Design Specification

Input voltage range:	90 VAC–240 VAC, 50/60 Hz
Outputs:	+5 VDC @ 1 A rated, 750 mA min.
	+12 VDC @ 1 A rated, 100 mA min.
	–12 VDC @ 1 A rated, 100 mA min.
	+24 VDC @ 1.5 A rated, 0.25 A min.
Output ripple voltage:	+5 V, +/– 12 V; 100 mVp-p(max)
	+24 V; 250 mVp-p(max)
Output Regulation:	+5 V, +/– 12 V; +/– 5% max.
	+ 24 V; +/– 10% max.
Cost Goal:	$25.00 BOM, 100 pcs

System protection and added features:

Low input inhibit:	The power supply and product is to be inhibited from operation any time the input voltage is below 85 VAC +/– 5%.
MPU power fail signal:	An open collector output is to be provided by the power system when ever the +5 V output is less than 4.6 V +/– 5%.

"Black box" predesign considerations (refer to Section 3.4)

1. Total output power: $P_o = (5\,V)(1\,A)+2(12\,V)(1\,A)+(24\,V)(1.5\,A)$
   $= 65$ watts
2. Estimated input power: $P_{in} = P_{o/effic} = 65\,W/.8 = 81.25\,W$
3. DC Input Voltages
   a. From 110 VAC line: $V_{in(L)} = (90\,VAC)(1.414) = 127\,VDC$
   $V_{in(H)} = (130\,VAC)(1.414) = 184\,VDC$
   b. From 220 VAC line: $V_{in(L)} = (185\,VAC)(1.414) = 262\,VDC$
   $V_{in(H)} = (240\,VAC)(1.414) = 340\,VDC$
4. Average input currents:
   a. Highest average I_{in}: $I_{in(max)} = P_{in/Vin(min)}$
   $= 81.25\,W/127\,VDC$
   $= 0.64$ Amps DC
   b. Min. average I_{in}: $I_{in(min)} = P_{in/Vin(max)}$
   $= 81.25\,W/340\,VDC$
   $= 0.24$ Amps DC
   Note: Make the AWG of the primary #20 AWG or equivalent.
5. Estimated peak current: $I_{pk} = 5.5P_{out}/V_{in(min)}$
   $= 5.5(65\,W)/127\,V = 2.81$ Amps
6. Heatsinking
   Rule of thumb for MOSFET-based flyback converters:
   35 percent of losses in MOSFETs and 60 percent of losses in rectifiers.
   Estimated losses = 16.25 W (at 80 percent efficient)
   a. MOSFETs: $P_D = (16.25\,W)(0.35) = 5.7$ Watts

b. Rectifiers:

$$P_{D(+5)} = (5/65)(16.25\,W)(0.6) = 0.75\,W$$
$$P_{D(+/-12)} = (12/65)(16.25)(0.6) = 1.8\,W$$
$$P_{D(+24)} = (36/65)(16.25\,W)(0.6) = 5.4\,W$$

Note: These dissipated heats are within the range of free-standing, screw-on heatsinks—call for samples of thermalloy heatsinks.

Predesign decisions

The topology is going to be an isolated, multiple output flyback converter that must meet the safety requirements of UL, CSA, and VDE. These considerations affect the design of the final packaging, transformer, and voltage feedback designs.

The controller to be used is to be the UC3843P current-mode controller IC running at a frequency of 50 kHz.

Transformer design (refer to Section 3.5.4)

The most common style of core is an E-E core in this application. For this power level, the size that is approximately appropriate is one with 1.1 inches (28 mm) per side. I will use the "F" material from Magnetics, Inc. (3C8 Ferroxcube).

The part numbers of the core parts (Magnetics, Inc.) are: core, F-43515-EC; and bobbin, PC-B3515-L1.

1. The minimum primary inductance will be

$$L_{pri} = \frac{V_{in(min)}\partial_{(max)}}{I_{pk} \cdot f} = \frac{(127\,V)(0.5)}{(2.81\,A)(50{,}000)} = 452\,\mu H$$

2. The air-gap required to prevent the core from entering saturation is

$$l_{gap} = \frac{(0.4\pi L_{pri} I_{pk})10^8}{A_c B_{max}^2} = \frac{.4(3.14)(.000451)(2.81)10^8}{(.904\,cm^2)(2{,}000\,G)^2} = 0.044\,cm = 17\,mils.$$

The closest standard gap for this core is 67 mils with an AL of 100 mH/1000 T. The final core part numbers are: gapped piece, F-43515-EC-02; ungapped piece, F-43515-EC-00.

3. The maximum number of turns required for the primary winding is

$$N_{pri} = 1000\sqrt{\frac{L_{pri}}{A_L}} = 1000\sqrt{\frac{.451\,mH}{100\,mH}} = 67.2 \text{ turns—round to 67 turns}$$

4. The number of turns required for the +5 V output are:

$$N_{sec} = \frac{N_{pri}(V_O + V_D)(1-\partial_{max})}{V_{in(min)} \cdot \partial_{max}} = \frac{(67\,T)(5\,V + .5)(1 - .5)}{(127\,V)(.5)} = 2.9 \text{ turms—rounds to 3.0 turns}$$

5. The number of turns for the remaining windings are

$$N_{sec2} = \frac{(V_{O2} + V_{D2})N_{sec2}}{(V_{O1} + V_{D1})}$$

+/−12 V:

$$N_{12} = \frac{(12\,\text{V} + .9\,\text{V})(3)}{(5\,\text{V} + .5\,\text{V})}$$
$$= 7.03 \text{ turns—round to } 7.0 \text{ turns}$$

+24 V:

$$N_{24} = \frac{(24\,\text{V} + .9\,\text{V})(3)}{(5\,\text{V} + .5\,\text{V})}$$
$$= 13.6 \text{ turns—round to } 14 \text{ turns}$$

Checking the errors in the respective output voltages:
+/−12 V: 11.93 V OK
+24 V: 24.76 V OK

Transformer winding technique

Because the transformer must operate in a safety regulated environment, I am going to use an interleaved method of winding as shown in Figure 1. Three layers of Mylar tape must be used between the primary and the secondary with a 2 mm space from the ends of the bobbin for the VDE creepage requirement as shown in Figure 3–68.

The wire gauges used in the respective winding are:

Primary: #24 AWG, 1 strand each section
+5 V: #24 AWG, 4 strands
+12 V: #20 AWG, 2 strands
−12 V: #22 AWG, 2 strands
+24 V: #22 AWG, 2 strands
V_{aux}: #26 AWG, 1 strand

The wiring arrangement of the windings is shown in Figure 3–69.

Designing the output-filter section (refer to Section 3.6)

The output rectifiers. +5 V output:

$$V_R > V_{out} + \frac{N_{sec}}{N_{pri}} V_{in(max)}$$
$$V_R > 5\,\text{V} + (3\,\text{T}/67\,\text{T})(340\,\text{V}) > 20.3\,\text{V}$$

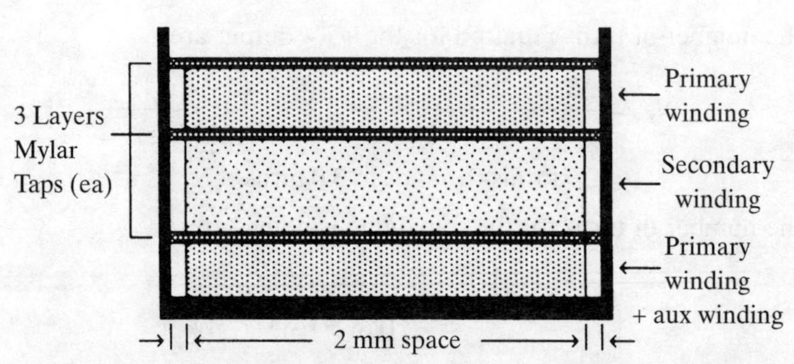

Figure 3–68 The transformer winding technique for design example 3.15.3.

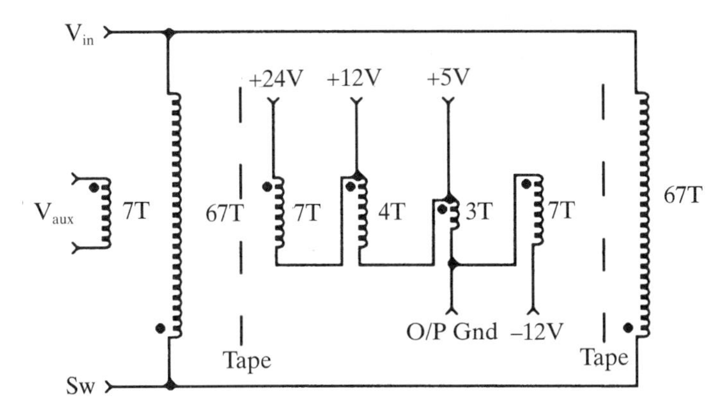

Figure 3–69 Transformer winding arrangement of design example 3.15.3.

IFWD: $I_F > I_{av} > 1$ A. Select Schottky rectifiers P/N MBR340.
+/−12 V: Use the same approach as above, Select MBR370.
+24 V: Select MUR420.

Determining the minimum values of output filter capacitors.
+5 V output:

$$C_{out(min)} = \frac{I_{out(max)} T_{off(max)}}{V_{ripple(desired)}}$$
$$= [(1.5\,\text{A})(18\,\mu\text{S})]/(100\,\text{mV})$$
$$= 270\,\mu\text{F make 2 ea } 150\,\mu\text{F @ } 10\,\text{V}$$

+/−12 V output:

$$C_{out} = 180\,\mu\text{F make 2 ea } 100\,\mu\text{F @ } 20\,\text{V}$$

+24 V Output:

$$C_{out} = 180\,\mu\text{F make 3 ea } 47\,\mu\text{F @ } 35\,\text{V}$$

Design of the controller-driver section
Selecting the power semiconductors (refer to Section 3.4). The power switch (power MOSFET)
The V_{DSS}:

$$V_{DSS} > V_{flbk} = V_{in(max)} + \frac{N_{pri}}{N_{sec}}(V_{out} + V_d)$$
$$V_{DSS} > 340\,\text{V} + (67\,\text{T}/3\,\text{T})(5\,\text{V} + 0.5\,\text{V}) > 462\,\text{V}$$

The I_D: approximately I_{pk} or > 3 A.
Select the IRF740.

Selecting the SMPS controller IC. The important factors within this application that affect the choice of switching power supply controller IC are MOSFET driver needed (totem-pole driver), single-ended output, 50 percent duty cycle limit desired, and current-mode control desired. The popular industry choice that meets these needs is the UC3845B.

Designing the voltage feedback loop (refer to Section 3.9)
The voltage feedback loop must be isolated from the input voltage line and the control IC. An optoisolator must be used. To minimize the drift effect of the optoisolator an error amplifier is desired on the secondary side. A TL431CP does this job nicely. The topology of the feedback circuitry is shown in Figure 3–70.

To improve the effects of poor output cross-regulation, sensing some current from each positive output greatly improves the behavior of each output in response to changes in individual loads.

The design of this section begins at the control IC. It is decided to bypass the error amplifier inside the UC3845. This means that the optoisolator must drive the same circuitry as the error amplifier itself. The error amplifier has a pull-up current source of 1.0 mA. The TL431 must draw 1.0 mA through the optoisolator LED in order to operate, and any control current must be added to this amount. If we arbitrarily assign a value of 1.0 mA/volt R_1 becomes

$$R_1 = \frac{5.0\,\text{V}}{5.0\,\text{mA}} = 1.0\,\text{K ohms}$$

R_2 (optoisolator LED bias resistor) is

$$R_2 = \frac{(5.0\,\text{V} - (2.5\,\text{V} + 1.4\,\text{V}))}{6.0\,\text{mA}} = 183\,\text{ohms, make } 180\,\text{ohms}$$

Let us assign a sense current of approximately 1 mA. Then for R_3

$$R_3 = \frac{2.5\,\text{V}}{1.0\,\text{mA}} = 2.5\text{K, make } 2.7\,\text{K}$$

The actual sense current becomes

$$I_{\text{sense}} = \frac{2.5\,\text{V}}{2.7\,\text{K}} = 0.926\,\text{mA}$$

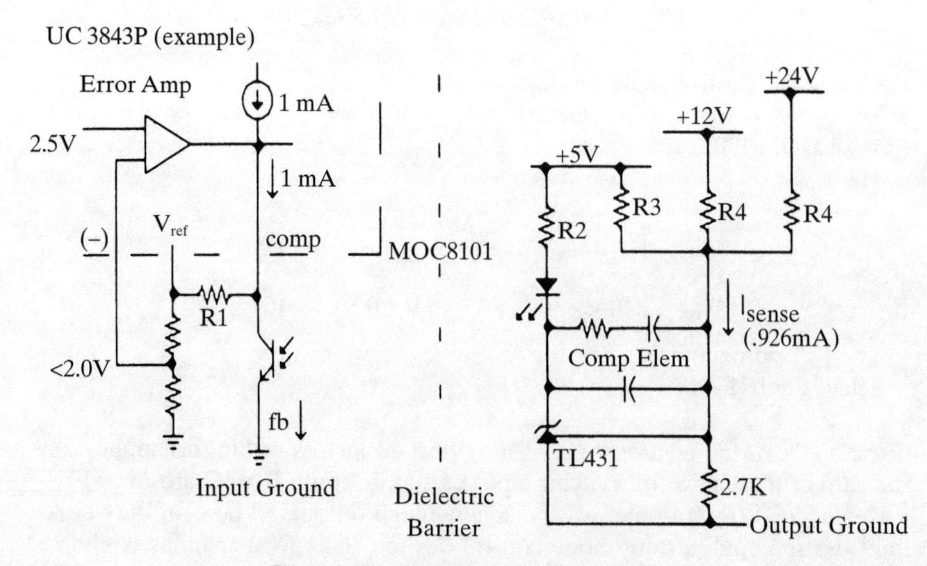

Figure 3–70 The voltage feedback circuit.

Now let us assign the degrees of sensing for each positive output in response to the requirements of the application. The +5 have an MCU and HCMOS logic and should be fairly tightly regulated to within 0.25 V. The +/– 12 V has operational amplifiers and an RSR232 driver, which are relatively insensitive to supply variations. The +24 V need only be regulated to within +/– 2 V. So assign sense current proportions as follows: +5 V, 70 percent; +12 V, 20 percent; +24 V, 10 percent.

The +5 V sense resistor (R_4) becomes

$$R_4 = \frac{(5.0\,\text{V} - 2.5\,\text{V})}{0.7(0.926\,\text{mA})} = 3,856\,\text{ohms, make } 3.9\,\text{K}$$

R_5 (+12 V)

$$R_5 = \frac{(12\,\text{V} - 2.5\,\text{V})}{0.2(0.926\,\text{mA})} = 51,295\,\text{ohms, make } 51\,\text{K}$$

R_6 (+24 V)

$$R6 = \frac{(24\,\text{V} - 2.5\,\text{V})}{0.1(0.926\,\text{mA})} = 232\,\text{K, make } 240\,\text{K}$$

The compensation elements will be designed later.

The current sense resistor

The current sense resistor in the source connection of the power MOSFET is approximately

$$R_{\text{sc}} = \frac{V_{\text{sc(max)}}}{I_{\text{pk}}} = \frac{0.7\,\text{V}}{2.81\,\text{Amps}} = 0.249\,\text{ohms}$$

This value usually must be lowered during the testing phase when it is sometimes determined that the supply cannot provide the full rated load at the minimum input voltage.

Designing the feedback loop compensation

This supply has the single-pole output filter characteristic found in all current-mode switching power supplies. Refer to Appendix B.

For the control-to-output characteristics, the lowest exhibited filter pole for the +5 V output (minimum load) is

$$f_{\text{fp}} = \frac{1}{2\pi(5\,\text{V}/.75\,\text{A})(300\,\mu\text{F})} = 79.6\,\text{Hz}$$

Since the +5 is the most sensed output but has only 5 W of the 65 W output, let us calculate the highest power output filter pole and use it for our compensation activities since its filter pole will be much lower in frequency and pushing its compensation lower in frequency will only boost the phase of the closed loop. This is erring in the safe direction.

$$f_{\text{fp(24)}} = \frac{1}{2\pi(24\,\text{V}/.25\,\text{A})(141\,\mu\text{F})} = 11.8\,\text{Hz}$$

The dc gain of the system is

$$A_{DC(max)} = \frac{(340\,V - 5.0\,V)^2\,3T}{(340\,V)(1\,V)67T} = 14.77$$

The gain expressed in dB is

$$G_{DC(max)} = 20\,Log(14.7) = 23.4\,dB$$

Assign the approximate location of the zero caused by the ESR of the output filter capacitor at 20 KHz.

Assign the location of the error amplifier compensating pole and zero. To compensate for the light-load output filter pole with a zero,

$$f_{ez} = f_{fp(light\ load)}$$
$$f_{ep} = f_{z(ESR)}$$

The bandwidth of the closed-loop system should be equal to or less than 10 KHz. The amount of gain needed to be added by the error amplifier in order to achieve this bandwidth is

$$G_{XO} = 20\,Log\left(\frac{10\,kHz}{11\,Hz}\right) - 23.4\,dB = 36.6\,dB$$

or an absolute gain of 63.

Calculate the values of the compensating elements using the component numbering system in the appendix:

$$C_1 = \frac{1}{2\pi(3.9\,K)(63)(20\,kHz)} = 32\,pF$$
$$R_2 = 3.9\,K(63) = 240\,K$$
$$C_2 = \frac{1}{2\pi(11.8\,K)(240\,K)} = 0.056\,\mu F$$

Designing the input EMI filter section (refer to Appendix E)

I will be using a second-order, common-mode filter. The main purpose of an EMI filter is to filter the switching noise and its harmonics from the input power line. One starts by estimating the amount of attenuation needed at the switching frequency.

A good starting point is to assume the need for 24 dB of attenuation at 50 kHz. That make the corner frequency of the common-mode filter

$$f_C = f_{sw}10^{\left(\frac{Att}{40}\right)}$$

where Att is the attenuation needed at the switching frequency in negative dB.

$$f_C = (50\,kHz)10^{\left(\frac{-24}{40}\right)} = 12.5\,kHz$$

Assume that a damping factor of 0.707 or greater is good and provides a −3 dB attenuation at the corner frequency and does not produce noise due to ringing. Also assume that the input line impedance is 50 ohms since the regulatory agencies use an LISN test which make the line impedance equal this value. Calculate the values needed in the common-mode inductor and "Y" capacitors:

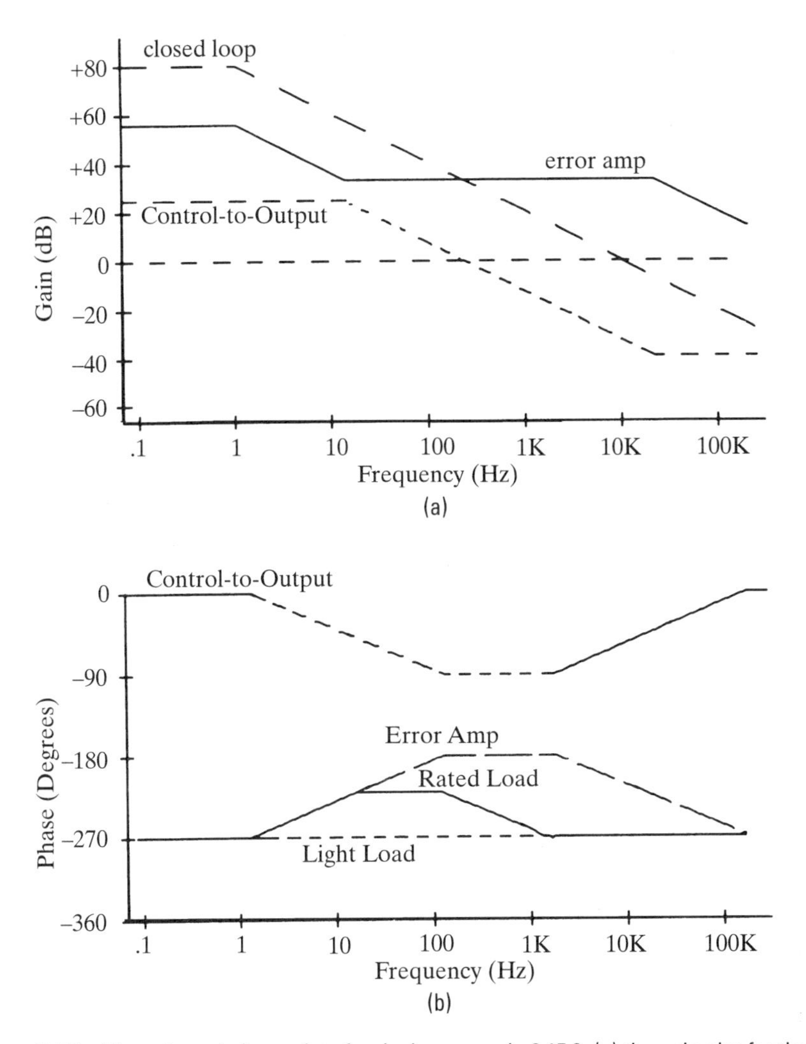

Figure 3–71 The gain and phase plots for design example 3.15.3: (a) the gain plot for the power supply; (b) the phase plot for the power supply.

$$L = \frac{R_L \cdot \zeta}{\pi \cdot f_C} = \frac{(50)(0.707)}{\pi(12.5\,\text{KHz})} = 900\,\mu\text{H}$$

$$C = \frac{1}{(2\pi f_C)^2 L} = \frac{1}{[2\pi(12.5\,\text{KHz})]^2(900\,\mu\text{H})}$$
$$= 0.18\,\mu\text{F}$$

Real world values do not allow a capacitor of this large a value. The largest value capacitor that will pass the ac leakage current test is 0.05 μF. This is 27 percent of the calculated capacitor value, so the inductor must be increased 360 percent in order to maintain the same corner frequency. The inductance then becomes 3.24 mH and the resultant damping factor is 2.5, which is acceptable.

Coilcraft offers off-the-shelf common-mode filter chokes (transformers); the part number closest to this value is E3493. With this filter design I can expect a minimum of –40 dB between the frequencies of 500 KHz and 10 MHz. If later

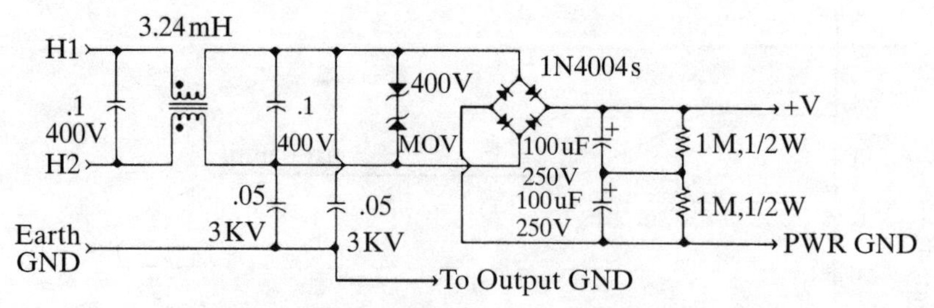

A. AC Input Section.

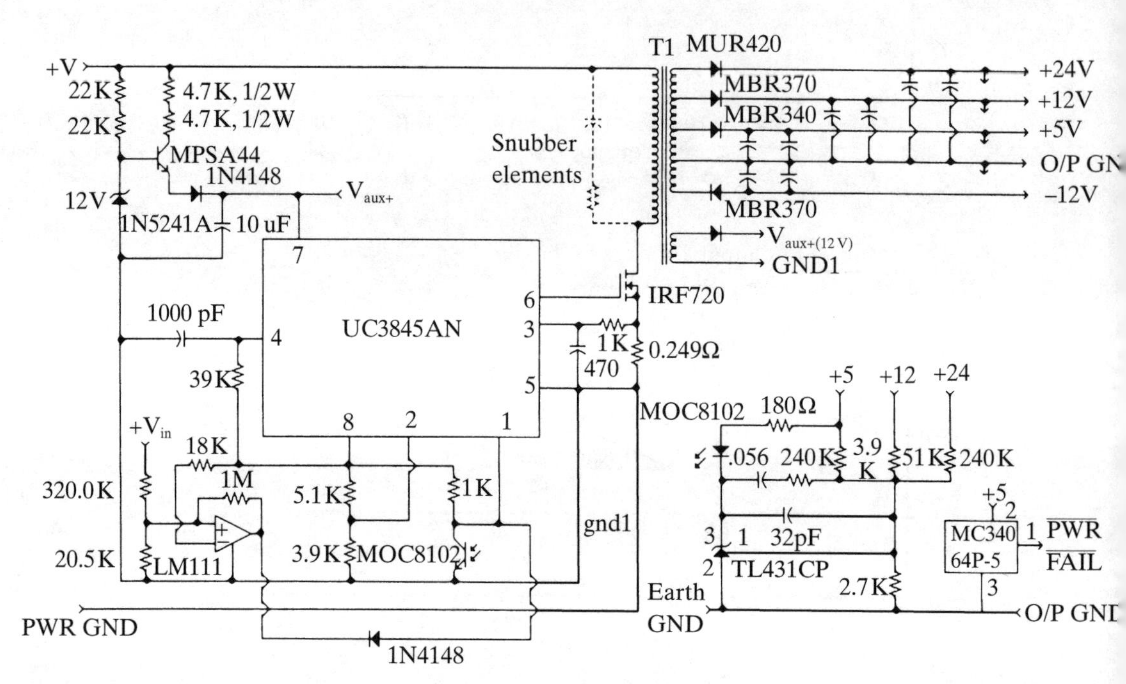

Figure 3–72 65 watt, off-line flyback converter.

during the EMI testing stage, I find I need additional filtering, I will add a third order to the filter design by using a differential-mode filter.

The final schematic can be seen in Figure 3–72.

3.15.4 A 280 Watt, Off-line, Half-Bridge Converter

This switching power supply is intended as a bulk power supply within a distributed power system. It will provide a "safe" bus voltage of 28 volts dc to the system. It has an ac input in which a tap must be changed in order to accommodate either a 110 VAC or a 240 VAC ac input power system. Please refer to Figure 3–77.

Design specification

Input voltage ranges:	90 to 130 VAC, 50/60 Hz
	200 to 240 VAC, 50/60 Hz
Output voltage:	+28 VDC at 10 A maximum rated,
	1 A minimum load
Output ripple voltage:	50 mVp-p
Output regulation:	+/− 2 percent
Cost target:	$50.00 US in 1000 piece quantities

"Black box" predesign considerations

1. Rated output power: $P_o = (28\,V)(10\,A) = 280\,W$
2. Estimated input power: $P_{in(est)} = 280\,W/0.8 = 350\,W$
3. DC input voltages (a voltage doubler is used for 110 VAC)
   a. From 110 VAC: $V_{in(low)} = 2(1.414)(90\,VAC) = 254\,VDC$
   $V_{in(hi)} = 2(1.414)(130\,VAC) = 368\,VDC$
   b. From 220 VAC: $V_{in(low)} = 1.414(185\,VAC) = 262\,VDC$
   $V_{in(hi)} = 1.414(270\,VAC) = 382\,VDC$
4. Average input currents (dc)
   a. Highest average: $I_{in(max)} = 350\,W/254\,VDC = 1.38\,A$
   b. Lowest average: $I_{in(min)} = 350\,W/382\,VDC = 0.92\,A$
5. Estimated maximum peak current $I_{pk} = 2.8(280\,W)/254\,VDC = 3.1\,A$

Design decisions

This is going to be a current-mode controlled half-bridge topology. It will have soft-start to minimize the start-up surge of the system. The power supply must meet UL, CSA, and VDE safety regulations.

It is decided that the power supply will operate at an operating frequency of 100 KHz, and the controller IC will be an MC34025P.

Transformer design (refer to Section 3.5.3)

I will be using an E-E core since it has the largest wire area of all the core styles. The large wire area will be needed for all the additional insulating layers required for the VDE certification. No air-gap is needed for bipolar forward-mode converters. The core material is going to be 3C8 (Ferroxcube) or "F" (Magnetics, Inc.) material. This material will yield reasonable core losses at this frequency.

The estimated core size for this application is approximately 1.3 inches (33 mm) on each side. The closest sized core is Magnetics part number F-43515. I will request that core and also F-44317, which is the next core size larger, just in case the windings grow beyond the limits of the window area.

For the F-43515 core. When calculating the number of turns for the primary, one needs to consider the initial start-up condition, which places the full input voltage across the primary winding for the first few milliseconds of operation. The designer must be assured that the transformer will not enter saturation during this period. The transformer design conditions become the maximum specified ambient temperature and the highest specified ac input. The number of primary turns needed for the primary winding will be

$$N_{pri} = \frac{(382\,V)\cdot 10^8}{4(100\,kHz)(2800\,G)(.904\,cm^2)} = 37.7\,T$$

make 38 turms.

Note: This would make the B_{max} during steady state operation at about 1300 to 1500 G.

$$N_{sec} = \frac{1.1(28\,V + 0.5\,V)(38\,T)}{(254\,V - 2\,V)(.95)} = 4.97\,T$$

Since there cannot be fractonal turns on an E-E core, round this up to 5.0 turns. This will yield a maximum duty cycle at the minimum input voltage of

$$\frac{4.95}{5.0} = \frac{X}{0.95}, \quad X = 94\%$$

This is acceptable.

For the auxiliary winding.

$$\frac{(5\text{T})(12.5\,\text{V})}{(28.5\,\text{V})} = 2.2\,\text{T}$$

Rounding the result to 2.0 turns would yield a secondary voltage of 11.4 V at low-line, which is acceptable.

Wire Gauges.

Primary: #19 AWG or equivalent
Secondary: #12 AWG or equivalent
Auxiliary: #28 AWG

Transformer winding technique

The transformer will be wound in an interleaved fashion. The primary winding will be made up of four strands of #22 AWG, and the secondary will be 5 mil thick foil, 0.5 inches (12 mm) width. Two strands of the primary will be wound first around the bobbin, along with the auxiliary winding, then three layers of 1 mil Mylar tape for insulation. The secondary then will be wound on the core. Then add another layer of Mylar tape and the remaining two strands of primary winding. The windings are then covered with at least two layers of Mylar tape. This process can be seen in Figure 3–73.

Selection of power semiconductors

1. Power switches: (refer to Sections 3.4 and 3.7)

$$V_{DSS} > V_{in} > 382\text{ VDC make } 500\text{ V}$$
$$I_D > I_{in(av)} > 2.75\text{ A, make} > 4\text{ A}$$

   Select IRF730
2. Output rectifiers:

$$V_R > 2V_{out} = 56\text{ VDC make} > 70\text{ VDC}$$
$$I_{FWD} > I_{out(max)} > 10\text{ A, make } 20\text{ A}$$

Select MBR20100CT

Design of the output filters

1. The minimum ac output filter inductance (refer to Section 3.5.5)

$$L_{o(min)} = \frac{(47\text{ V} - 28\text{ V})(4.25\,\mu\text{s})}{1.4(1\text{ Amp})} = 57.6\,\mu\text{H}$$

   Using the LI^2 method of determining the size of an MPP toroid, one determines the Magnetics core part number P/N 55930A2. The number of turns should be

$$N_{Lo} = 1000\sqrt{\frac{.576}{157}} = 19.2\,\text{T make } 20\,\text{T}$$

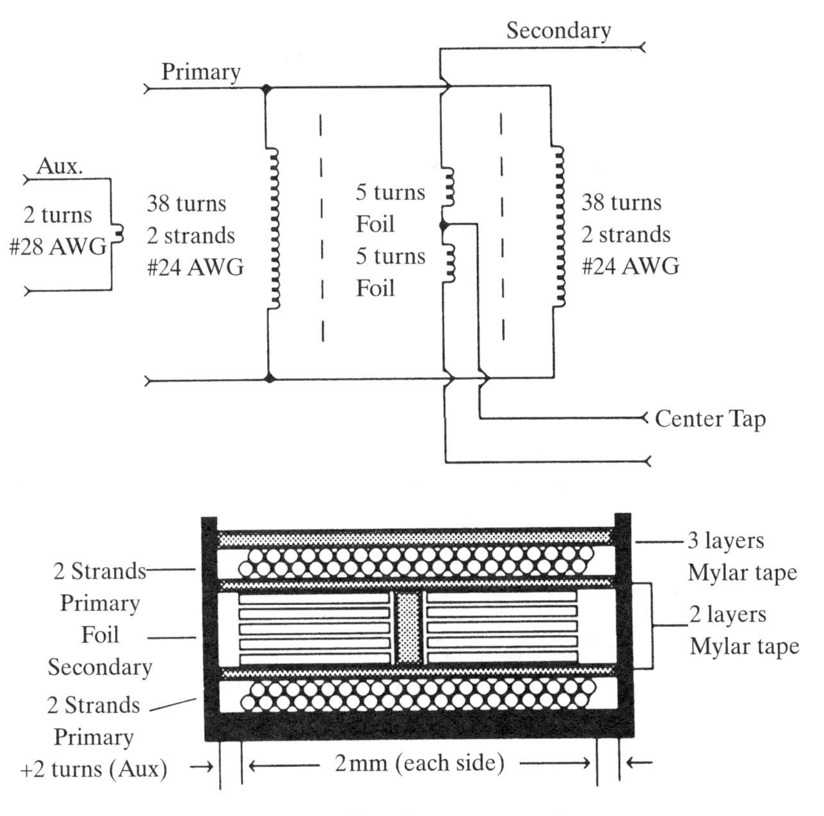

Figure 3–73 Transformer construction.

The toroid should have an overall wire gauge of #12AWG. I will use 100 strand Litz wire to minimize the skin effects.

2. The minimum output filter capacitance (refer to Section 3.6):

$$C_{o(\min)} = \frac{(10\,\text{A})(4.25\,\mu\text{S})}{.05\,\text{V}_{\text{p-p}}} = 850\,\mu\text{F}$$

Make this capacitor four paralleled aluminum electrolytic capacitors of 220µF in value. This will keep the RMS ripple current experienced by each capacitor to less than 3A.

3. Design the output dc filter choke (refer to Section 3.5.7). Referring to the permeability versus dc bias graph (Figure 3–22) and selecting a permeability that does not excessively degrade at a reasonable dc bias level, I choose a 60mu core at an "H" level of 40 Oersteds.

Using the same core size as above and solving Equation 3.35 one gets:

$$N = \frac{(400)(6.35\,\text{cm})}{0.4\pi(10\,\text{A})} = 20.2\text{ turns, make 20 turns}$$

We still need #12AWG wire, and since Litz wire is easier to wind onto the core, we will use that. Otherwise, in this application, it is unnecessary.

Designing the gate drive transformer (refer to Section 3.5.8)

This procedure is identical to that of designing a forward-mode power transformer. I will use a small E-E core and will separate the primary and the secondary windings with several layers of insulating Mylar tape. The gate drive transformer has the same voltage stresses as the main transformer and therefore should be insulated accordingly. The tape will not allow a MOSFET failure to propagate to the control circuitry.

I will use Magnetics E-E core P/N F-41808EC without an air-gap.

1. Solve for the number of turns on the primary winding. (Use Equation 3.20 a, b, or c. This transformer is designed the same way as a forward-mode power transformer).

$$N_{pri} = \frac{(18\,V)10^8}{4(100\,kHz)(1800\,G)(.228\,cm^2)} = 11\ turns$$

2. Since the input control IC bias voltage is about 15 V, we need to make the turns ratio about 1:1. So the secondary turns will be 11 turns.

The primary winding will be wound first, then the two layers of Mylar tape. The secondaries will be bifilar-wound next and then covered with two layers of Mylar tape. The wire gauges will be #30 AWG for each winding.

Designing the bootstrap start-up circuit (refer to Section 3.10)

I will use a similar boostrap start-up circuit as the previous design example. For control ICs that have a small voltage hysteresis, such as the MC34025, the start-up circuit must sustain the full operating current of the IC and MOSFET drive current during overcurrent conditions and start-up. By providing an auxiliary voltage from the main transformer that is higher than the "regulated" output of the start-up stage, it cuts-off any current flow through the highly dissipative collector resistors during operation. This will save a few watts during normal operation.

The transistor is operating as a starved (highly current-limited) linear regulator. Here the collector resistors dissipate the bulk of the power. The transistor, though, should be able to handle about one watt of dissipation at an ambient temperature of +50°C. This dictates that a TO-220 package should be used. It must also handle 400 VDC in breakdown voltage. A TIP50 would be more than sufficient for this purpose.

The total resistance of the collector resistors (two resistors in series for voltage breakdown reasons) should have an approximate resistance of

$$R_{Coll} = \frac{254\,V}{15\,mA} = 16.9\,K, \text{ make 2 each } 8.2\,K$$

The power dissipated by these resistors is

$$P_{D(max)} = \frac{(382\,V)^2}{16.4\,K} = 8.8\,W$$

If we place two 8.2 K ohm, 5 W resistors in series, we can spit the dissipation between two resistors and also guarantee that the resistors will not reach their breakdown voltage.

For the base resistor(s) design,

$$R_{base} = \frac{254\,V}{.50\,mA} = 508\,K, \text{ make } 510\,K\ \text{ohms}$$

Once again, place two resistors in series to avoid resistor voltage breakdown problems (250 V for 1/4 W resistors). So two 240 K, 1/4 W resistors will be used.

Design of the controller circuitry

The overall control strategy will be current-mode control. The industry standard controller is the UC3525N or the MC34025P. The IC can be configured as either a current-mode or voltage-mode controller. I will use the current-mode control configuration.

The oscillator frequency is set by referring to a timing graph. In order to provide a 100 kHz operating frequency, the values for RT and CT are

RT = 7.5 K ohms
CT = 2200 pF

The design of the current sensing circuit

I have decided to use a current transformer to sense the primary current waveform, since resistive methods are impractical in the half-bridge topology. Several transformer manufacturers make current transformers for such a purpose wound on a toroidal core. Coilcraft makes current transformers with 50, 100, and 200 turns on their secondaries. The secondary voltage must be determined in order to have representative current waveforms of the level to work with the control IC. The voltage needed on the output of the current transformer is

$$V_{CT(sec)} \approx V_{sc} + 2\mathrm{V}_{fwd} = 1.0\,\mathrm{V} + 2(0.65) = 2.3\,\mathrm{V}$$

Selecting a 100:1 ratio current transformer, the secondary current is given by

$$I_{sec} = (N_{pri}/N_{sec})I_{pri} = (3.1\,\mathrm{Amps})/100 = 31\,\mathrm{mA}$$

The resistor needed to convert this current into the needed voltage is

$$R_{sc} = 2.3\,\mathrm{V}/31\,\mathrm{mA} = 75\,\mathrm{ohms}$$

In order to improve the slope compensation circuitry that depends upon a resistor to ground all the time, I will split this resistor between the secondary winding of the current transformer and after the rectifiers. I will double the value of the two resistors (150 ohms each), so that when the diodes are conducting, the net value is the same.

It is necessary to add a small leading edge spike filter to the current sensing output. To keep the time delay to a reasonable value, I will use a 1 K ohm resistor and a 470 pF capacitor.

Slope compensation

Every current-mode control application that exceeds 50 percent duty cycle must have slope compensation on the current ramp waveform. Otherwise an instability will occur whenever the duty cycle exceeds 50 percent. This is typically done by summing into the current waveform some of the oscillator ramp waveform. This will increase the slope of the current waveform and therefore trip the current sense comparator earlier. A common problem is the inadvertent loading of the oscillator, so I will use a PNP emitter-follower to buffer the oscillator. The circuit configuration can be seen in Figure 3–74.

The design of the slope compensation circuit is almost fairly qualitative and may eventually need to be adjusted at the breadboard stage. To estimate how much additional ramp voltage is needed to keep the power supply stable, one performs the following equation. A_i is the gain or step-down influences of the transformers between the output and the current sense pin.

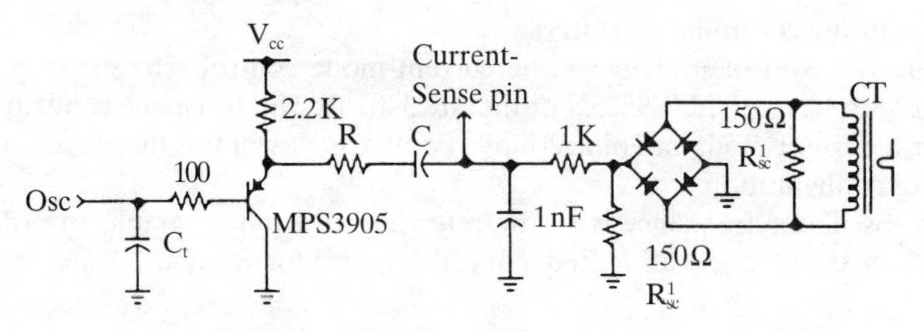

Figure 3–74 Current slope compensation pin.

$$S_e = \frac{V_{sec}(\partial_{max} - 0.18)A_i}{L_o} = \frac{V_{sec}(\partial_{max} - 0.18)N_{sec} \cdot N_{1CT} R_{sc}}{L_o \cdot N_{pri} \cdot N_{2CT}}$$
$$= \frac{32\,V(0.94 - 0.18)(5\,T)(1\,T)(75\,\Omega)}{(58\,\mu H)(38\,T)(100\,T)}$$
$$= 4.1 \times 10^4\,V/\mu Sec$$

The actual level of the ramp voltage needed to be summed into the current ramp signal at the end of the maximum on-time is

$$\Delta Vr = (4.1 \times 104\,v/\mu S)(4.25\,\mu S) = 0.174\,\text{Volts}$$

The components connected between the emitter-follower and the current-sensing filter capacitor can be thought of as a resistor divider. An additional 0.17 V needs to appear at pin 7 (through a 1 K resistor) so the amount of current that must be contributed to that node is 0.17 V/1 K which is 170 μA. The capacitive coupling of the PNP to pin 7 essentially centers the oscillator waveform upon the current ramp. So,

$$R_{sc} = \frac{V_{osc}}{2 \cdot I_{sc}} = \frac{(4.5\,V - 2.3\,V)}{2(170\,\mu A)} = 6.47\,K \text{ make } 6.2\,K$$

Designing the voltage feedback loop

The voltage feedback loop must be isolated from the primary to the secondary. I am choosing an optoisolated method. The voltage feedback circuit will be the arrangement shown in Figure 3–75.

The error amplifier within the MC34025 has a totem-pole output circuit, which means that its output is not easily overriden. It will be used as a simple voltage follower and the error amplifier function will take place completely within the TL431 on the secondary side of the power supply.

Starting on the secondary side of the power supply, assign the sense current through the voltage sensing resistor divider to be approximately 1 mA (or 1 K per volt). Using the closest resistor value for the lower resistor a 2.7 K resistor yields and actual sense current of 0.926 mA. One can immediately calculate the value of the upper resistor (R3):

$$R_{upper} = \frac{(28\,V - 2.5\,V)}{0.926\,mA} = 27.54\,K$$

Make this 27 K

The value of the resistor that would provide the bias current through the optoisolator and the TL431 is set by the minimum operating current require-

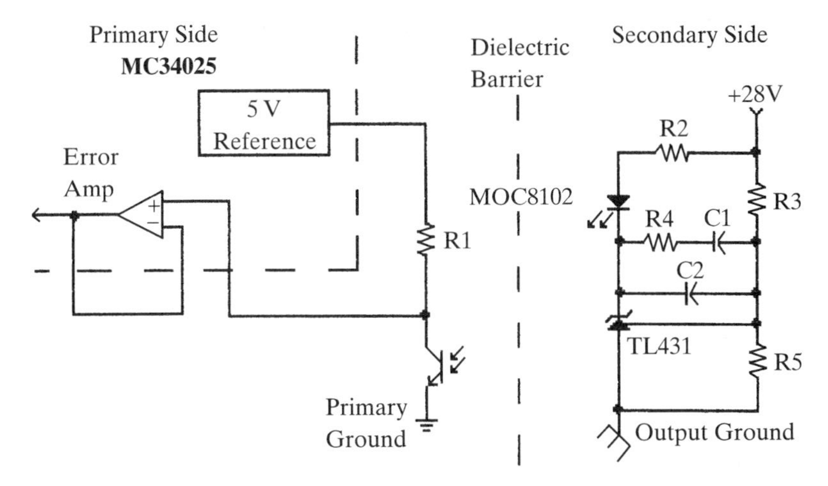

Figure 3–75 Voltage feedback arrangement.

ments of the TL431. This is a minimum of 1 mA. Let us assign the maximum current through the branch to be 6 mA. That makes the bias resistor value (R5)

$$R_{\text{bias}} = \frac{(28\,\text{V} - (2.5\,\text{V} + 1.4\,\text{V}))}{6\,\text{mA}}$$

$$= 4016\,\text{ohms, make this } 3.9\,\text{K}, \frac{1}{4}\,\text{W}$$

On the primary side of the power supply, the transistor output of the optoisolator will be a simple common-emitter amplifier. The MOC8102 has a typical current transfer ratio of 100 percent with a +/– 25 percent tolerance. When the TL431 is full-on, 6 mA will be drawn from the transistor within the MOC8102. The transistor should be in a saturated state at that time, so its collector resistor (R1) must be

$$R_{\text{collector}} = \frac{(5\,\text{V} - 0.3\,\text{V})}{6\,\text{mA}} = 783\,\text{ohms}$$

make this 820 ohms

This completes the design of the uncompensated voltage feedback circuit.

Designing the voltage feedback compensation (refer to Appendix B)

Current-mode controlled, forward-mode converters have a one pole filter characteristic. The optimum compensation is the single-pole, single-zero method of compensation.

One starts this by calculating the approximate elements of the control-to-output characteristic.

The intrinsic gain of the open loop power supply is

$$A_{\text{DC}} = \frac{V_{\text{in}}}{\Delta V_{\text{e}}} \cdot \frac{N_{\text{sec}}}{N_{\text{pri}}}$$

$$= \frac{(382\,\text{V})(5\,\text{T})}{(1\,\text{V})(38\,\text{T})} = 50.2$$

In terms of dB the gain at dc of the power supply is

$$G_{DC} = 20\,\text{Log}(A_{DC}) = 34\,\text{dB}$$

The location of the lowest manifestation of the output filter pole occurs at the lightest anticipated output load on the power supply. The equivalent light-load resistance is 28 V/1 A or 28 ohms. It is found from

$$\begin{aligned} f_{fp} &= \frac{1}{2\pi R_L C_o} \\ &= \frac{1}{2\pi(28\,\Omega)(880\,\mu\text{F})} \\ &= 6.5\,\text{Hz} \end{aligned}$$

The zero in the control-to-output characteristic caused by the ESR of the output filter capacitor can be found by two methods: if the value of the actual ESR is known from the capacitor's data sheet, then the location of the zero can be calculated, if not, it can be grossly estimated. Using four aluminum electrolytic capacitors in parallel should cut the total ESR to one-quarter that exhibited by each. I will estimate the zero to be at 10 KHz.

Using the procedure given in Appendix B, the location of the pole and zero within the error amplifier's compensation network are

$$\begin{aligned} f_{ez} &= f_{fp} = 6.5\,\text{Hz} \\ f_{ep} &= f_{z(\text{ESR})} = 10\,\text{kHz} \end{aligned}$$

I am selecting the bandwidth of the closed loop (fxo) power supply to be 6 kHz. It could be as high as 15 to 20 kHz, but there is a double pole at one-half the switching frequency which, if approached too closely, would ruin the phase and gain margin of the closed loop.

The error amplifier must add gain to the eventual closed loop function in order to achieve the desired closed loop bandwidth. This gain is

$$\begin{aligned} G_{XO} &= 20\,\text{Log}\left(\frac{f_{XO}}{f_{fp}}\right) - G_{DC} \\ &= 20\,\text{Log}\left(\frac{6000}{6.5}\right) - 34\,\text{dB} = 25.3\,\text{dB} \end{aligned}$$

Converting this back to absolute terms:

$$A_{XO} = 10^{\left(\frac{G_{XO}}{20}\right)} = 18.4$$

Knowing or assigning the values to the critical points in the closed-loop characteristics, one can now calculate the actual component values. (Refer to Figure 3–69 for the component designations.)

$$\begin{aligned} C_2 &= \frac{1}{2\pi \cdot A_{XO} \cdot R_3 \cdot f_{ep}} \\ &= \frac{1}{2\pi(18.4)(27\,\text{K})(4000)} = 80\,\text{pF} \\ R_4 &= A_{XO} \cdot R3 \\ &= (18.4)(27\,\text{K}) = 496\,\text{K} \end{aligned}$$

Make 510 K

$$C_2 = \frac{1}{2\pi \cdot R4 \cdot f_{ez}}$$
$$= \frac{1}{2\pi(510\text{K})(6.5\text{Hz})} = 0.048\,\mu\text{F}$$

Make $0.05\,\mu\text{F}$

Designing the rectifier and input filter circuit (refer to Section 3.12)
The approximate value of the bulk input filter capacitor is found from

$$C_{in} = \frac{I_{in(av)}}{8 \cdot f \cdot V_{ripple(p-p)}}$$
$$= \frac{(1.38\,\text{A})}{8(120\,\text{Hz})(20\,\text{V})} = 72\,\mu\text{F}$$

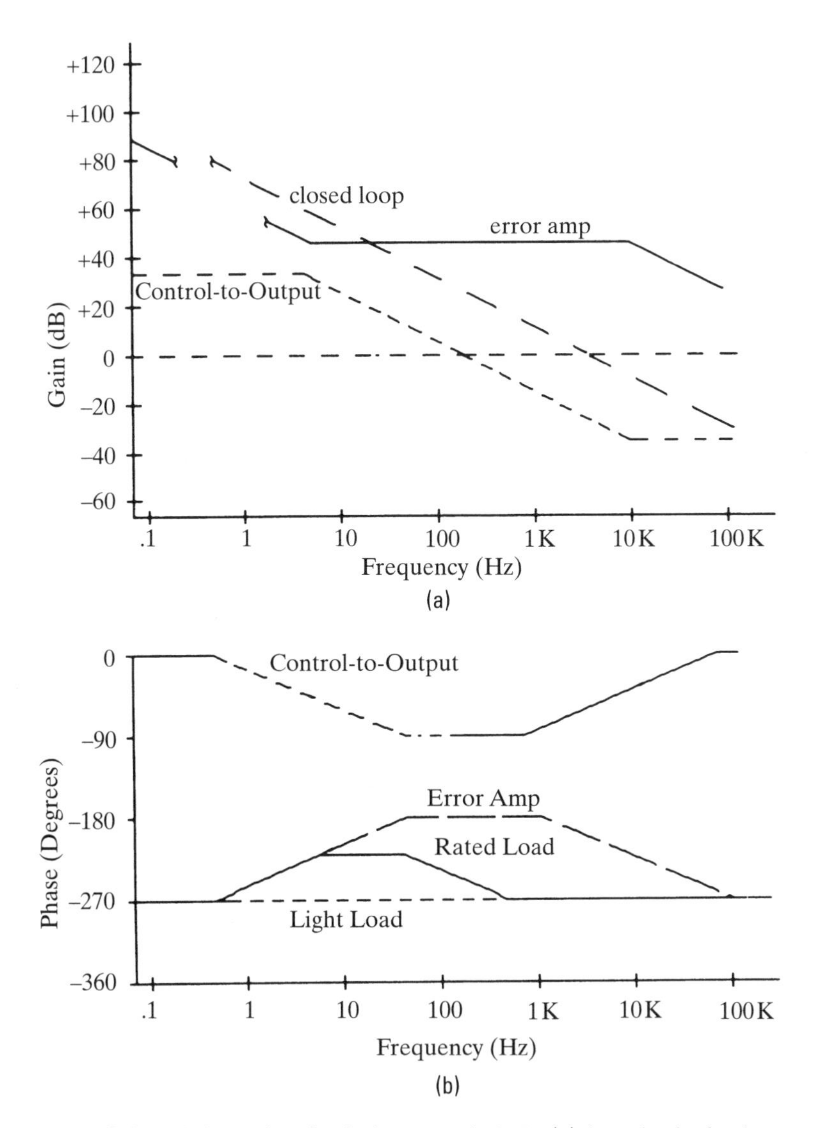

Figure 3–76 Gain and phase plots for design example 3.15.4 (a) the gain plot for the power supply; (b) the phase plot for the poser supply.

The standard recovery rectifiers used in the rectifier bridge must accommodate the highest average current expected which occurs at the ac low-line voltage. This has been calculated during the predesign estimates (black box considerations). The rated forward current should be more than 2 A and minimum blocking voltage should be twice the maximum high ac line crest voltage. This is more than 764 V. A 1N5406 would be a good choice.

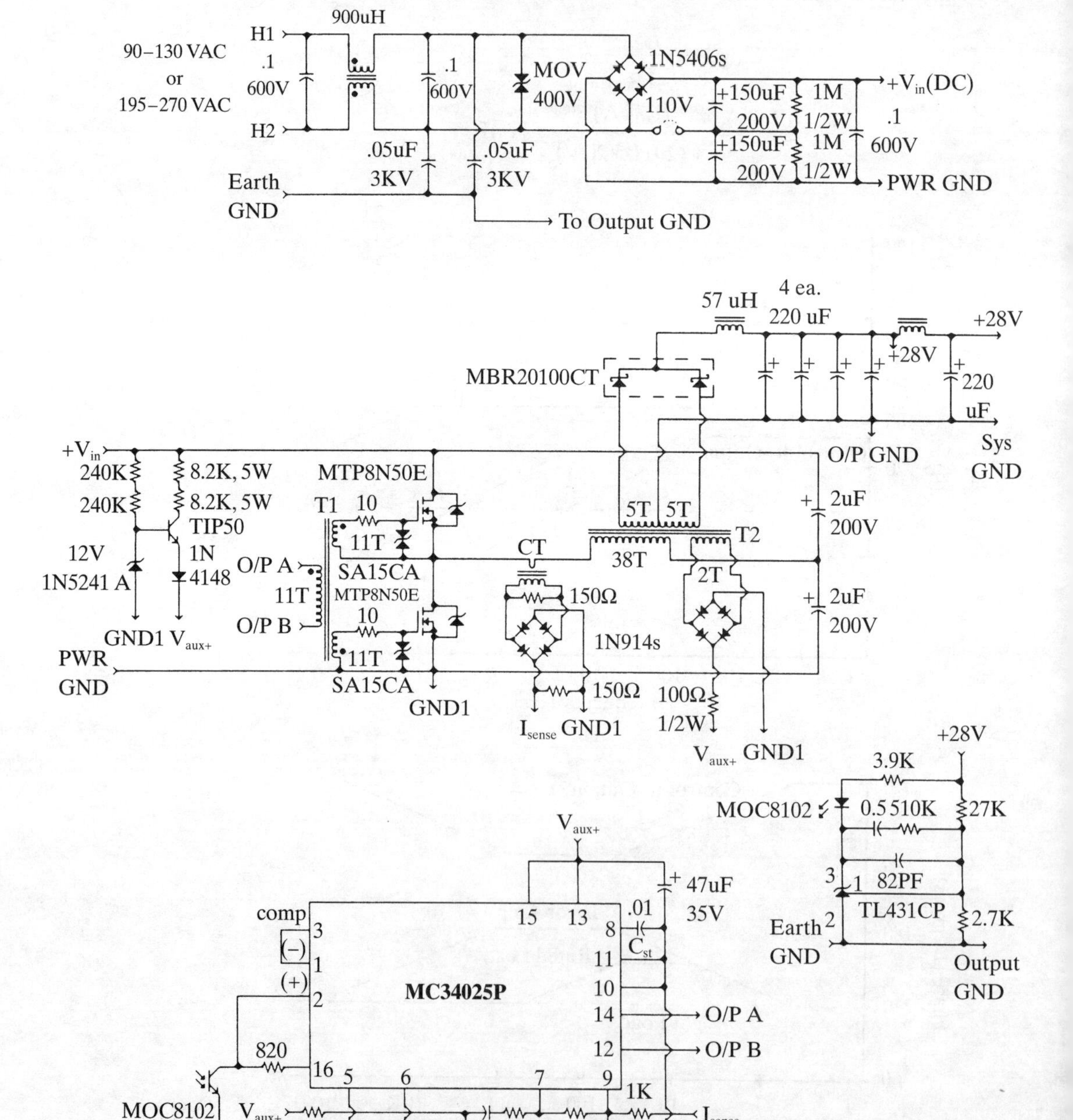

Figure 3–77 Schematic of a 100 kHz, 280 W, half-bridge converter.

Designing the input EMI filter (refer to Appendix E)
I will use a second-order, common-mode filter. The frequency of operation is 100 KHz. A good starting point is to assume the need for 24 dB of attenuation at 100 KHz. That make the corner frequency of the common-mode filter

$$f_C = f_{SW} \cdot 10^{\left(\frac{Att}{40}\right)}$$

where Att is the attenuation needed at the switching frequency in negative dB.

$$f_c = (100\,\text{kHz})10^{\left(\frac{-24}{40}\right)} = 25\,\text{kHz}$$

Assume that a damping factor of 0.707 or greater is good and provides a –3 dB attenuation at the corner frequency and does not produce noise due to ringing. Also assume that the input line impedance is 50 ohms since the regulatory agencies use an LISN test that make the line impedance equal this value. Calculating the values needed in the common-mode inductor and "Y" capacitors:

$$L = \frac{R_L \cdot \zeta}{\pi \cdot f_c} = \frac{(50)(0.707)}{\pi(25\,\text{kHz})} = 450\,\mu\text{H}$$

$$C = \frac{1}{(2\pi f_c)^2 L} = \frac{1}{[2\pi(25\,\text{kHz})]^2(450\,\mu\text{H})}$$

$$= 09\,\mu\text{F make } 1\,\mu\text{F}$$

Real world values do not allow a capacitor of this large a value. The largest value capacitor that will pass the ac leakage current test is 0.05 μF. This is 50 percent of the calculated capacitor value, so the inductor must be increased 200 percent in order to maintain the same corner frequency. The inductance then becomes 900 μH and the resultant damping factor is 2.5 which is acceptable. The resulting schematic is shown in Figure 3–77.

4. Waveshaping Techniques to Improve Switching Power Supply Efficiency

Much research and work has been done in the past two decades to improve the efficiency of the basic PWM switching power supply. During the 1980s, the improvements largely took the form of improved semiconductor devices and ferrite materials. Their contributions allowed the switching frequencies to rise and their efficiency to improve about another +5 to +10 percent over the bipolar transistor-based designs. The most recent techniques include the use of resonant and charge redirection techniques. These modifications along with the use of synchronous rectifiers (where applicable), allowed switching power supplies to routinely exceed 90 percent efficiency.

The ultimate goal of the recent circuit techniques is to reduce or eliminate the voltage-current product, primarily during the switching transitions. This "tuning" of the waveforms inside the basic PWM converter can add about +5 to +10 percent efficiency to the supply. The tuning process, though, can add a significant amount of time to the development process. Care must be taken so that more sources of EMI are not created by the rapid switching of charge within the power sections of the supply. The printed circuit board design also becomes a significant factor in the overall design of the supply.

Any energy that has been redirected away from the power stages for the purpose of improving efficiency must be reinserted back into the power section in a place where the energy can be recovered. Otherwise there will be no improvement in efficiency. So, a good understanding of both the supply's operation and the tuning circuit being placed into it is necessary to take advantage of the benefits.

4.1 Major Losses within the PWM Switching Power Supply

To improve the efficiency of a switching power supply, one must be able to identify and roughly quantify the various losses. Losses within a switching power supply roughly fall into four categories: switching, conduction, quiescent, and resistive losses. These losses usually occur in combination within any lossy component and are treated separately.

Section 3.4, Table 3–3, provides some insight as to where the major losses occur and to what degree. The losses highlighted in that table are for the basic PWM switching power supply without much effort placed into making them more efficient. These efficiency numbers, therefore, can be seen as the baseline efficiency of a particular topology. The areas where the major losses occur can be viewed on the ac nodes within the power section. From the single or multiple nodes, depending upon whether the topology is transformer-isolated or not, one can view the switching transitions and the conduction conditions of the power switch(es) and rectifier(s). The most informative ac node is the drain, or collector, of the power switch. The second most important ac node is the anode of the output rectifier(s). These nodes will be the primary focus of the majority of our work in improving the efficiency of the switching power supply.

Losses associated with the power switch

The power switch is one of two of the most prominent sources for loss within the typical switching power supply. The losses basically fall into two categories: *conduction losses* and *switching losses*. The conduction loss is where the power switch is in the ON state after the drive and switching waveforms have stabilized. Switching losses occur when the power switch has been driven into a new state of operation. The drive and switched waveforms are in a state of transition. These periods and their typical waveforms can be viewed in Figure 4–1.

The conduction loss (t_2) is measured as the product of the switch terminal voltage and current waveforms. These waveforms are typically quite linear and the power loss during this period is given in Equation 4.1.

$$P_{D(\text{conduct})} = (V_{\text{sat}})(I_{\text{sat}}) \tag{4.1}$$

To control this loss one typically attempts to minimize the voltage drop across the power switch during its on-time. To do this, the designer must operate the switch in a saturated state. These conditions are given in Equations 4.2a and b. This is identified by overdriving the base or gate such that the collector or drain current is controlled by the external elements and not by the power switch itself.

$$\text{BJT:} \quad P_{D(\text{condut})} = (V_{CE})(I_C) \tag{4.2a}$$

$$\text{MOS:} \quad P_{D(\text{conduct})} = (I_D)^2(R_{DS(\text{on})}) \tag{4.2b}$$

The switching losses during the transition times of the power switch are more complicated, both in their nature and their associated contributors. The waveforms that show the losses can only be viewed with the use of an oscilloscope with a voltage probe across the drain and source (collector and emitter) connections, and an ac current probe measuring the drain, or collector, current. The method of quantifying the amount of loss during each switching transition must be carefully viewed with probes with shielded cables and short connection leads. This is because any long, unshielded wires tend to pick-up noise radiated from other portions of the supply, thus presenting an inaccurate representation of the true waveform. Once good waveforms are received, one can approximate the areas under both curves as a piecewise summation of the areas of simple triangles and rectangles. For example, the turn-on switching loss of Figure 4–1 can be written as Equation 4.3.

$$P_{D(\text{turn-on})} = f_{sw}\left[V_1/2(I_1 + (I_2 - I_1)/2)\right] t_{\text{turn-on}} \tag{4.3}$$

This results in watts for the loss seen only during the turn-on transitions of the power switch. One would add the turn-off and conduction loss to this amount to arrive at the total loss within the power switch.

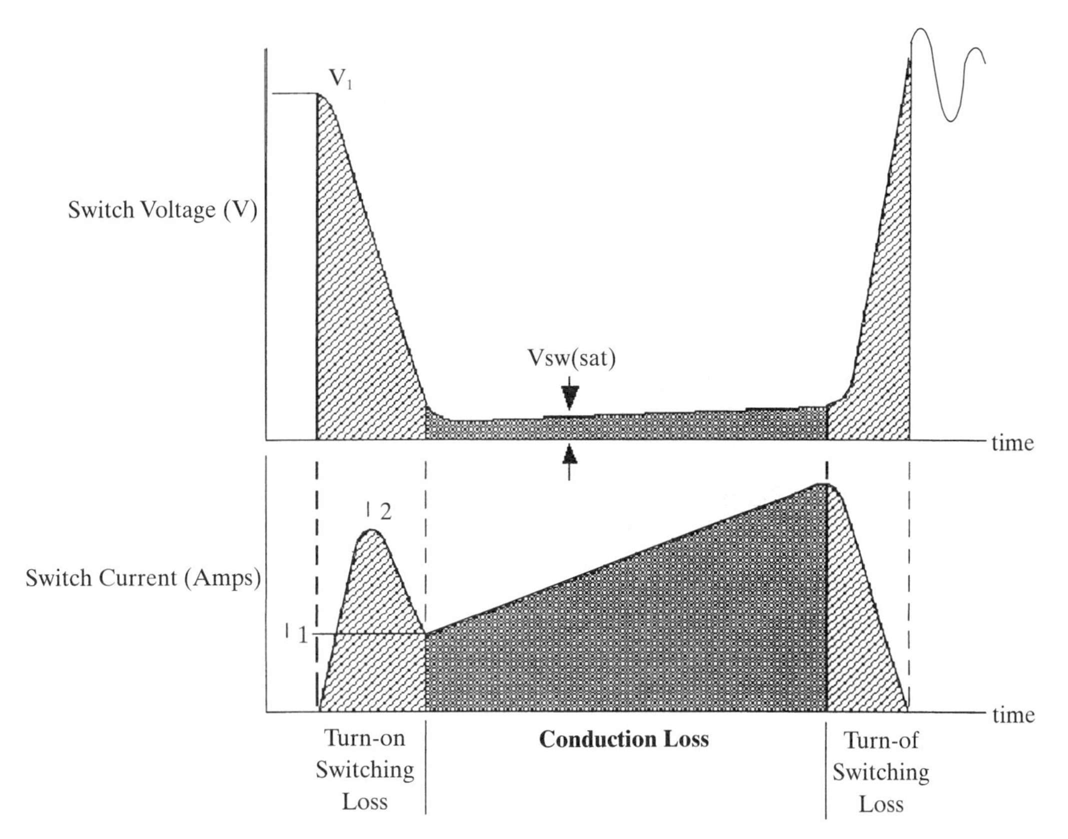

Figure 4–1 Power switch losses.

Losses associated with the output rectifier

The output rectifier represents between 40 and 65 percent of the total losses within the typical nonsynchronous rectified switching power supply. So, it is very important to understand this section. The waveforms associated with the output rectifier can be seen in Figure 4–2.

The losses can once again be broken into three periods: the turn-on loss, the conduction loss, and the turn-off loss.

The conduction loss of a rectifier is when the current and voltage waveforms have stabilized when the rectifier is conducting. Its loss is controlled by selecting a rectifier with the lowest forward voltage drop for the operating current. P-N diodes have a more flat V-I characteristic in the forward direction, but have a fairly high voltage drop (0.7 to 1.1 V). Schottky diodes have a lower "knee" voltage (0.3 to 0.6 V), but have a more resistive voltage-current characteristic. That means that the forward voltage increases more significantly with higher currents as compared to the P-N diode. The loss can be quantified in a manner as shown in Equation 4.3 by breaking the transitional sections of the waveform into piecewise rectangle and triangular areas which are then used to calculate the loss for that period.

Analyzing the switching loss of an output rectifier is much more complicated. The inherent behavior of the rectifier itself causes problems within the local circuits.

During turn-on, the transition is controlled by the forward recovery characteristic of the selected rectifier. The *forward recovery time* (t_{frr}) is the time it takes

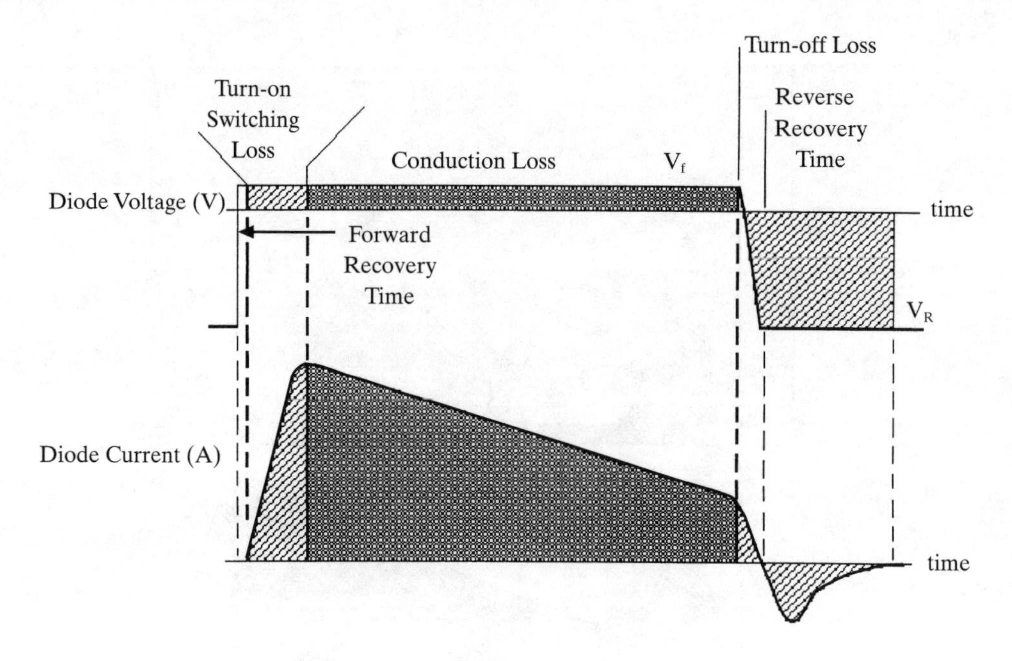

Figure 4–2 Rectifier losses.

for the diode to begin conducting forward current after a forward voltage is placed across its terminals. In P-N diodes, this period can be 5 to 15 nS for ultrafast rectifiers. Schottky rectifiers can sometimes exhibit a longer forward recovery-like characteristic because of their inherently higher junction capacitances. Although this is not a significant loss, it can cause other problems inside the supply. During the forward recovery period, the inductor or transformer has no significant load impedance because the power switch is open and the rectifier still appears open-circuited. This allows any stored energy to create ringing in its waveforms until the rectifier finally begins to conduct forward current and it clamps the power signal.

During the turn-off transition, the reverse recovery characteristic dominates its behavior. The reverse recovery characteristic within P-N diodes is caused by the carriers trapped within the junction when the reverse voltage is applied to the diode's terminals. These carriers, which have limited mobility, need to reverse direction and exit the junction from the direction they had originally come. This appears to be a reverse current flowing through the diode, just after the reverse voltage is applied. The loss associated with this can be significant, because the reverse voltage can rapidly climb to very high levels before the charge has completely emptied from the junction region. The reverse current can also be reflected through any power transformer and add to the loss within the power switch during its turn-on transition. This can be seen, for example, in Figure 4–1 during the turn-on period as a current peak.

Reverse recovery-like behavior can also occur in high voltage Schottky rectifiers. This behavior is caused not by carriers, but the higher junction capacitance exhibited by this type of Schottky diode. High-voltage Schottky diodes are those that have a reverse breakdown voltage of greater than 60 V.

Losses associated with the filter capacitors

The input and output filter capacitors do not contribute a significant source of loss to a switching power supply, although they can greatly affect the operating life of the supply. The input capacitor(s), if chosen incorrectly, can make the supply appear like it is operating less efficiently than it actually is.

Every capacitor has a small resistance and inductance in series with the specified capacitance of the capacitor. The *equivalent series resistance (ESR)* and *equivalent series inductance (ESL)* are parasitic elements caused by the construction of the capacitor. Both tend to isolate the internal capacitance from the signal on its terminals. Hence a capacitor will have its best characteristics at dc, but may behave more poorly at the switching frequency of the supply.

The input and output capacitors are the only source (or sink) of the high frequency currents created by the power switch or the output rectifier. So, by viewing these current waveforms, one can reasonably determine the currents flowing through the ESR of these capacitors. This unavoidably creates heating within the capacitor. The major design activity surrounding filter capacitors is to assure that the internal heating of the capacitor is kept low enough to assure the product life specified for the supply.

Calculating the real power loss created by the ESR of the capacitor is given in Equation 4.4.

$$P_{\mathrm{D(esr)}} \approx (I_{\mathrm{sw}})^2 (R_{\mathrm{ESR}}) \quad \text{[input capacitor]} \tag{4.4a}$$

$$\text{or } (I_{\mathrm{D}})^2 (R_{\mathrm{ESR}}) \quad \text{[output capacitor-boost]} \tag{4.4b}$$

Here not only does the resistive portion of the capacitor model cause problems, but if the PCB is laid out asymmetrically between paralleled capacitors, the trace inductance causes unbalanced heating within the capacitors, thus shortening the life of the hottest capacitor.

The quiescent losses

Quiescent losses are associated with all of the functions required to operate the power circuits. That is all of the circuits associated with the controller IC and any feedback circuits elsewhere in the supply. These losses are typically small compared to the other losses within the supply, but can be analyzed to see if any improvement can be made.

The first circuit that can burn up a significant amount of power is the start-up circuit. Here dc current is taken from the input voltage so that the control IC and driver circuits have enough power to start the power supply. If the type of start-up circuit does not cutoff its current flow after a successful start-up has been done, then up to 3 W can be continuously dissipated within the circuit depending upon the input voltage.

The second significant area is the power switch driver circuit. For bipolar power transistors used as power switches, the base drive current must be greater than the peak drain current divided by the gain (h_{fe}) of the transistor. A typical gain of a power transistor is between 5 and 15, which means, for example, a peak current of 10 A would require a base current of between 0.66 and 2 A. The base-emitter has a drive voltage of 0.7 V and if this current is not derived from a voltage very close to this voltage then a significant loss will result.

Power MOSFETs can be driven more efficiently than bipolar power transistors. The gate of a MOSFET has two equivalent capacitors connected to the terminal, the gate-to-source capacitor (C_{iss}) and the drain-to-source capacitance (C_{rss}). The loss exhibited by the gate drive of the MOSFET is created by charging the gate capacitances from the auxiliary voltage to turn the MOSFET ON, and this discharging the capacitances to ground when turning the MOSFET OFF. The equation for the gate drive loss is given in Equation 4.5.

$$P_{D(gate)} \approx 0.5 f_{sw}\left[C_{iss}\left(V_{drive}\right)^2 + C_{rss}\left(V_D\right)^2\right] \tag{4.5}$$

There is not much that can be done about this loss except to select a MOSFET with lower values for C_{iss} and C_{rss} and to possibly slightly lower the maximum gate drive voltage.

Losses associated with the magnetic components

This area is very complicated for the average design engineer because of the unusual nature of magnetic terminology. The losses described next are typically presented in graphical form by the core manufacturers, which are very easy to use. These losses are being presented here for one's appreciation of the nature of the losses.

There are three major losses associated with transformers and inductors: hysteresis loss, eddy current loss, and resistive loss. These losses are controlled during the transformer or inductor's design and construction.

The hysteresis loss is caused by the amount of turns placed upon the core by the driving circuit. This dictates how large an area within the B-H curve is swept out during each cycle of operation (see Figure D-3 in Appendix D). The area swept out by the operating *minor-loop* is the amount of work that is required to apply force to the magnetic domains within the core and some of them remain reoriented (residual flux density). The larger the area swept out, the more hysteresis loss. The loss is given by Equation 4.6.

$$P_{hyst} \approx k_h V_c f_{sw}\left(B_{max}\right)^2 \tag{4.6}$$

where: k_h is the hysteresis loss constant for the material.
V_c is the volume of the core (Cm^2).
f_{sw} is the switching frequency (Hz).
B_{max} is the maximum excursion of the operating flux density.

As seen in the equation, the loss is proportional to the operating frequency and to the square of the peak operating flux density (B_{max}). Although this loss is not as significant as the losses seen within the power switch and rectifiers, it can be a problem if it is not addressed properly. At 100 kHz, B_{max} should be set to about 50 percent of the material's saturation flux density (B_{sat}). At 500 kHz, the B_{max} should be no more than 25 percent of B_{sat}, and at 1 MHz, the B_{max} should be about 10 percent of B_{sat}. This is based upon the behavior ferrite material typically used in switching power supplies (3C8, etc.).

The eddy current loss is a much smaller loss than the hysteresis loss, but increases significantly with the operating frequency. It is shown in Equation 4.7.

$$P_{eddy} \approx k_e V_e f_{sw}^2\left(B_{max}\right)^2 \tag{4.7}$$

where: k_e is the eddy current loss constant for the material.

Eddy currents are circular currents induced in wide areas within the core in the surrounding wires and structures by the presence of high magnetic fields. There is not much that a typical designer can do to reduce these losses.

Resistive losses are those losses associated with the resistance of the windings contained within the transformer or inductor. There are two forms of resistive losses, *dc resistance losses* and *skin effect resistive losses*. The dc resistive losses are simply caused by the dc resistance exhibited by the length of wire used in the winding multiplied by the square of the RMS value of the current waveform. The skin effect is the effective increase in the resistance of the wire caused by the "pushing" of the current flow from the center of the wire to the surface by the intense ac magnetic fields within the wire. The current flows over a smaller cross-sectional area, thus appearing to be a smaller diameter wire. The equation that combines the two losses in one expression is given in Equation 4.8.

$$mr \approx r_{\mathrm{DC}} \sqrt{\frac{8\pi^2 \times 10^{-7} f_{\mathrm{sw}} u_{\mathrm{r}}}{r_{\mathrm{m}}}} \tag{4.8}$$

where: mr is the ratio of the ac resistance to the dc resistance.
r_{DC} is the dc resistance of the wire (ohms).
f_{sw} is the switching frequency (Hz).
u_{r} is the relative permeability of the wire material.
r_{m} is the resistivity of the conductor material.

The solid wire can be replaced by a tubular wire which has a thickness given by Equation 4.9.

$$dw - \frac{1}{2\pi} \sqrt{\frac{10^7 r_{\mathrm{m}}}{f_{\mathrm{sw}} u_r}} \quad \text{meters} \tag{4.9}$$

The *leakage inductance* (represented by a small inductor in series with a winding) causes some flux not to couple with the core, but to escape into the surrounding air and materials. Its behavior is not governed by its associated transformer or inductor, hence any reflected impedance to the winding in question does not affect the behavior of the leakage inductor.

The leakage inductance causes a problem because it traps energy which is not passed onto the load but causes ringing energy within the surrounding components. The physical design of the transformer or inductor controls the amount of leakage inductance a winding will exhibit. Its value varies from unit to unit, but falls close to a nominal value.

Some general rules of thumb: Things that decrease the leakage inductance exhibited by a winding are: a longer physical winding length, closer physical distance to the core, close coupling techniques between windings, and turns ratios that are similar (i.e., close to 1:1). For the typical E-E core used in dc-dc converters, the amount of leakage inductance one can expect is between three and five percent of the inductance of the winding. In off-line converters, the leakage exhibited by the primary winding may go as high as 12 percent of winding inductance value if the transformer must meet stringent safety regulatory requirements. The tape needed to isolate the windings, make the windings shorter, and space the windings away from the core and from the other windings.

The parasitic loss caused by the leakage inductance can be harnessed as seen later.

In dc magnetic applications, an air-gap is usually required somewhere along the magnetic path of the core. In ferrite cores, the gap is placed in the centerleg of the core. The flux leaves one end of the core and flows towards the opposing end. The flux, though, repels itself and causes the flux lines to "bulge" out away from the centerline of the core. The presence of an air-gap creates an area of intense magnetic intensity, which can cause eddy currents to flow within the neighboring wires or within metallic structures close to the gap. This loss is typically not very large and it is difficult to identify.

4.1.1 An Overview of the Major Parasitic Elements within a Switching Power Supply

Parasitic elements are the unintended electrical behavior of physical elements within the circuit. They typically store energy and react with existing elements that cause the creation of noise and losses. Identifying, quantifying, and minimizing or harnessing their effects is a significant challenge to the designer. Parasitic behavior is accentuated under the influence of ac waveforms. There are two major nodes within the typical switching power supply that have large levels of ac. The first node is the collector or drain of the power switch, and the second is the anode of the output rectifier. Much attention is given to these particular nodes.

Major parasitic elements within all converters

There are common parasitic elements that exists in all switching power supplies. Their influence is typically noticed when one views the waveforms at the major ac nodes within the converter. Some are even characterized within the datasheets of the physical components, such as the intrinsic capcitances of a MOSFET. Some of the major parasitic elements for two types of common converters can be seen in Figure 4–3.

Some parasites are well defined, such as the MOSFET capacitances. Others may be distributed over many sources and are lumped together to make any modeling effort easier. Attempting to assign a value to these less-defined parasitic elements is very difficult and usually they are just treated as an empirical value. That is, when it is time to redirect energy, the components chosen are the ones that provide the best desired results. Placing the parasitic elements in the proper locations within the schematic is important because some electrical branches are only active during a portion of the converter's operation. For example, the junction capacitance of a rectifier is only significant when the rectifier is reverse-biased and disappears when the diode becomes forward-biased. Table 4–1 lists some of the more easily identified parasitic elements, the components that cause them, and the general range of values. Some specific parasitic values may be obtained from the particular component's data sheet.

The printed circuit board (PCB) contributes parasitic influences distributed almost everywhere. Many of these influences want to be minimized though the use of good PCB layout rules (refer to Section 3.14). Traces that carry high peak currents are susceptible to inductance and resistance contributed by any PCB trace. These traces must be short and wide. PCB pads (or islands) that have high ac voltages, such as the drain or collector of the power switch or the anode of rectifier, are susceptible to capacitive coupling to neighboring traces. This couples ac noise into the quieter surrounding traces. This can be avoided by making all of the layers beneath the ac trace the same signal as the ac node by connecting them through "vias" or feedthru holes. The parasitic effects that remain are typically summed into neighboring parasitic elements.

Understanding the nature of the parasitic elements contributed by each of the components that make up a typical converter will aide in the specifying of magnetic components, designing PCBs, designing filters for EMI control, etc. This is the *black-magic* portion of all switching power supply designs.

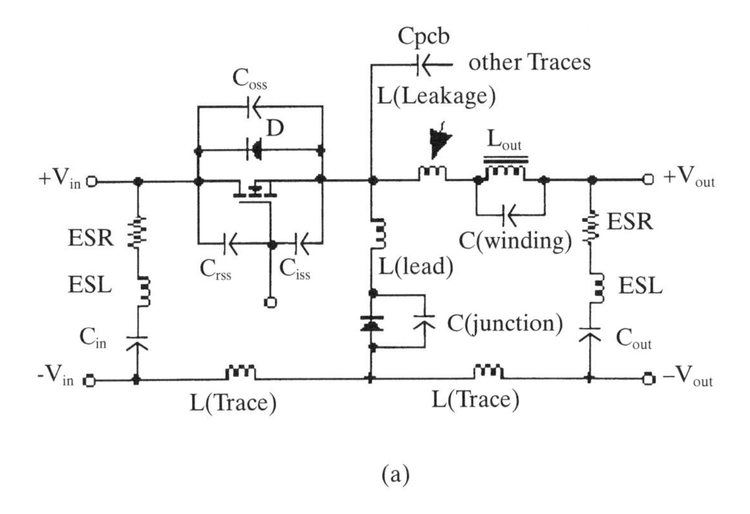

(a)

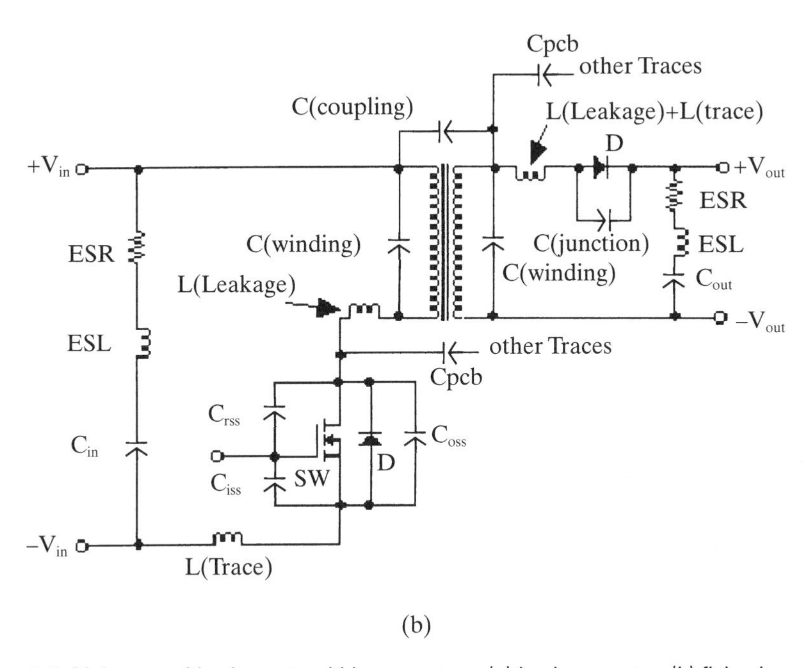

(b)

Figure 4–3 Major parasitic elements within converters: (a) buck converter; (b) flyback converter.

4.2 Techniques for Reducing the Major Losses

A PWM switching power supply that is designed with no extraordinary loss-control methods will exhibit efficiencies as seen in Table 3–3. For switching power supplies that have no problem in getting rid of the heat, such as some off-line applications, the aforementioned efficiencies may be satisfactory. For portable applications and equipment that must be small in size much better efficiencies must be sought. To improve the overall efficiency of a power supply, several techniques can be used.

Table 4–1 Major Sources of Parasitic Behavior

Source	Name	Typical Range
MOSFET	C_{oss}	100–800 pF*
	C_{iss}	200–3000 pF*
	C_{rss}	20–400 pF*
	Intrinsic diode	0.2–1.0 I_D
	Lead inductance	2–10 nH
Rectifier	Lead inductance	2–10 nH
	Junction capacitance	20–400 pF#
Capacitor	ESR	0.05–10 Ω
	ESL	10–100 nH
Inductor	Leakage inductance	1–8% L(winding)
	Winding capacitance	~1.75 ln(T) [pF]
Transformer	Leakage inductances	1–8% L(winding)
	Winding capacitances	~1.75 ln(T) [pF]
	Core losses	See Appendix D
	Coupling capacitance	1–100 pF

* Low voltage trench MOSFETs (ID = 1–20 A).

Schottky and P–N diodes measured at rated reverse voltage.

As seen in Section 4.1, the major types of losses are the conduction and switching losses. Conduction losses are addressed by selecting a better power switch or rectifier with a lower conduction voltage. The synchronous rectifier can be used to reduce the conduction loss of a rectifier, but it can only be used for forward-mode topologies, and excludes the discontinuous boost-mode converters. The synchronous rectifier will improve the efficiency of a power supply about one to six percent depending upon the average operating duty cycle of the supply. For further improvements, other techniques must be pursued.

Switching losses are more significant within a switching power supply when the input or output voltage(s) are high (greater than 20 VDC). The rectifier forward voltage drops are less significant when compared to the input or output voltage. The instantaneous product of the voltage and current in switching transitions are proportional to both voltage and current.

Switching losses occur at two equivalent nodes within every switching power supply: the drain (or collector) of the power switch(es), and the anode of the output rectifier(s). These are the only ac nodes within each type of PWM switching power supply. Within the nontransformer isolated topologies, these nodes are physically one node where the collector (or drain) of the power switch is directly connected to the anode of the output rectifier. Within transformer-isolated topologies, these two nodes are separated by the transformer and the two nodes are treated slightly differently.

The waveforms of the typical PWM switching power supply can be seen in Figures 4–1 and 4–2 of Section 4.1. As one may observe in these figures, the values of voltage and current are significant values during the transitions, and voltage or current "spikes" can also occur during these periods, making the loss even higher.

There are four goals one wants to accomplish at these two nodes:

1. Lower the voltage and current cross-over point during all turn-on and off transitions.

2. Minimize the effects of the reverse recovery of any P-N rectifiers.
3. Remove any spikes created by the parasitic elements.
4. Recover as much of this "loss" energy as possible and return it to the power flow of the power supply.

The designer cannot accomplish all of the goals completely, but improving the conditions can add another +3 to +9 percent in overall power supply efficiency.

One additional consideration when working on these circuits is to make all efforts to band-limit the waveforms in order to reduce any radiated EMI. The transitions of a switching power supply create the majority of the EMI energy radiated into the surrounding environment. Typically, by adding a small inductance to the current path that returns the energy to the supply, one can greatly improve the EMI performance.

To accomplish this, one normally uses additional reactive elements with diodes or MOSFETs to control the effects. The types of modifications to the standard PWM topologies fall into three categories:

1. Lossless snubbers.
2. Active clamps.
3. Quasi-resonant modifications.

Lossless snubbers and active clamps produce PWM waveforms with "soft" edges.

For the power switch ac node, the voltage wants to be delayed on the turn-off transition. This provides loading of the magnetic element during the forward recovery time of the output rectifier. For the output rectifier ac node, the current wants to be delayed at its turn-off. This limits the reflected current spike caused by the reverse recovery period of the rectifier. These techniques are shown in the following sections.

4.3 Snubbers

Snubbers are passive networks that delay the risetime of the voltage waveform. Historically, snubbers have been used to keep power devices within their forward- and reverse-biased safe operating areas (FBSOA and RBSOA) or to control RF emissions from the power supply. They are essentially lossy tank circuits (L-C circuits with R). Using them offered more of an advantage than the loss incurred. Semiconductor components are more rugged today and the traditional need for the RLC snubber for protection has lessened, but occasionally a snubber is still needed.

In modern, high-efficiency switching power supplies, the trend is to recover the energy from any waveshaping circuits and to return it to the power circuits for use. Here the *lossless snubbers* are used. The challenge in the modern switching power supplies is to minimize the losses everywhere in the circuit. For completeness, though, the design of both is offered within this section.

4.3.1 Design of the Traditional Snubber

The traditional snubber has been the approach used to keep bipolar power transistors away from second breakdown conditions. It is also useful in the reduction of radiated EMI by controlling the dv/dt of rectifiers with abrupt reverse

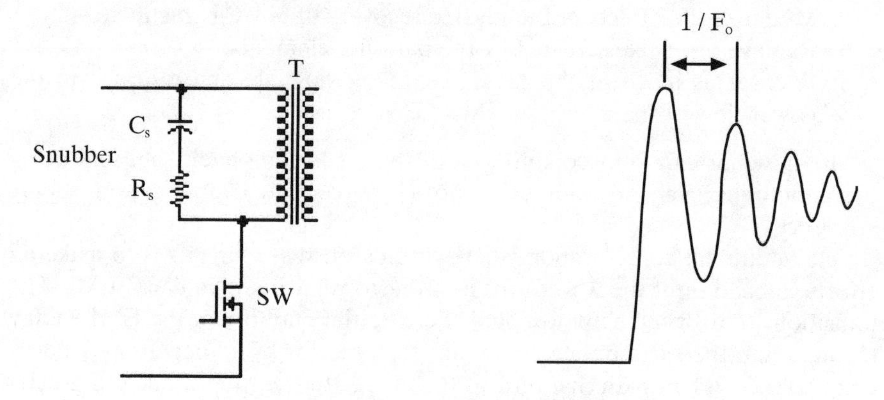

Figure 4–4 The traditional snubber.

recovery characteristics. Its design is very important because if designed improperly, it will loose more power than is required. Its design, though, is based more upon an empirical process than a theoretical approach. This is because the waveforms that the snubber needs to affect are generated, in large part, by the parasitic elements existing in the circuit. The snubber should be designed after the circuits have been physically built. That is, one has to build the first prototype from the final PC board, transformer, power switch, and rectifiers so that the parasitic elements will be most like the final product. The traditional snubber can be seen in Figure 4–4.

The empirical design process of the traditional snubber is as follows:

1. Measure the period of the ringing that is exhibited on the unsnubbed signal (I/F_o).
2. Place a high-frequency capacitor (ceramic or film) across the primary winding of the transformer, rectifier, or the element to be snubbed. Determine the capacitor value that produces an oscillation period which is three times the original period (C_o).
3. The resistor that should be placed in series with the capacitor will approximately have the value of:

$$R - \frac{1}{2\pi F_o C_o} \tag{4.10}$$

There is more than one combination of R and C that will produce a satisfactory waveform, but the above values of R and C should yield the least lossy and most effective values. If one were to change the values, larger resistance and smaller capacitance produces less loss.

NOTE: Do not use snubber values or snubber elements intended for silicon-controlled rectifier (SCR) circuits in switching power supplies. The impedances and parasitic values of these circuits are much lower than within switching power supplies. They will create far too much loss in switching power supply circuits.

4.3.2 The Passive Lossless Snubber

The lossless snubber is a means of redirecting energy that is contained within a spike or an edge and possibly restoring the energy to either the input or the output. It is very important to redeposit the energy in a place where it is going to be used, otherwise the energy becomes a loss.

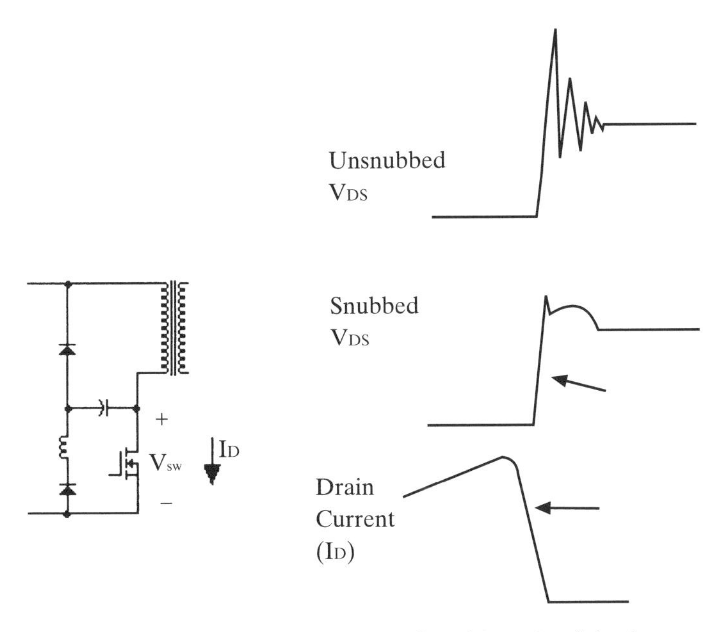

Figure 4–5 Lossless snubber for a one-transistor forward or flyback converter.

The power switch turn-off transition creates significant spikes. The major cause of the turn-off spike is the energy stored in the parasitic and actual magnetic components prior to the output rectifier turning-on. During the forward recovery period of the rectifier, the rectifier is open-circuited and the stored magnetic energy creates large voltage swings on all of the ac nodes.

The passive lossless snubber uses a diode to select the desired edge. The rapid dv/dt is gated into a discharged capacitor. The amount of slowing of the edge is determined by the value of this snubber capacitor. After the capacitor's voltage has equalized with the post-transition voltage, the energy trapped by the capacitor must hopefully be deposited in a useful place prior to the next cycle. Lossless snubbers are specifically designed for each topology and may be implemented in many forms.

One form of lossless snubber for a one-transistor, transformer-isolated converter can be seen in Figure 4–5. The initial condition of the snubber capacitor is zero volts. When the power switch drain or collector voltage exceeds the input voltage during the turn-off transition, the capacitor then begins to charge, taking energy from what is typically a spike. The capacitor remains charged until the drain or collector voltage returns to ground where the capacitor is discharged through the diode and inductor. The cycle is then repeated. The loss from this type of lossless snubber resembles more the gate drive loss of a power MOSFET. The loss is diverted to the ground which bypasses the power sections of the circuit.

In boost converters the lossless snubber shown in Figure 4–6 can be used.

The reset should be resonant to reduce the amount of noise that could be generated by the rapid discharge of the clamp capacitor. An inductor of about 2 to 3 uH is useful. An etched printed circuit board spiral inductor can also be used for this purpose, if the space can be spared. Since the average current is small, the trace need be only about 10 to 20 mils (0.25 to 0.5 mm) in width. Three spiraled turns produce about the inductance needed.

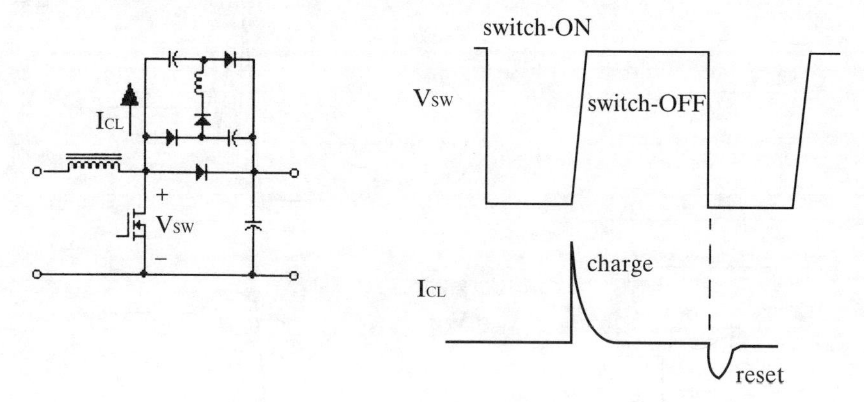

Figure 4–6 Lossless snubber for a boost converter.

4.4 The Active Clamp

An active clamp is a reactive circuit connected to an ac node whose affects are directly controlled by the control IC. Active clamps are used to slow the transition time of the voltage waveform for only one edge of an ac power waveform. By slowing the transition time of the voltage waveform, the switching loss during that transition can be significantly reduced by lowering the instantaneous voltage value during the cross-over period.

A typical active clamp can be seen in Figure 4–7 along with its effect upon the ac waveform. During turn-off, a discharged clamp capacitor is placed in parallel with the power switch. The capacitor then forms a resonant tank circuit with the primary inductance plus its leakage inductance. The capacitor's voltage then increases until it reaches the value of the reflected transformer voltage. The series MOSFET clamp switch must be turned off prior to the next turn-on period of the power switch. The clamp switch must then be turned back on just prior to the next turn-off of the power switch with sufficient time to allow the capacitor to completely discharge. It is important to not allow the clamp capacitor to be electrically attached to the drain during its turn-on transition. The cycle is then repeated.

Only a few controller ICs available at the time of this publication directly support an active clamp drive (for example, Texas Instruments UCC3580). There will be more since its function seems to increase the efficiency of a switching power supply by several percentages.

Making an active clamp from a controller IC that does not support its direct drive depends upon which signals are brought out from the IC. Figure 4–8 gives a couple of ideas.

At the present, active clamps are still not broadly used because of the difficulty in implementation. This will change as the industry evolves.

4.5 Saturable Inductors to Limit Rectifier Reverse Recovery Current

Continuous-mode converters, both forward and boost, suffer from one common problem. The output rectifiers have forward current flowing through them just

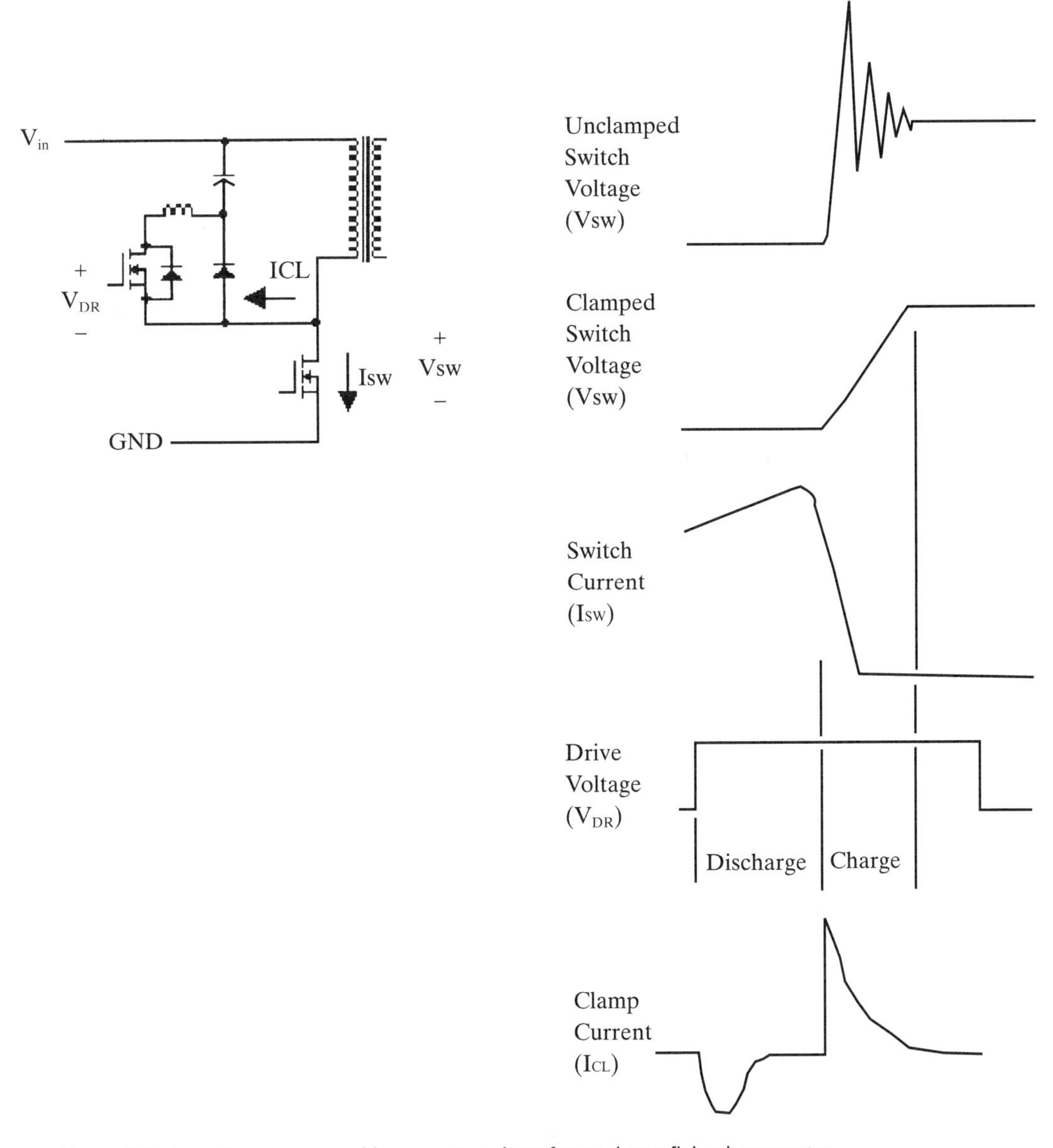

Figure 4–7 An active clamp used in a one-transistor forward or a flyback converter.

prior to an instant voltage reversal across their terminals. This causes a sizeable spike of reverse current to flow through the diode. This spike usually flows through the power switch at its turn-on transition. This is true for both the nontransformer-isolated and transformer-isolated topologies. If the input voltage of the supply is high, this added switching loss can be much more than the conduction loss of the power switch. This current spike caused by the reverse recovery of the output rectifier at the turn-on of the power switch can be seen in Figure 4–1.

One method to reduce this phenomenon is to add a saturable inductor in series with the output rectifier. A saturable inductor is an inductor whose core exhibits a very square B-H curve as exhibited by core materials, such as magnisil and Orthanol. Inductors made with these core materials have very high

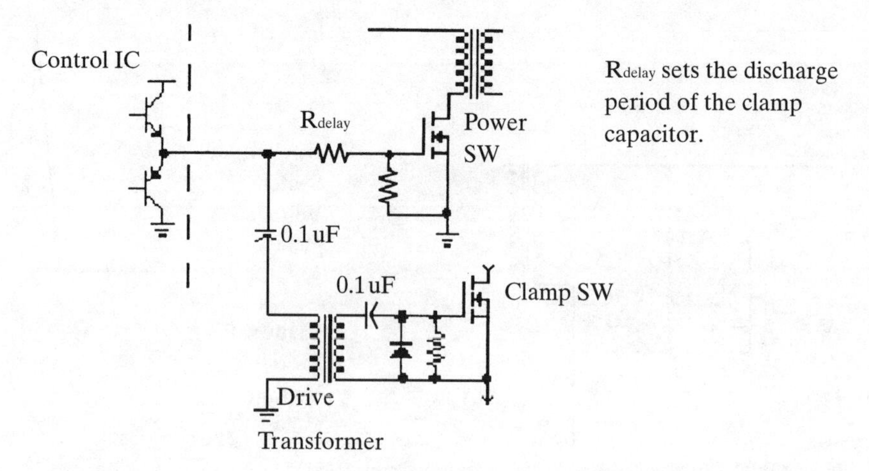

Figure 4–8 Designing an active clamp from an IC that does not drive an active clamp.

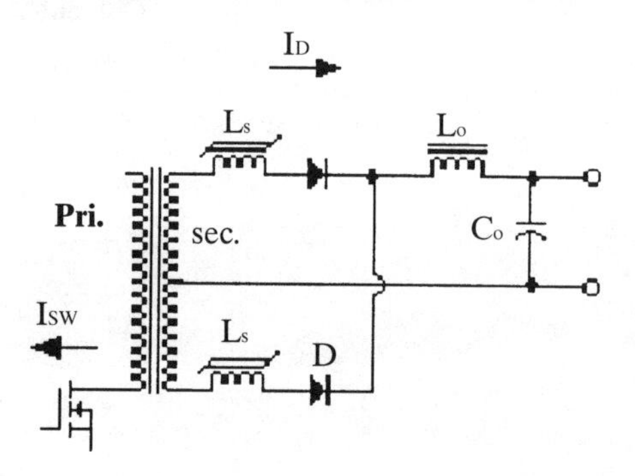

Figure 4–9 Typical use of saturable inductors in reducing reverse recovery losses.

permeabilities and quickly enter saturation, but do pass through a period of linear inductance behavior. Essentially, anytime their current is reversed, they behave as an inductor for the amount of time dictated by the value of the stepped current, the turns, and the permeability. Then they effectively short-circuit across their terminals when they enters saturation. A common implementation can be seen in Figure 4–9.

The saturable inductor should be a small size because one only needs to exhibit an inductive characteristic for as long as the expected reverse recovery time of the associated output rectifier (20 to 50 nS). Any longer, and the output power will be needlessly reduced. Also the same inductive behavior will affect the forward recovery period of the rectifier, which may cause more ringing on the opposite edge. Design of the saturable reactor is best done with the core manufacturer's design data.

One normally must check the saturable inductor's affect upon the circuit. One should verify its performance by measuring the overall efficiency of the power supply versus every value of the saturable inductor's value. Although one can make an educated guess from the output rectifier's specified reverse recovery time, the process ends-up being an iterative "tuning" process.

4.6 Quasi-resonant Converters

Quasi-resonant converters are a separate class of switching power supplies that tune the ac power waveforms to reduce or eliminate the switching loss within the supply. This is done by placing resonant tank circuits within the ac current paths to create pseudo-sinusoidal voltage or current waveforms. Because the tank circuits have one resonant frequency, the method of control needs to be modified to a variable frequency control where the resonant period is fixed and the control varies the period of the non-resonant period. The quasi-resonant converters usually operate in the 300 kHz to 2 MHz frequency range.

The advantages of a quasi-resonant converter over a classic PWM converter are smaller size and typically higher efficiencies. Although when a smaller size is pursued by increasing its operating frequency, the improvement in efficiency is sacrificed due to other frequency dependent losses.

The disadvantage of the quasi-resonant converter compared to the newer lossless snubber and active clamp techniques in addition to the basic PWM converters, is the voltage or current stresses placed upon the power components. The peak voltage or current values that exist within quasi-resonant converters can be two to three times higher than in PWM converters. This forces the designer to use higher-rated power switches and rectifiers which may not have as good conduction characteristics.

In general, it is probably better for the typical application to use the lossless snubbing and active clamp techniques than the quasi-resonant techniques, because of design time and end product cost.

4.6.1 Quasi-resonant Converter Fundamentals

Quasi-resonant converters force the voltage or current waveform into a haversine waveshape. If the power switch(es) are switched at the right moments, then there are no switching losses experienced. Also because of the controlled rates of change for the voltage or current waveforms, much better RFI/EMI performance is realized. Most of the basic topologies that exist within the PWM family are also in the quasi-resonant family.

Quasi-resonant converters utilize an L-C tank circuit, which "rings" at its natural resonance frequency in response to a step change in its terminal voltage or current. The tank circuit is placed between the power switch and the transformer and/or the transformer and the output filter.

Control is accomplished by controlling the nonresonant portion of the waveform and with a fixed period allowed for the resonant portion of the waveform.

4.6.1.1 The Zero-current Switching Quasi-resonant Converter

The zero current switching (ZCS) quasi-resonant (QR) switching power supply forces the current through the power switch to be sinusoidal. The transistor is always switched when the current through the power switch is zero. To understand the operation of a ZCS QR switching power supply, it is best to study in detail the operation of its most elementary topology—the ZCS QR buck converter (and its waveforms) as seen in Figure 4–10.

As one can see, there is the familiar choke input filter (L-C) on the output, which is characteristic of the buck and all forward-mode converters. The configuration shown in Figure 4–10 is called a parallel resonant topology because the load impedance (the L-C filter acting as a damping impedance) is placed in parallel to the resonant capacitor. The input to the L-C filter stage

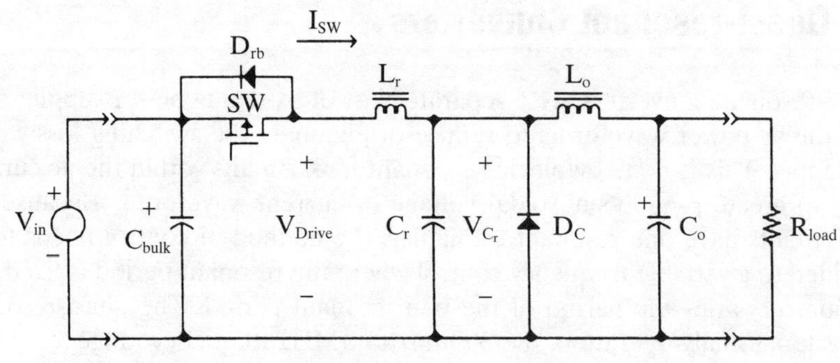

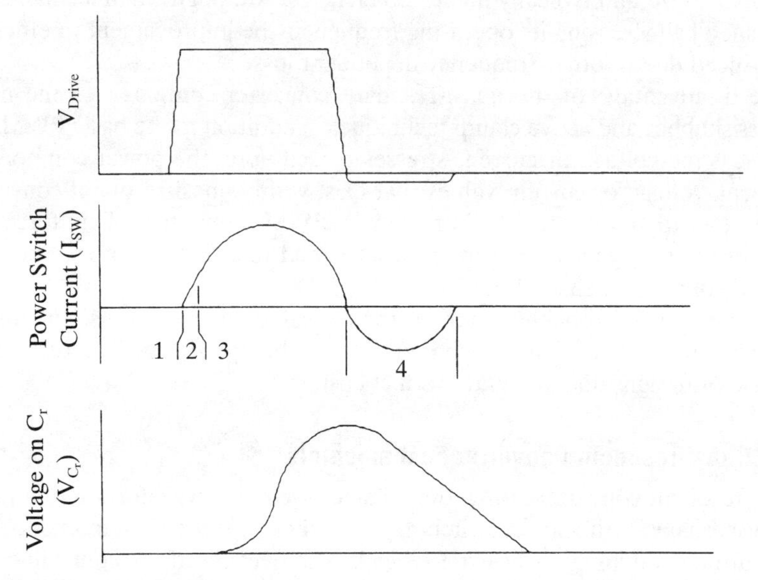

Figure 4–10 The schematic and waveforms of a ZCS quasi-resonant buck converter.

represents a high ac impedance at the tank circuit's resonance frequency. Otherwise, it would lower the Q of the tank circuit and adversely affect its ability to "ring."

The operation of the ZCS QR buck converter can be broken into four periods. Period 1 is its quiescent or initial state, where the tank elements are "uncharged," the power switch is off, and the catch diode is conducting the load current through the inductor as in a PWM buck converter. Period 2 starts when the power switch turns on and the voltage across the power switch makes a step change. With the catch diode conducting, the resonant capacitor is effectively shorted to ground. The power switch only sees the inductance of the resonant inductor. The switch current is at zero when the voltage has switched since current cannot change instantaneously through the resonant inductor. The switch current begins a linear ramp from zero amps with a positive slope of $+V_{in}/L_r$. This continues until the amount of current flowing through the switch and resonant inductor exceeds the load current being conducted by the catch diode. The catch diode then turns off in a zero current fashion by the load current being displaced by the power switch.

Period 3 begins when the resonant capacitor is released. The current waveform now assumes a sinusoidal waveshape, proceeds over its crest, and decreases until it passes through zero amps. The resonant inductor's current then begins to flow backwards through the antiparallel diode of the power switch. The power switch may turn off at any time during the "ringback" period with no switching loss since any current during this time is flowing through the antiparallel diode. Any excess inductor energy is returned to the input filter capacitor. During period 3, the resonant capacitor's voltage exhibits a sinusoidal waveform, but is lagging the current waveform by 90 degrees. So, when the inductor's current passes through zero, the resonant capacitor's voltage has reached its peak and starts its decline.

During period 4, the input to the buck filter inductor (L_o) represents a current sink and creates a downward linear voltage ramp that discharges the rest of the resonant capacitor's energy into the L-C filter. When the voltage ramp reaches zero, the quiescent state of the tank circuit is again reached and it awaits the next conduction period of the power switch.

As one can see, the power switch switches at zero current on both switching edges. The commutation diode switches at zero current because of the current displacement by the power switch at its turn-on, and a linearly declining voltage ramp at its turn-off (zero voltage turn-off). The net result is no switching losses in both power semiconductors including the ringback diode.

The "on" period of the power switch must be set to the resonant period of the tank circuit. The power transferred to the load is done by varying the amount of on-times per second of the power switch. So the ZCS quasi-resonant converter needs a fixed on-time, variable off-time method of control. The control ICs presently available on the market perform just this function. The control equation is given by

$$\frac{V_o}{V_i} = \frac{f_s}{f_r} \tag{4.11}$$

Care must be taken to specify a low and high limit to the output power of the converter. At light loads, the frequency can drop drastically. If the frequency comes within a decade of the output L-C filter pole, too much ripple will be allowed to pass to the load. At heavy loads, the frequency can become too high, and the power switch conduction cycles will run together, thus producing a nonzero current switching situation. So,

$$10.f_{fp} \text{ " } f_{sw} \text{ " } 1.1.f_r \tag{4.12}$$

where: f_{fp} is the L–C filter pole frequency (see Equation B.8).
f_{sw} is the switching frequency of the supply.
f_r is the resonant frequency of the tank circuit (see Equation 4.13).

Frequency limiting is also provided on all the resonant controllers on the market today. The resonant frequency of the tank circuit is given by

$$f_r = \frac{1}{2\pi\sqrt{L_r C_r}} \tag{4.13}$$

One-half of the reciprocal of this equation is the positive half period of the tank circuit waveform. Since the tank circuit has been partially emptied of its energy, the ringback period is shorter than a half period. On the average, an additional 75 percent of the above period should be reserved for the ringback period.

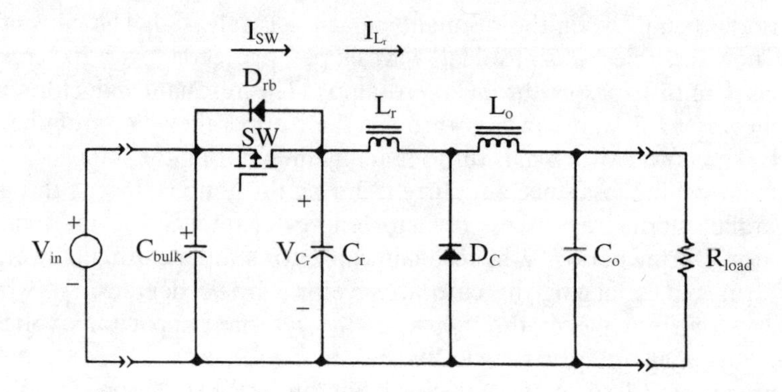

A. Schematic of a ZCS Quasi-Resonant Buck Converter

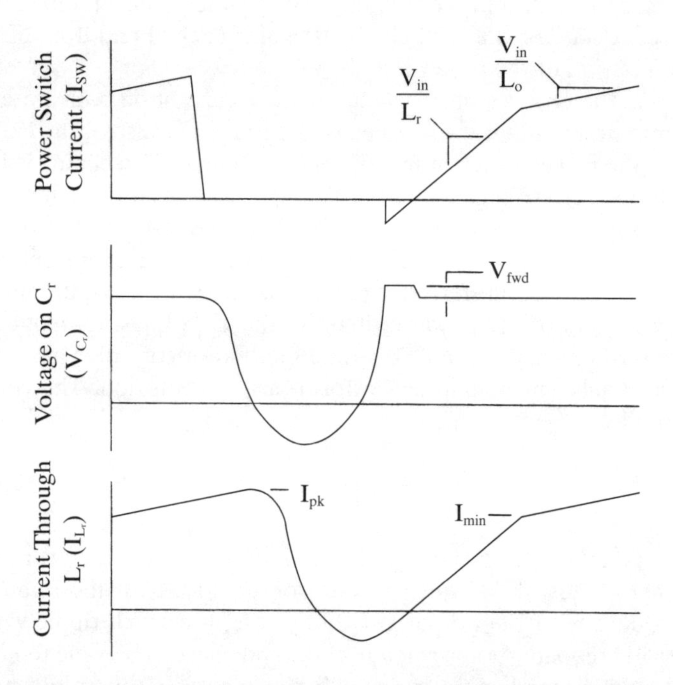

Figure 4–11 The schematic and waveforms of a ZVS quasi-resonant buck converter.

4.6.1.2 The Zero-voltage Switching Quasi-resonant Converter

A second type of quasi-resonant converter is the zero voltage switching (ZVS) quasi-resonant family. A ZVS QR buck converter and its waveforms are shown in Figure 4–11. Here the power switch remains "on" most of the time and performs resonant off periods to decrease the output power. Actually, the ZCS and the ZVS families mirror one another. If you were to compare the switch voltage and current waveforms between the two families, and if one inverts both the voltage and current waveforms in order to reference them to the power switch, the waveforms would have a striking resemblance to one another.

The time segments are mirrored oppositely in the ZVS family as compared to the ZCS family. The commutating diode conducts during the resonant off-period. The power switch is "on" between the resonant off-periods, and the power switch's current during these on-periods is the familiar current ramp from a forward-mode PWM converter whose slope is $(V_{in} - V_{out})/L_{out}$. This period is also the quiescent time of the tank circuit. During this time, the resonant inductor is saturated and is effectively a short-circuit. The resonant capacitor has V_{in} across its terminals.

A resonant period is initiated by the power switch turning off. The resonant capacitor's voltage cannot change instantaneously, so the power switch's terminal voltage remains at V_{sat} while the current is being cut off. The resonant capacitor's voltage then begins to fall. The resonant inductor's current has already reached zero. The catch diode then begins contributing load current when the resonant inductor's current falls below the value of the load current. The resonant inductor falls out of saturation and the forward-biased catch diode shunts the filter-end of the resonant inductor to ground which allows the resonant tank to resonate. The power switch end of the resonant inductor, the resonant capacitor's voltage, rings through a half-sinusoidal waveform. Then the resonant capacitor's voltage rings back above the input voltage and current flows through the antiparallel diode of the power switch. Once again, during the ringback period, the power switch can turn back on. After the power switch turns back on, the catch diode is on, and the switch sees the impedance of the resonant inductor to ground. So, a current slope of $+V_{in}$/Lr is seen. When this current exceeds the load current being conducted by the catch diode, the diode turns off. Then the resonant inductor can enter saturation and the current ramp reverts to a slope of $+(V_{in} - V_{out})/L_{out}$. This completes the periods of operation of the ZVS quasi-resonant buck converter.

The method of control of the ZVS QR converter is inverse from the ZCS. At light loads, the control frequency is high, thus performing many off-times. When the load is heavy, the number of off-times decreases. So its control relationship is

$$\frac{V_o}{V_i} = \frac{f_r}{f_s} \qquad (4.14)$$

Once again, high and low frequency limiting must be introduced on the control IC in order to minimize output ripple voltage and zero switching loss conditions.

ZVS QR topologies appear to be the more popular of two methods of quasi-resonant technologies. This is mainly due to two reasons: first, its typical variation in frequency over its input and load variations is 4:1 as opposed to 10:1 for the ZCS topologies; secondly, it has a better heavy load performance. Also, some of the more troublesome parasitic elements within the circuit can be more easily harnessed.

4.6.2 Quasi-resonant Switching Power Supply Topologies

As with PWM switching power supplies, there are comparable topologies within the zero-current switching (ZCS) and zero-voltage switching (ZVS) quasi-resonant families. You'll immediately recognize the family members upon seeing them.

Because current or voltage waveforms within QR converters are sinusoidal, the peak values are higher than those equivalent parameters found in PWM switching power supplies where the waveforms are typically rectangular or trapezoidal. One can expect the peak values to be about 1.5 or more times higher than PWM topologies. The ZCS QR supplies present a high current stress upon the power switches, and the ZVS QR supplies present a high voltage

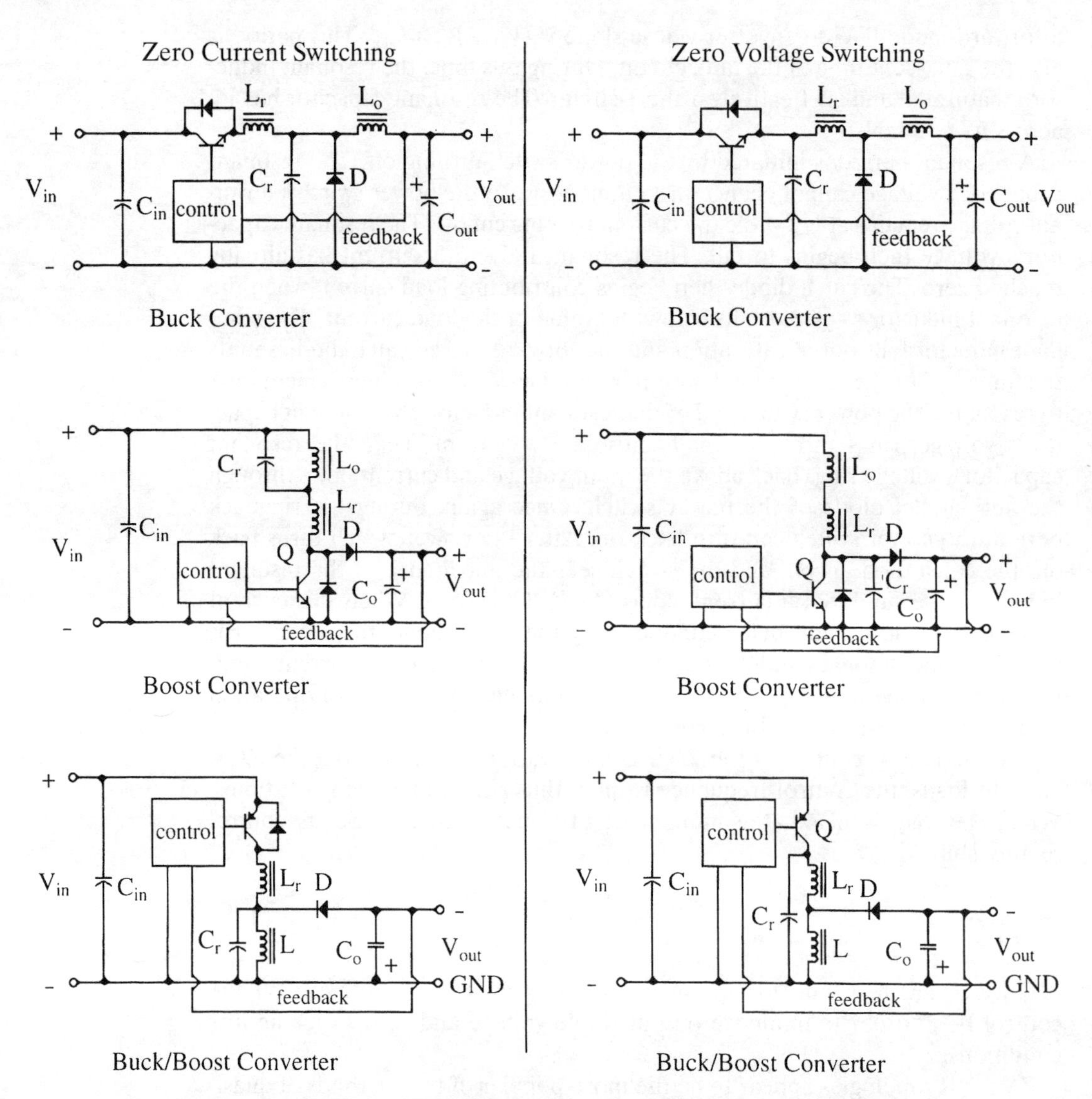

Figure 4–12 Nontransformer-isolated quasi-resonant topologies.

stress. This stress can vary with the input voltage and the output loading, making some topologies better suited for certain ranges of input voltages and output powers. For instance, the ZCS QR supplies are good at high input voltages, but poor at high output powers. Conversely ZVS QR converters are good at high output powers but poor at high input voltages.

Most zero-current quasi-resonant applications are no more than 300 W, but zero-voltage quasi-resonant applications can provide many kilowatts in output power. Two power switch topologies will distribute the losses in two packages thus making them better for the higher output powers. The topologies are summarized in Figures 4–12 through 4–14.

4.6.3 Designing the Resonant Tank Circuit

The resonant tank circuit gives the quasi-resonant switching power supply its unique functionality. Because switching power supplies are made up of many

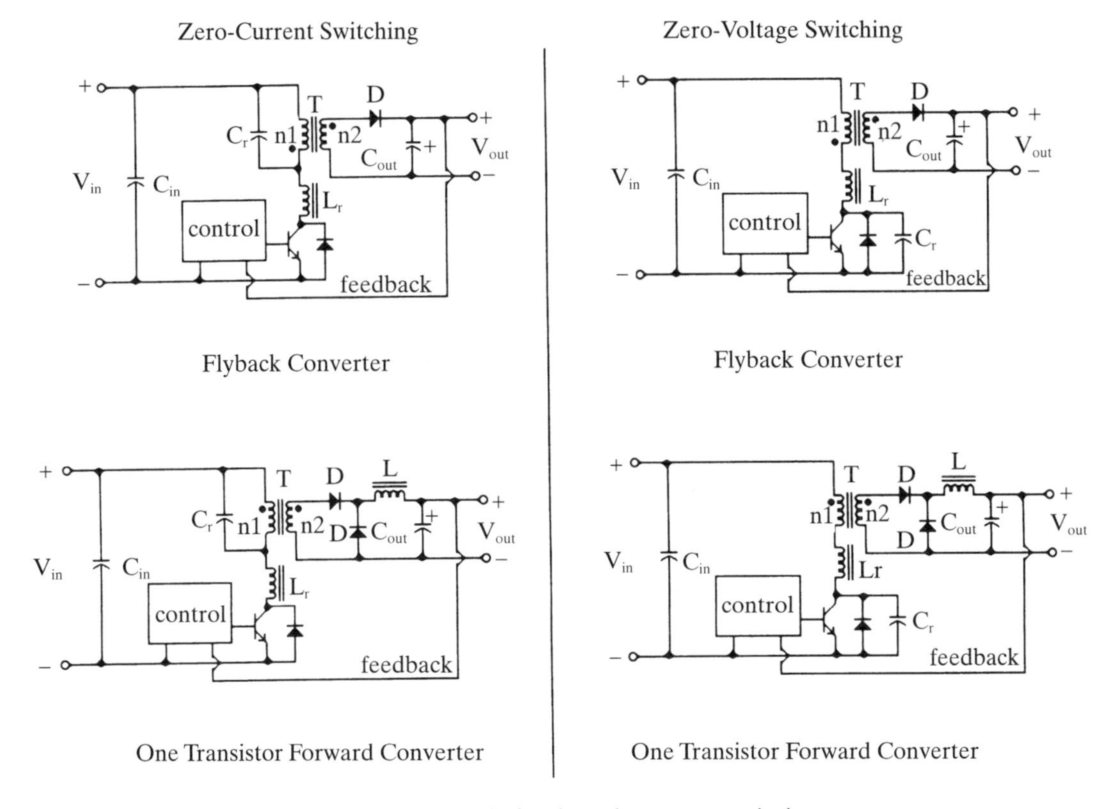

Figure 4–13 One transistor, transformer-isolated quasi-resonant topologies.

power elements that exhibit some other parasitic characteristics, the resonant topologies can actually turn these annoying parasitic characteristics into useful functions within the power supply. The skill of the designer is to know where parasitic behaviors are located and how to best arrange the L-C tank circuit to take advantage of them.

A tank circuit is made up of reactive elements with no loss elements (resistive). It yields the lowest branch impedance when the ac impedances of the inductor and the capacitor equal one another, or the net phase shift equals zero degrees. This point is called *its natural resonance frequency*. If the tank circuit is subjected to a signal whose frequency is equal to its resonance, it will actually produce a voltage much greater than the exciting signal. The degree of this "amplification" is referred to as its Q. It is proportional to the amount of resistive damping or loss that exists within the tank circuit. A tank circuit, if subjected to a transient spike or a step function, will "ring" at its natural resonance frequency, thus forcing the input signal into a sinusoidal wave shape as it passes through the circuit.

Let us examine how the quasi-resonant switching power supplies employ tank circuits within their topologies. Any energy taken from the tank circuit by the output is a loss of its stored energy to the tank circuit. The problem is to harness this energy within the tank circuit, but not load the tank circuit too much as to ruin its Q. There are two ways to remove energy from an L-C tank circuit: by placing relatively high impedance in parallel with the capacitor (parallel loading); or by placing a relatively low impedance in series with the inductor

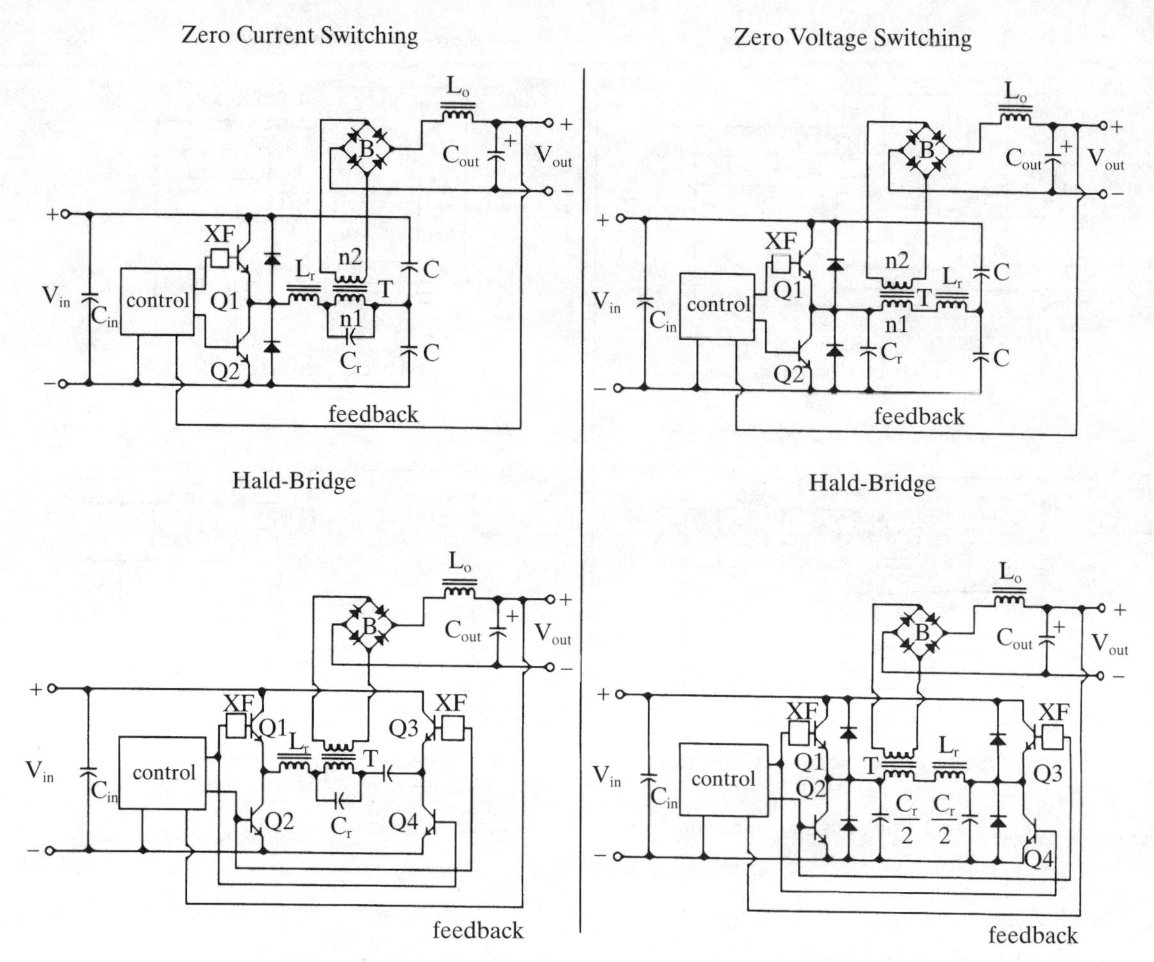

Figure 4–14 Two or more transistor, transformer-isolated quasi-resonant converters.

(series loading). This leads to quasi-resonant "subspecies": series-resonant, zero current (or voltage) quasi-resonant switching power supplies and parallel-resonant zero current (or voltage) quasi-resonant switching power supplies. They differ in how they load the resonant tank circuit and can be seen in Figure 4–15.

Let us examine how one determines the values of the inductor and capacitor. Several assumptions have to be made at the beginning of the design process since several of the tank circuit's characteristics are variable within the application. The first is to assume a value for the Q of the tank circuit. In the application, the Q varies greatly with the amount of load placed on the output of the supply. So, a good value to start with is

$$Q_{est} = 5$$

Next, decide the natural resonance frequency of the tank circuit. For the available quasi-resonant controller ICs on the market, the range is between 1 and 2 MHz. This limit should be considered the maximum limit within conventional QR designs. So 1 to 1.5 MHz is a typical choice. Lower frequencies can be used and some efficiencies can be gained. The equation for the resonance frequency is

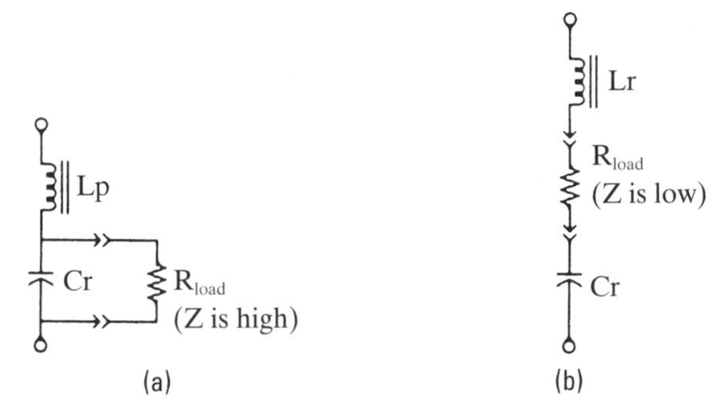

Figure 4–15 The two methods of loading a resonant tank circuit.

$$f_r = \frac{1}{2\pi\sqrt{LC}} \tag{4.15}$$

where: L is the value of the resonant inductance in Henries.
C is the value of the resonant capacitor in Farads.

Here we have one equation and two unknowns, obviously another relationship is needed. Many values will satisfy the above equations.

The typical assumption is to make the energies balance between the L and the C. This is done by

$$E_{tank} = \frac{1}{2} L_r \cdot i_L^2 = \frac{1}{2} C_r \cdot V_{pri}^2 \tag{4.16}$$

Substituting $E_{tank} = P_{out/fr}$ and solving for the value of the capacitor from the energy stored within the capacitor (second equation) is

$$C_r = \frac{2P_{out}}{V_{pri}^2 f_r} \tag{4.17}$$

Solve this equation by using the parameters known from your application. Then substituting this relationship for the resonant capacitor into the resonance equation (Equation 4.8), one gets

$$L_r = \frac{1}{C_r (2\pi f_r)^2} \tag{4.18}$$

The results of this exercise should be considered a starting point. This is because the parasitic elements in the power supply in the area of the tank circuit are of comparable value. The particular parasitic elements are the leakage inductance(s) of the transformer which would sum into the above value, and the output capacitance of a power MOSFET (if used) which also sums into the resonance capacitor's value. Some applications actually use these parasitic elements as the resonance elements, but much caution must be used, because these are elements resulting from the physical construction of the transformer and MOSFET. These can vary greatly from part to part.

What if the above relationship between the resonant inductor and capacitor were "pushed" one direction or the other. One could use a "large" L and a "small" C or vice versa. The relative size of the two components does affect some behaviors of the circuit. For instance, if a "small" C is used in a ZVS QR

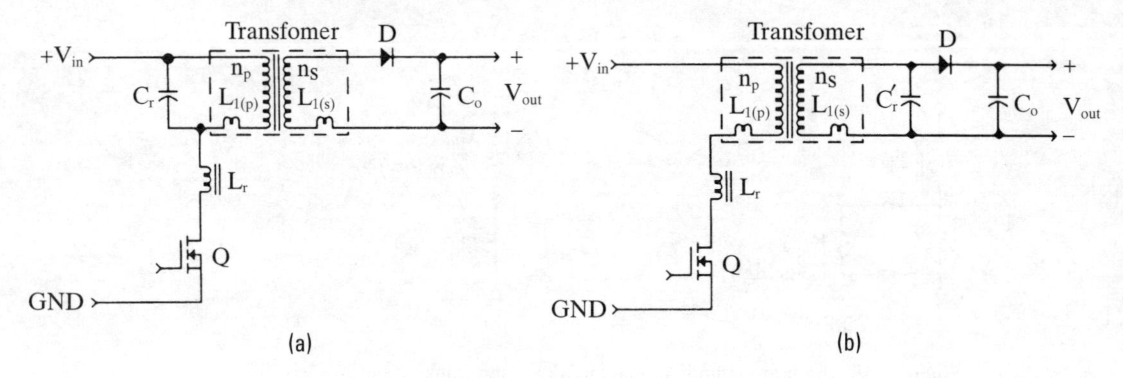

Figure 4–16 "Pushing" the resonant capacitor through a transformer in ZCS QR supplies: (a) primary-side resonance; (b) secondary-side resonance.

supply the peak ring voltage will increase thus producing a higher voltage stress on the power switch. Inserting the desired limiting voltage value into Equation 4.17 and solving will adjust the estimated capacitor value to achieve this maximum voltage. In typical 1 to 1.5 MHz QR applications the range of values for the tank elements are

$$C_r \approx 7{,}000 \text{ to } 10{,}000\,\text{pF}$$

$$L_r \approx 2 \text{ to } 5\,\text{uH}$$

Another arrangement for the tank circuit in ZCS QR supplies takes advantage of the leakage inductances of the primary and secondary(ies). The designer can "push" the resonant capacitor through the transformer and place it on the secondary as seen in Figure 4–16. This is done by utilizing the equation that describes the reflection of impedances through a transformer.

$$Z_{sec} = \left(\frac{n_{sec}}{n_{pri}}\right)^2 \cdot Z_{pri} \tag{4.19}$$

Substituting the value of the resonant capacitor (across the primary winding) for Z_{pri} one can determine the value needed on the secondary of the transformer. Without "pushing" the capacitor through the transformer, inductive voltage spikes still are present on the primary during transitions. When the resonant capacitor "pushed" through the transformer, the leakage inductances are summed into the resonant inductor and their dv/dts are now controlled by the tank circuit and the voltage spikes disappear. This eliminates the need for snubbing across the primary winding and will improve the RFI behavior of the supply.

If the transformer has multiple secondaries, the ZCS resonant capacitor can be split and placed on each secondary or can be placed across the entire secondary winding if an autotransformer secondary arrangement is used as shown in Figure 4–17.

If multiple isolated secondaries are used, one might split the values of the secondary resonant capacitors in proportion to the respective output powers. That is

$$C_{r(n)} \approx \left(\frac{P_{0(n)}}{P_{0(total)}}\right)\left(\frac{n_{sec(n)}}{n_{pri}}\right)^2 \cdot C_{r(pri)} \tag{4.20}$$

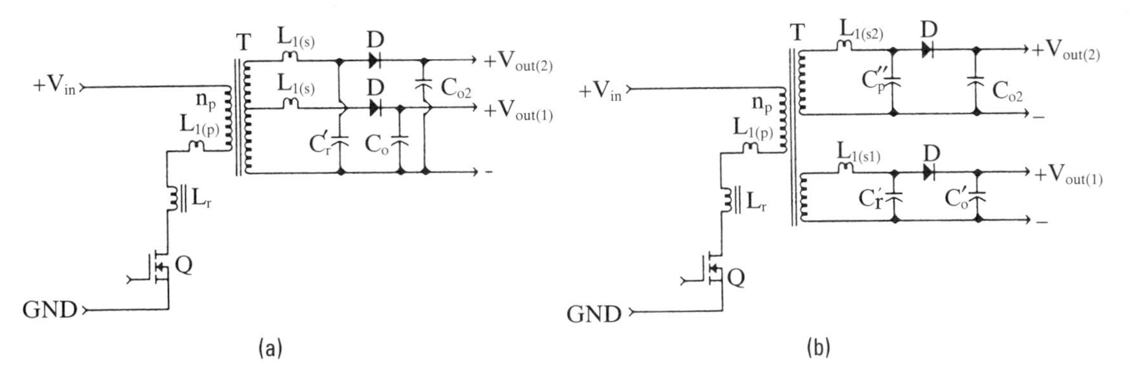

Figure 4–17 Second-side resonance arrangements for complex secondary designs in ZCS QR supplies: (a) autotransformer secondaries; (b) isolated secondaries.

One caution: Whenever the resonant capacitor is placed on a lower voltage output, the ripple current entering the capacitor increases. So, carefully check the ratings against the application.

In ZVS QR supplies that use power MOSFETs, the resonant capacitor (placed across the drain and source pins) will be summed with the output capacitance (C_{OSS}) of the MOSFET during its off-time. Some designers use only the C_{OSS} of the MOSFET. Caution must be used with this approach since this capacitance is highly nonlinear and varies with voltage and from part to part. But, by doing this, the designer reduces what is called charge dumping. During the turn-on of the MOSFET, the drain must shunt the charge developed across the C_{OSS} to ground.

Identifying and harnessing these parasitic losses into the operation of the QR power supply is an interesting challenge that should be undertaken. So, carefully analyze its operation and play with the arrangement of the tank circuits.

4.6.4 Phase Modulated PWM Full-bridge Converters

Since full-bridge converters occupy the highest output power region of the topologies, their power switch losses pose a particularly significant problem. In the normal PWM full-bridge converter, the power switches in opposing corners are switched simultaneously. This practice completely releases the primary winding from any low impedance ac ground. This results in large spikes and ringing caused by the leakage inductance of the primary and any residual magnetization inductance. Historically, this noise can only be reduced by the use of a lossy snubber network.

By changing the power switch control strategy so that only one corner power switch is first turned off, the opposing end of the primary winding remains connected to an ac ground. This allows the unloaded end of the primary winding to be "tuned" with a resonant capacitance and produce a controlled dv/dt. The value of the leakage inductance and the output capacitance of the power MOSFET contribute to the resonant network. The open-end terminal of the primary winding voltage then rings to the opposing input voltage rail where that MOSFET is then turned-on in a ZVS fashion. The MOSFET on the lagging end of the primary winding can then be turned off and its end of the primary can ring to its opposing input voltage and the opposing MOSFET can then be turned on. The waveforms can be seen in Figure 4–18.

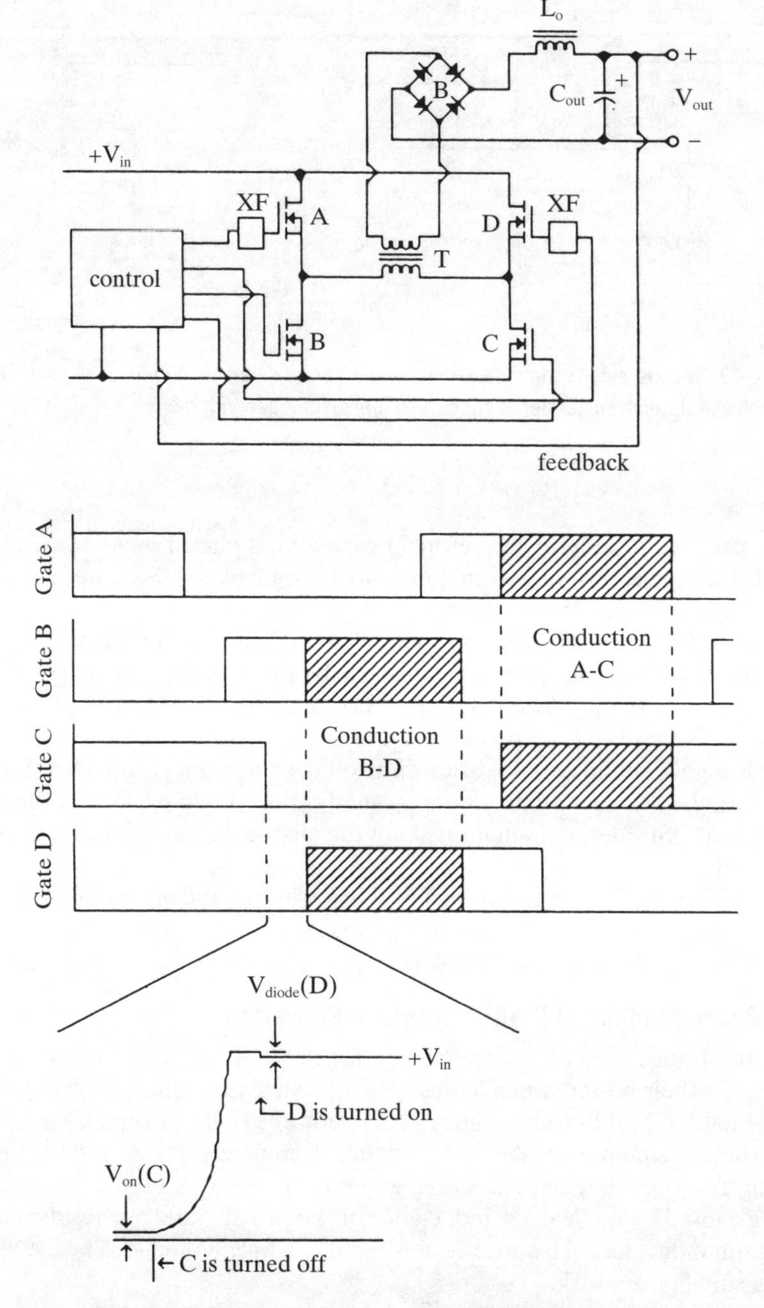

Figure 4–18 Phase modulation of a PWM full-bridge converter.

Since both ends of the primary winding have a single-ended loaded winding during their respective turn-off transitions, each of the MOSFETs accomplish ZVS turn-off. The output rectifiers gain some efficiency since their current transitions appear more zero-current switching in nature.

There are control ICs presently in the market. This is one technique to increase the frequency of full-bridge converters and gain three to five percent in efficiency over the normal PWM full-bridge converters. The cost of the supply does increase since more gate drive transformers are required.

4.7 High Efficiency Design Examples

The following design examples contain various techniques to improve the overall efficiency of common designs. Most of the designs of active clamps and lossless snubbers are empirical, so they are not included within these design examples.

Care must be taken in using the new techniques because of the patents that cover many of their uses. There may be a licensing fee that needs to be paid directly to the patent holder or a small per-piece fee that is paid when buying the control IC. Please investigate this before introducing these power supplies to production.

4.7.1 A 10 Watt Synchronous Buck Converter

Application

This design example is a redesign of the PWM design example 1 that shows how one would include synchronous rectifiers (Section 3.15.1).

In designing a synchronous switching power supply, one must be very careful when choosing the control IC. For the sake of maximizing efficiency and minimizing space, the average synchronous controllers take many liberties in system operation, which make them only good for the IC supplier's mentioned application. Many subtleties of operation cannot be determined unless the datasheet is read completely. For instance, every time I attempt to design a synchronous converter and try to use off-the-shelf ICs, I throw out three to four designs. This is because I encounter an unexpected operational mode that cannot be overcome or modified because the function is not brought out to a pin. Needless to say, it is a frustrating process if one wants something different from the "cookie cutter" solution (refer to Figure 4–20 for the schematic).

Design specification

Input voltage range:	+10 V to +14 VDC
Output voltage:	+5.0 VDC
Rated output current:	2.0 A
Overcurrent limit:	3.0 A
Output ripple voltage:	+30 mV p-p
Output regulation:	+/− 1%
Maximum operating temperature:	+40°C

"Black box" predesign estimates

Output power:	(+5.0 V)(2 A) = 10.0 W (max)
Input power:	P_{out}/Est. Effic. = 10.0 W/0.90 = 11.1 W
Power switch loss:	(11.1 − 10 W)(0.5) = 0.5 W
Catch diode loss:	(11.1 − 10 W)(0.5) = 0.5 W

Input average currents

Low input:	(11.1 W)/(10 V) = 1.11 A
High input:	(11.1 W)/(14 V) = 0.8 A
Estimated peak current:	1.4 ($I_{out(rated)}$)
	1.4(2.0 A) = 2.8 A

The desired frequency of operation is 300 kHz.

Inductor design (refer to Section 3.5.5)
The worst case operating condition is at the high input voltage.

$$L_{\text{min}} - \frac{(V_{\text{in(max)}} - V_{\text{out}})(1 - V_{\text{out}}/V_{\text{in(max)}})}{1.4 I_{\text{out(min)}} f_{\text{sw}}}$$

$$L_{\text{min}} - \frac{(14\,\text{V} - 5\,\text{V})(1 - 5\,\text{V}/14\,\text{V})}{1.4(0.5\,\text{A})(300\,\text{kHz})} = 27.5\,\mu\text{H}$$

where: $V_{\text{in(max)}}$ is the maximum possible input voltage.
V_{out} is the output voltage.
$I_{\text{out(min)}}$ is the lightest expected load current.
f_{sw} is the operating frequency.

The inductor is to be a surface mount toroid on a J-leaded plastic mounting board. There are standard surface mount inductors available from many suppliers. I have chosen the part DO3340P-333 from Coilcraft (33 μH).

Selection of the power switch and synchronous rectifier MOSFETs

Power switch. The power switch is going to be a transformer-coupled N-channel power MOSFET. I plan to use a dual N-Channel MOSFET in an SO-8 package to help save on PCB space. The maximum input voltage is 14 VDC. Therefore, a V_{DSS} rating of +30 VDC or higher will be satisfactory. The peak current is 2.8 A.

The first step in the selection process is to determine the maximum $R_{\text{DS(on)}}$ for the desired MOSFETs. This is done by examining the thermal model (see Appendix A). The maximum $R_{\text{DS(on)}}$ is found by

$$R_{\text{DS(on)(max)}} = \left[T_{\text{j(max)}} - T_{\text{amb(max)}}\right] \Big/ \left[(I_{\text{D}})^2 (R_{\text{JA}})\right]$$

It is also desired to keep the heat dissipation less than 1 W for the device so the estimated $R_{\text{DS(on)}}$ should be less than

$$R_{\text{DS(on-max)}} = P_{\text{D(est)}} \big/ (I_{\text{pk(est)}})^2$$

$$= (1\,\text{W})\big/(2.8\,\text{A})^2 < 127 \text{ milliohms (max. each MOSFET)}$$

I will select the FDS6912A dual N-channel MOSFET with an on-resistance of 28 milliohms at a VGS of 10 V packaged in an SO8 package.

Synchronous Diode. A Schottky diode about 30 percent of the rating of the continuous rating of the synchronous MOSFET needs to be placed in parallel with the MOSFET's intrinsic diode. This would be about 0.66 A at 30 V. I will use an MBRS130. This diode produces a 0.35 V forward voltage drop at 0.66 A.

Alternate choice. Fairchild Semiconductor, at the time of writing this book, released an integrated Schottky diode and MOSFET where the parallel Schottky diode is placed directly on the MOSFET silicon die (SyncFET). The SyncFET has a 40 milliohm N-channel MOSFET, copackaged with a 28 milliohm SyncFET. The part number is FDS6982S.

The output capacitor (refer to Section 3.6)
The value for the output capacitor is determined by the following equation"

$$C_{out(min)} - \frac{I_{out(max)}\left(1 - DC_{(min)}\right)}{f_{sw}\left(V_{ripple(p\text{-}p)}\right)}$$

$$C_{out(min)} - \frac{(2\,A)(1 - 5\,V/14\,V)}{(300\,kHz)(30\,mV)} = 142\,\mu F$$

The major concern of both output and input filter capacitors is the ripple current entering the capacitor. In this application, the ripple current is identical to the inductor ac current. The maximum limits of the inductor current is 2.8 A for I peak and about one-half the maximum output current or 1.0 A. So the ripple current is 1.8 A p-p or an estimated RMS value of 0.6 A (about one-third of the p-p value).

I will be using a surface mount tantalum capacitor because they typically exhibit about 10 to 20 percent of the ESR of electrolytic capacitors. I will also derate the rating of the candidate capacitors by 30 percent at +85°C ambient temperature.

The best candidate capacitors are from AVX. They have very low ESR and thus can handle very high levels of ripple current. These capacitors are exceptional. I will place two pieces of the following parts in parallel on the output.

AVX:
TPSE107M01R0150 100 μF (20%), 10 V, 150 mohm, 0.894 A_{rms}
TPSE107M01R0125 100 μF (20%), 10 V, 125 mohm, 0.980 A_{rms}

Nichicon:
F751A107MD 100 μF (20%), 10 V, 120 mohms, 0.92 A_{rms}

The input filter capacitors (Sections C.1 and C.2)

This capacitor experiences the same current waveform at the power switch, which is a trapezoid with an initial current of about 1 A rising to 2.8 A with very sharp edges. This capacitor has much more rigorous operating conditions than the output filter capacitor. I will estimate the RMS value of the trapezoidal current waveform as a piecewise superposition of two waveforms: a rectangular 1 A peak waveform and a triangular waveform with a 1.8 A peak. This yields an estimated RMS value of 1.1 A.

The value of the capacitor is then calculated as

$$C_{in} = \frac{P_{in}}{f_{sw}\left(V_{ripple(p\text{-}p)}\right)^2} = \frac{11.1\,W}{(300\,kHz)(0.5)^2}$$

Capacitors at the higher voltages have smaller capacitance values. There will be two 100 µF capacitors in parallel. The candidates are

AVX (two required per system):
TPS107M020R0085 100 $\mu\mu$F (20%), 20 V, 85 mohms, 1.534 A_{rms}
TPS107M020R0085 100 μF (20%), 20 V, 200 mohms, 1.0 A_{rms}

Selecting the controller IC (U1)

The desired features one might look for in a buck controller IC are:

1. Ability to operate directly from the input voltage.
2. Pulse-to-pulse overcurrent limiting.
3. Totem-pole MOSFET drivers.

4. Some control over the delay between the power switch and the synchronous rectifier MOSFETs.

There are very few synchronous buck controllers on the market that are not targeted to the +5 to 1.8 V microprocessor local regulator application (i.e., V_{dd} of +12 V, and V_{in} of +5 V). Also ICs that bring out enough functions to pins so that one may tailor its performance to the application. I threw out two entire product offerings from two major California companies (not to throw stones) and found only one part that fits my needs, the UC3580–3 from Unitrode/TI.

The internal reference presented to the voltage error amplifier is 2.5 V +/– 2.5 percent.

Setting the frequency of operation (R7, R8, and C8)

R8 charges the timing capacitor (C8) and R7 discharges the timing capacitor. First, one decides the maximum duty cycle of the converter. Since the output voltage is about 50 percent of the lowest input voltage, I will choose a 60 percent maximum duty cycle. From the data sheet:

$$\text{max duty cycle} = R8/(R8 + 1.25R7)$$

or

$$R8 = 1.875 \cdot R7$$

The charge time is 0.6/300 kHz or 2 μS maximum on time. The parameter tables use a timing capacitor value of 100 pF which is fairly small and will not dissipate much energy. I will use this value. The value of R8, therefore, is

$$R8 = (2.0\,\mu\text{S})/(100\,\text{pF}) = 20\,\text{k ohms}$$

$$R7 = (20\text{k})/1.875 = 10.66\,\text{k make } 12\,\text{k}$$

The volt-second limiter (R4 and C5)

This IC has a method of feed-forward maximum pulsewidth limiting. As the input voltage increases, the expected operating pulsewidth of the buck converter decreases. An R-C oscillator is directly connected to the input voltage and its timeout is inversely proportional to the input voltage. Its timeout period is set to about 30 percent longer than the excepted operating pulsewidth. If the pass unit is still conducting at the moment that the volt-second oscillator times out, the pass unit is cutoff.

I am selecting C5 to also be 100 pF since its timing is about the same frequency as the oscillator. That makes R4 around 47 Kohms.

Setting the deadtime between the pass unit and the synchronous rectifier MOSFETs

One could perform the turn-on and turn-off delay calculations presented in Section 3.7.2 and still have to adjust the value of the deadtime delay-setting resistor (R6) at the breadboard stage. A starting value of 100 nS is good. The typical MOSFET turn-on delay is about 60 nS. The 100 nS will assure that there is no push-through current.

The IC produces an asymmetrical deadtime delay. From the graph in the datasheet a value of 100 kohms will produce a pass unit turn-on delay of about 110 nS and a turn-off delay of 180 nS.

I am fully planning to reduce these delays at the breadboard stage. Delays this long cause the diodes to conduct too long, thus causing the losses to be high. But this is operating on the safe side.

Design of the gate drive transformer (T1)

The gate drive transformer is a very simple 1 : 1 turns ratio forward-mode transformer. There are not extraordinary demands being placed on the transformer since it is a very low power, ac-coupled (bipolar flux), 300 kHz transformer.

A ferrite toroid about 0.4 inches (10 mm) is sufficient, such as TDK part number K_5T10x2.5x5 (B_{sat} is 3300 G), or Philips part number 266T125–3D3 (B_{sat} is 3,800 G).

From Section 3.5.3, the number of turns to produce 1,000 G (0.1 T) or $0.3B_{sat}$ is

$$N_{pri} = \frac{V_{in(nom)} 10^8}{4 \cdot f \cdot B_{max} \cdot A_C}$$

$$N = \frac{(12\,V)10^8}{4(300\,kHz)(1,000\,G)(0.06\,cm)^2}$$

$$= 16.6 \text{ turns round up to 17 turns}$$

The gate drive transformer will be bifilar wound where two identical wires (about #30 AWG) will be wound simultaneously with equal turns. For convenience, the transformer will be mounted on a four-pin, gull wing, surface mount header.

Current sense resistor (R15) and voltage sensing resistor divider (R11 and R13)

The IC provides only a shutdown pin with an activation threshold of 0.4 V minimum. I plan to enter a hiccup mode of overcurrent protection as a form of back-up protection. To minimize the size of the current sensing resistor, I will employ a variation of the current foldback sensing circuit. Here 0.35 V will be contributed by a resistor in the voltage sensing resistor divider (R14). The value of R15 is then

$$R15 = 0.05\,V/3\,A = 16.6 \text{ mohms make 20 mohms}$$

The Dale resistor is WSL-2010-.02–05.

I will assign a sense current of about 1.0 mA through the voltage sensing resistor divider. This make the resistance represented by the sum of R13 and R14

$$R_{sum} = 2.5\,V/1.0\,mA = 2.5k$$

R14 is then

$$R15 = 0.35\,V/1.0\,mA = 350 \text{ ohms make 360 ohms}$$

R13 then becomes

$$2.5\,k - 360\,ohms = 2.14\,k\,ohms$$
$$\text{make } 2.15\,k\,ohms \text{ 1 percent tolerance.}$$

R11 is then

$$R11 = (5.0\,V - 2.5\,V)/1\,mA = 2.5k$$
$$\text{make } 2.49k \text{ ohms } 1\%$$

Compensating the voltage feedback loop (refer to Appendix B)

This is a voltage-mode, forward converter. To produce the best transient response time, a two-pole, two-zero method of compensation is going to be used.

Defining the control-to-output characteristic

The output filter pole is determined by the filter inductor and capacitor and is a –40 dB/decade rolloff. Its nominal corner frequency is

$$f_p = \frac{1}{2\check{s}\sqrt{L_o C_o}}$$

$$f_p = \frac{1}{2\pi\sqrt{(33\,\mu\text{H})(200\,\mu\text{F})}} = 1{,}959\,\text{Hz} \tag{3.58}$$

The zero caused by the output filter capacitor(s) (ESR is two 150 mohms in parallel) is

$$f_{esr} = \frac{1}{2\pi R_{esr} C_o} = \frac{1}{2\pi(75\,\text{m})(200\,\mu\text{F})} = 10{,}610\,\text{Hz}$$

The intrinsic absolute gain of the power circuit at dc is

$$A_{DC} \approx V_{in}/\Delta V_{error} = (14\,\text{V})/(2.9\,\text{V}) = 4.8$$

$$G_{DC} = 20\,\text{Log}(A_{DC}) = 13.6\,\text{dB}$$

Locating the compensating poles and zeros of the error amplifier

I shall use a gain cross-over frequency of 15 kHz, which is quite satisfactory for the majority of applications. This yields a transient response time around 200 μS.

$$f_{xo} = 15\,\text{kHz}$$

First one assumes that the final closed loop compensation network will have a continuous –20 dB/decade slope. To achieve a 15 kHz cross-over frequency the amplifier must add gain to the input signal and "push-up" the gain curve of the Bode plot.

$$G_{xo} = 20\,\text{Log}(f_{xo}/f_{fp}) - G_{DC} = 20\,\text{Log}(15\,\text{kHz}/1{,}959\,\text{Hz}) - 13.6\,\text{dB}$$

$$G_{xo} = G2 = +4.1\,\text{dB}$$

$$A_{xo} = A2 = 1.6\ (\text{absolute gain})$$

This is the gain needed at the midband plateau (G2) to achieve the desired cross-over frequency.

The gain exhibited at the first set of compensating zeros is

$$G1 = G2 + 20\,\text{Log}(f_{ez2}/f_{ep1}) = +4.1\,\text{dB} + 20\,\text{Log}(980\,\text{Hz}/10{,}610\,\text{Hz})$$

$$G1 = -16.5\,\text{dB}$$

$$A1 = 0.15\ (\text{absolute gain})$$

To compensate for the double filter pole, I will place two zeros at one-half the filter pole frequency:

$$f_{ez1} = f_{ez2} = 980\,\text{Hz}$$

The first compensating pole will be located at the capacitor's ESR frequency (4,020 Hz):

$$f_{ep1} = 10{,}610\,\text{Hz}$$

The second compensating pole is just used to maintain high-frequency stability by depressing the gain above the cross-over frequency:

$$f_{ep2} = 1.5 f_{xo} = 22.5\,\text{kHz}$$

Now one can begin calculating the component values within the error amplifier. Refer to Figure 4–19.

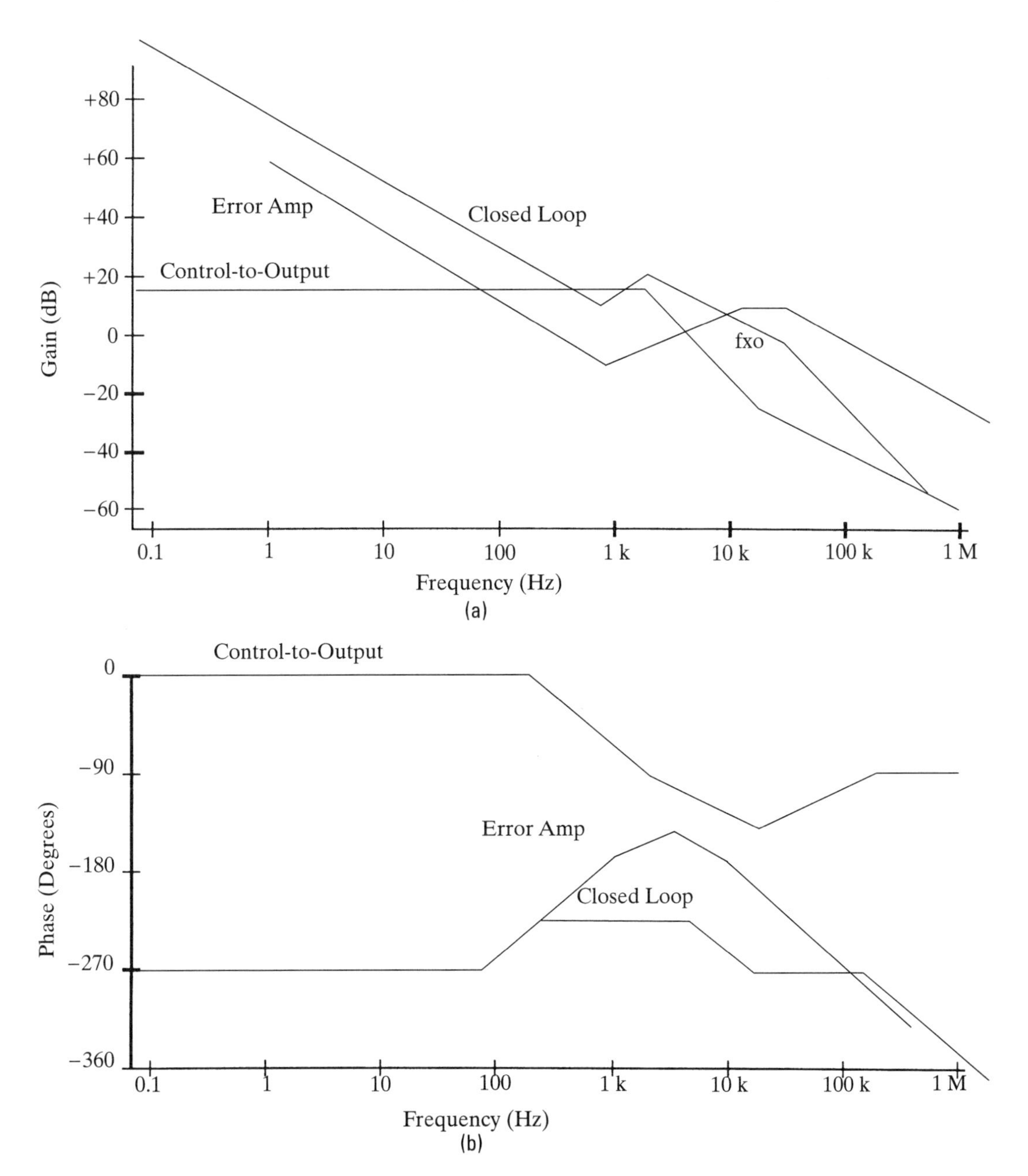

Figure 4–19 The gain and phase Bode plots for design example 4.7.1: (a) the gain plot for the synchronous buck converter; (b) the phase plot for the buck synchronous converter.

$$C13 = \frac{1}{2\pi(f_{xo})(A2)(R11)} = \frac{1}{2\pi(15\,\text{kHz})(1.6)(2.49\,\text{kž})}$$

$$C13 = 0.0026\,\mu\text{F make } 0.0027\,\mu\text{F}$$

$$R10 = (A1)(R11) = (0.15)(2.59\,\text{k}\Omega) = 373\,\Omega \text{ make } 360\,\Omega$$

$$C12 = \frac{1}{2\pi(f_{ez1})(R10)} = \frac{1}{2\pi(10{,}610\,\text{Hz})(360\,\text{ž})}$$

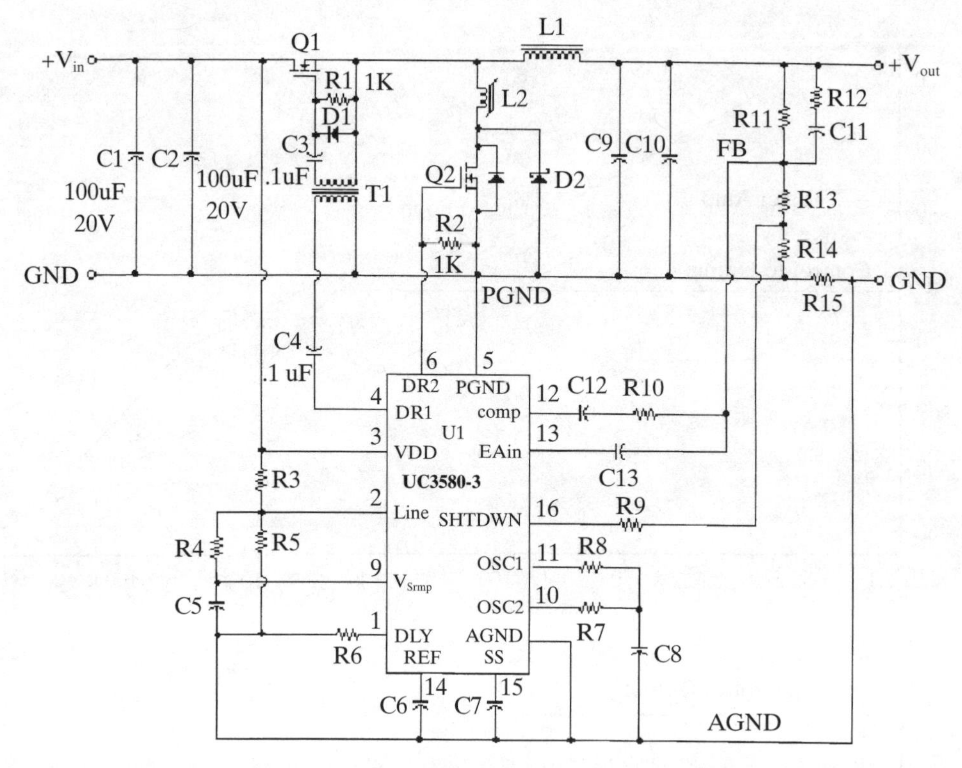

Figure 4–20 A 10 watt synchronous buck converter.

$$C12 = 0.042\,\mu\text{F make } 0.05\,\mu\text{F}$$

$$R12 = R10/A2 = (360\,\Omega)/(1.6) = 225\,\Omega \text{ make } 220\,\Omega$$

$$C10 = \frac{1}{2\check{s}\,(f_{ez2})(R12)} = \frac{1}{2\check{s}\,(22.5\text{kHz})(220\,\check{z}\,)}$$

$$C10 = 0.31\,\mu\text{F make } 0.33\,\mu\text{F}$$

4.7.2 A 15 Watt, ZVS Quasi-resonant, Current-mode Controlled Flyback Converter

This design is a novel turn about to the traditional voltage-mode controlled ZVS quasi-resonant converters. By modifying a conventional, non-duty cycle limited current-mode control IC into fixed off-time, current-mode controlled on-time method of control, a ZVS topology can be created. The added advantage is having the overcurrent protection and responsiveness of current-mode while having the reduced switching losses of a resonant technology. Although its frequency of operation may not exceed 1 MHz, it does offer the advantages of no switching losses and reduced EMI radiation. Please refer to Figure 4–22 for the schematic.

Design specification

Input voltage range:	18 V to 32 VDC, +24 VDC (Nominal)
Output voltage:	+15 VDC @ 0.5 A to 1 A
Low voltage "no start" voltage:	8.0 V +/− 1.0 V

Predesign "black box" considerations

Output power: $V_{out} \cdot I_{out} = (15\,V)(1\,A) = 15$ Watts
Maximum peak current:

$$I_{pk} \approx \frac{5.5 \cdot P_{out}}{V_{in(min)}} = \frac{5.5(15\,W)}{18\,V} = 4.6 \text{ Amps}$$

Input average currents:

$$I_{in(av)} \approx \frac{P_{out}}{eff \cdot V_{in(nom)}} = \frac{15\,W}{9(24\,V)} = 0.7 \text{ Amps}$$

$$I_{in(av-hi)} \approx \frac{P_{out}}{eff \cdot V_{in(low)}} = \frac{15\,W}{9(18\,V)} = 0.926 \text{ Amps}$$

Determining the gauge wire needed on the primary, the supply is required to operate at 18 V at the rated load current, so the primary wire gauge needs to be equivalent to #20 AWG.

Designing the flyback transformer
The transformer is going to be the only nonsurface mount component inside the power supply because there is no surface mount core style large enough to support 15 W. A toroid will work but I will be using a low profile E-E core from TDK. The "EPC" style of low profile cores will do.

Determining the core material. The power supply will be operating over the frequency range of around 150 kHz to 500 kHz, which is an area where two types of core material could be used. The "F", "3C8", and "H_7P_4" (similar materials made by different manufacturers) are useable up to about 800 kHz. The "N", "3C85", and "H_7P_{40}" are useable into the MHz region, with slightly lower core losses in this operational region. For this application, I will use the H_7P_{40} material (TDK).

Determining the core size. TDK rates its cores by the amount of power that can be handled by the core in a one-transistor forward converter. Its volume requirements are very similar to a flyback converter. The EPC core that rated at 15 W or greater is the EPC 17 core size. The part numbers for this assembly are: core, PC40EPC17-Z; bobbin,BER17–1111CPH; and clamp, FEPC17-A.

Determining the primary inductance. I am assigning a maximum on-time of 7 uS, which would occur at the minimum input voltage. The primary inductance would then be

$$L_{pri} = \frac{V_{in(min)} T_{on(max)}}{I_{pk}}$$
$$= \frac{(18\,V)(7\,\mu S)}{(4.6\,A)} = 27.3\,\mu H$$

The length of the air-gap should be approximately

$$l_{gap} \approx \frac{0.4\pi L I_{pk}^2 10^8}{A_c B_{max}^2}$$
$$= \frac{0.4\pi (27\,\mu H)(4.6\,A)^2 10^8}{(0.22\,cm^2)(1800)^2} = .101\,cm$$

The A_L of the core with this air-gap is about 55 nH/N^2. This A_L is the version used by TDK and uses the following equation to determine the number of turns:

$$N_{pri} = \sqrt{\frac{L_{pri}}{A_L}} = \sqrt{\frac{27\,\mu\text{H}}{55\,\mu\text{H}}} = 22.2 \text{ turns, make 22 turns}$$

The secondary winding's inductance controls how quickly the core would empty itself of the stored energy for the discontinuous-mode of operation. Since the input and output voltages are very close in magnitude, a 1 : 1 turns ratio would be possible. This would result in a 3 μS off-time within a comparable PWM system. Let us use a 1 : 1 turns ratio; it could be bifilar wound to promote the tightest coupling.

$$N_{sec} = 22 \text{ turns}$$

Wire gauges

Primary: #20 AWG or equivalent—make 3 strands of #24 AWG
Secondary: #20 AWG or equivalent—make 3 strands of #24 AWG

To reduce confusion, I will use two differently colored wires.

Transformer winding technique

The primary and secondary windings will be twisted together prior to winding them onto the bobbin. The windings will be separated on each end (distinguished by the colors) and soldered to the assigned terminals. A layer of Mylar tape will be placed on the exterior surface for aesthetics and safety.

Designing the resonant tank circuit

This is a first-pass estimate for the values of the tank circuit since at this time it is impossible to predict the influences of all the parasitic elements that would occur in the physical circuit. Readjustment of calculated tank values and the off-time setting on the controller IC will be necessary at the breadboard stage.

First one assumes that the power stored within the L-C tank circuit will be shared equally, or that

$$\frac{P_{out}}{f_{op}} = \frac{C_r V_c^2}{2} = \frac{L_r i_L^2}{2}$$

Rearranging and solving for the resonance capacitor value,

$$C_r = \frac{2P_{out}}{V_C^2 \cdot f_{op}}$$

I wish to limit the peak voltage across the resonant capacitor (C7) to less than 100 V and solve

$$C_r = \frac{2(15\,\text{W})}{(70\,\text{V})^2\,(250\,\text{kHz})} = 0.024\,\mu\text{F}, \text{make } .02\,\mu\text{F}$$

I have chosen the maximum operating frequency of the power supply to be 250 kHz. At light loads, the maximum on-time should be in the order of 10 to 15 percent. So, the resonance frequency will be around 250 kHz as well. Solving for the resonant inductor,

$$L_r = \frac{1}{C_r(2\pi f_r)^2}$$

$$= \frac{1}{(.02\,\mu\text{F})(2\pi(250\,\text{kHz}))^2} = 20\,\mu\text{H}$$

Designing the output rectifier/filter stage

Selection of the output rectifier

$$V_r = V_{out} + \frac{N_{sec}}{N_{pri}}(V_{in(max)})$$

$$= 15\,\text{V} + 32\,\text{V} = 47\,\text{V}$$

Let us select the MBR360 for D4.

Calculating the needed output filter capacitor

$$C_o = \frac{I_{out(max)}T_{off}}{V_{ripple}}$$

$$= \frac{(1\,\text{A})(2\,\mu\text{S})}{50\,\text{mV}} = 40\,\mu\text{F}$$

make C8 equal 47 μF at 25 VDC. I will use a very good grade tantalum capacitor and place in parallel a 0.5 μF ceramic capacitor.

Designing the bootstrap start-up section

I will be using the current limited linear regulator type of start-up circuit.

For R1, the base bias resistor,

$$R1 = (18\,\text{V} - 12\,\text{V})/(0.5\,\text{mA}) = 12\,\text{K}$$

For R2, the collector current limiting resistor,

$$R2 = (18\,\text{V} - 13\,\text{V})/(10\,\text{mA}) = 500\text{ ohms make }510\text{ ohms}$$

Design of the controller section

I will be using a garden variety UC3842 current-mode control IC. The IC selection is important because it cannot have 50 percent duty cycle limiting. The oscillator will be operating as a fixed off-time one-shot, which means shorting the timing capacitor to ground and the on-time is then only controlled by the current sense input pin. When appropriate current is achieved, then the timing capacitor will be release and the oscillator will operate like a one-shot timer for the off-time and restart the next on cycle.

Modifying the controller to fixed off-time. This is done by placing a small signal N-channel MOSFET across the timing capacitor. The gate is connected to the gate of the main power MOSFET. A couple of good choices for the small signal MOSFET is the BS170 or 2N7002. By referring to the data sheet for the timing elements and needing an off-time of approximately 2 μS, means that the timing components will be approximately 15 Kohms for the timing resistor and 220 pF for the timing capacitor. These values will be adjusted to match the resonant half-period of the tank circuit during the breadboard stage.

Design of the voltage feedback loop

I am going to use a sense current of 1.0mA. This makes the lower resistor in the voltage sensing resistor divider (R9) a 2.49K 1 percent resistor.

The upper resistor (R8) is

$$R8 = (15\,\text{V} - 2.5\,\text{V})/1\,\text{mA} = 12.5\text{K ohms}$$

make 12.4K 1 percent.

Designing the feedback loop compensation

Zero voltage quasi-resonant power supplies usually vary in frequency 4:1 with varying input line and load. That variation would make the estimated minimum switching frequency at 80kHz. We will need to estimate this for thc compensation.

The control-to-output characteristic curves for a current-mode controlled flyback-mode converter, even though it is operating in variable frequency, are of a single-pole nature. So a single pole-zero method of compensation should be used. The placement of the filter pole, ESR zero, and dc gain are

$$A_{\text{DC}} = \frac{(28\,\text{V} - 15\,\text{V})^2}{28\,\text{V} \cdot 2.5\,\text{V}}$$
$$= 2.41$$

$$G_{\text{DC}} = 20\,\text{Log}(2.41) = 7.7\,\text{dB}$$

$$f_{\text{fp(hi)}} = \frac{1}{2\pi(15\,\text{V}/1\,\text{A})(47\,\mu\text{F})}$$
$$= 225\,\text{Hz (at rated load (1A))}$$

$$f_{\text{fp(low)}} = \frac{1}{2\pi(15\,\text{V}/0.5\,\text{A})(47\,\mu\text{F})}$$
$$= 112\,\text{Hz (at light load (0.5A))}$$

The control-to-output curves can be seen in Figure 4–21.

The gain cross-over frequency should be less than $f_{sw}/5$ or

$$f_{xo} < 80\,\text{kHz}/5 < 16\,\text{kHz}$$

I'll settle for 10kHz.

To determine the amount of gain needed to boost the closed loop function to 0dB at the gain cross-over frequency (see Equation B.24):

$$G_{xo} = 20\log(f_{xo}/f_{\text{fp(hi)}}) - G_{\text{DC}}$$
$$= 20\log(10{,}000/225) - 7.7\,\text{dB}$$
$$= 25.2\,\text{dB (needed for Bode Plot only)}$$

$$A_{\text{XO}} = 18.3\ \text{(scaler gain—needed later)}$$

Locate the compensating error amplifier zero at the location of the lowest manifestation of the filter pole or

$$f_{ez} = f_{\text{fp(low)}} = 112\,\text{Hz}$$

Locate the compensating error amplifier pole at the lowest anticipated zero frequency caused by the ESR of the capacitor or

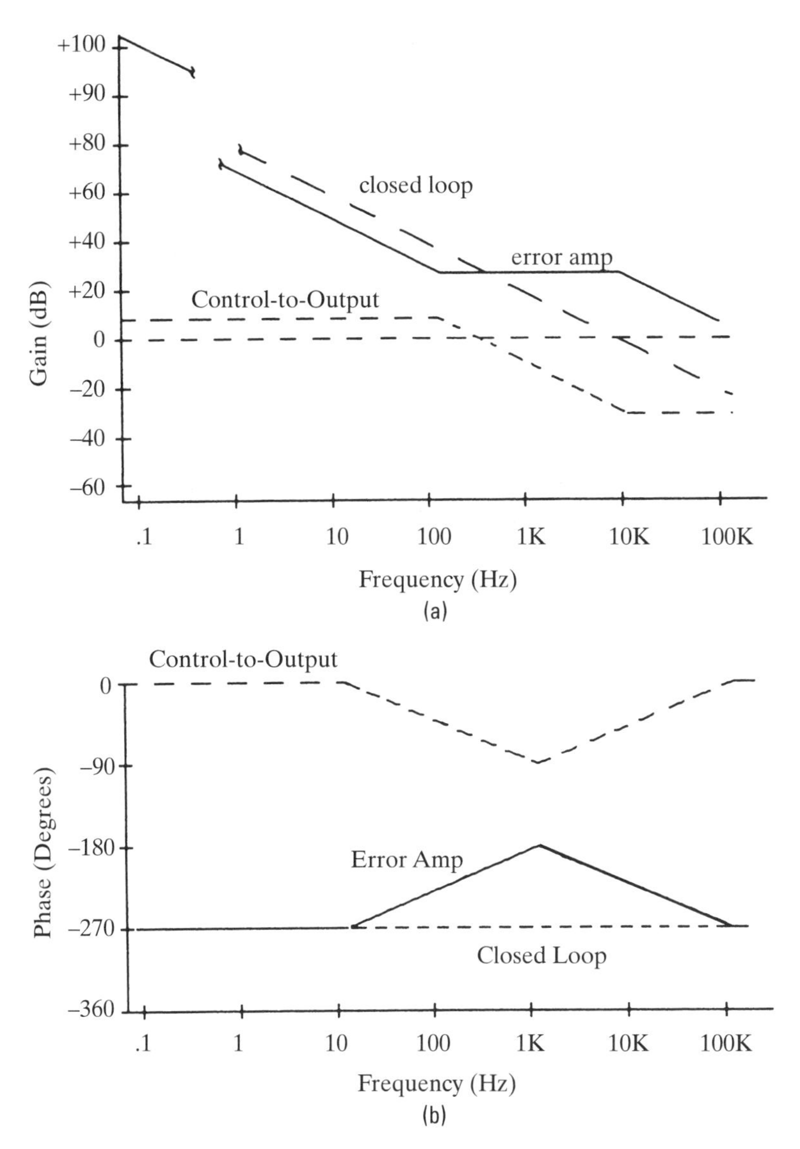

Figure 4–21 The gain and phase Bode plots for design example 4.7.1 (compensation design): (a) the gain plot for the power supply; (b) the phase plot for the power supply.

$$f_{ep} = f_{p(ESR)} = 10\,\text{kHz (approximated)}$$

Knowing the upper resistor in the +5 V voltage sensing divider (12.4 K),

$$\begin{aligned} C3 &= \frac{1}{2\pi \cdot A_{xo} \cdot R1 \cdot f_{xo}} \\ &= 1/2\pi(18.3)(12.4\,\text{K})(10\,\text{KHz}) \\ &= 70\,\text{pF make } 68\,\text{pF} \end{aligned}$$

$$\begin{aligned} R3 &= A_{xo} R1 \\ &= 18.3(12.4\,\text{K}) \\ &= 227\,\text{K} - \text{round to } 220\,\text{K} \end{aligned}$$

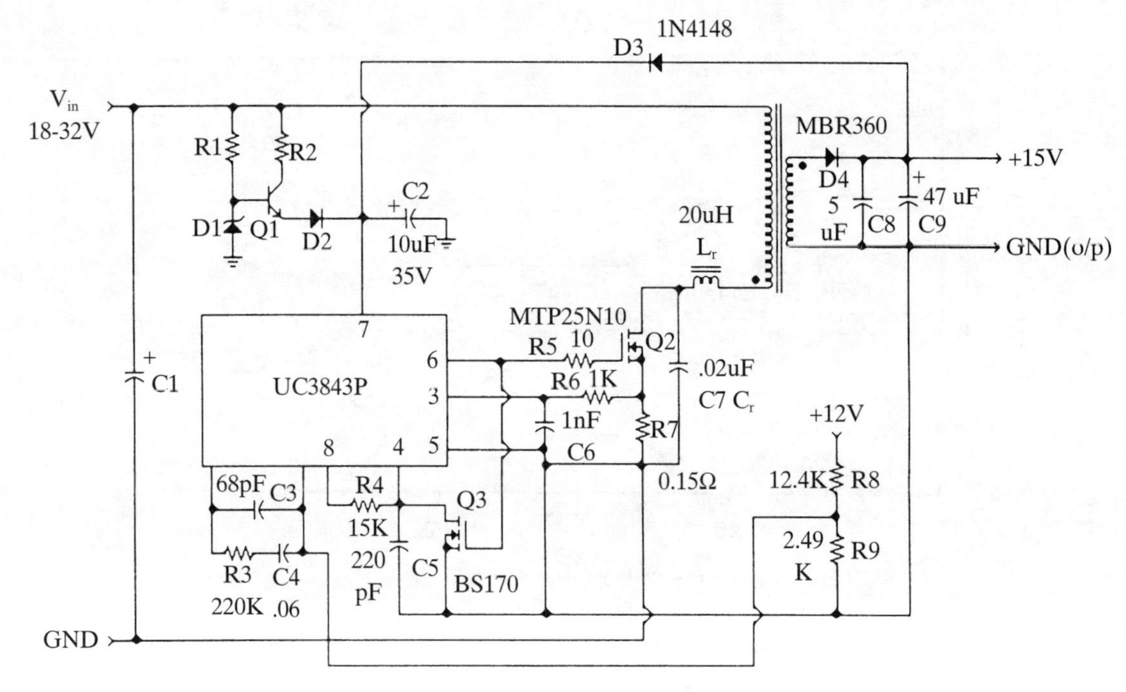

Figure 4–22 A ZVS quasi-resonant current-mode flyback converter.

$$
\begin{aligned}
C4 &= \frac{1}{2\pi f_{ez} R2} \\
&= 1/2\pi(112\,\text{Hz})(220\text{K}) \\
&= 0.065 - \text{round to } 0.06\,\mu\text{F}
\end{aligned}
$$

4.7.3 A Zero Voltage Switched Quasi-resonant Off-line Half-bridge Converter

This converter is intended to function as a bulk power supply for a distributed system. It has only one +28 VDC output at 10 A. This is going to be a classic ZVS quasi-resonant half-bridge converter; that is, variable frequency, voltage-mode controlled with averaging overcurrent protection. It is representative of the designs using the available control ICs on the market today.

This design example is the PWM design example (Section 3.15.4), modified to a quasi-resonant topology. Please refer to Figure 4–25.

Design specification

Output voltage:	+28 VDC +/– 0.5 V
Output rated current:	10 A rated, minimum Load 1 A
Input ac voltage range:	105 VAC–130 VAC
	208 VAC–240 VAC
Output ripple voltage:	50 mV p-p
Output regulation:	+/– 2 percent

Predesign considerations

1. Rated output power: (28 V)(10 A) = 280 W
2. Estimated input power: $P_{in(est)} = 280\,\text{W}/0.8 = 350\,\text{W}$

3. DC input voltages (a voltage doubler is used for 110 VAC)
   a. From 110 VAC: $V_{in(low)} = 2(1.414)(90\,VAC) = 254\,VDC$
      $V_{in(hi)} = 2(1.414)(130\,VAC) = 368\,VDC$
   b. From 220 VAC: $V_{in(low)} = 1.414(185\,VAC) = 262\,VDC$
      $V_{in(hi)} = 1.414(270\,VAC) = 382\,VDC$
4. Average input currents (dc):
   a. Highest average: $I_{in(max)} = 350\,W/254\,VDC = 1.38\,A$
   b. Lowest average: $I_{in(min)} = 350\,W/382\,VDC = 0.92\,A$
5. Estimated maximum peak current: $I_{pk} = 2.8(280\,W)/254\,VDC = 3.1\,A$

This converter is going to be able to pass UL, CSA, and VDE safety requirements, which will affect the design approach that is taken. It will operate between the frequencies of 1 MHz at the minimum specified load and no less than 200 kHz at the maximum specified load.

Design of the transformer

I will be using an E-E core with the necessary Mylar tape insulating layers to meet the isolation requirements of the safety approval agencies. At these operating frequencies, a high-frequency core material will have to be used to minimize the core losses at 1 MHz. The material should be the "K" material from Magnetics, Inc, "3C85" material from Phillips, or "N67" material from Siemans.

One consideration is the variation in the operating flux density over the operating range of the supply. At minimum load (1 A) the frequency will be at its highest (1 MHz), and the B_{max} should be about $0.1\,B_{sat}$. At maximum load (10 A) the operating frequency is no less than 200 kHz, so its B_{max} should be less than $0.3B_{sat}$. Let us start by assigning the B_{max} at 1 MHz at 200 G and use this as the reference operation point.

The size of the core will be approximately 1.3 inches (41 mm) on a side. This corresponds to the K-43515 core from Magnetics, Inc. In designing the transformer, one must keep in mind that the first pulse issued in starting up is at the full input voltage and the subsequent pulses will be at or closer to one-half of the input voltage. To be positive that the core will not enter saturation during that first pulse, one limits the maximum B_{max} at a higher than normal level:

$$N_{pri} = \frac{(382\,V)\cdot 10^8}{4(200\,KHz)(2{,}200\,G)(.904\,cm^2)} = 24\,T$$

Checking the maximum operating flux densities using Faraday's law, at light load, the frequency is approximately 1 MHz, the B_{max} is 211 G. At its heavy load (200 kHz) the B_{max} is 1,100 G. This seems to be exactly what is required.

In calculating the number of secondary turns needed, one needs to consider that the resonant transition period is approximately 0.5 μSec within a 2.5 μSec maximum period per MOSFET at 200 kHz. That corresponds to an 84 percent duty cycle. So,

$$N_{sec} = \frac{1.1(2.8\,V + 0.5\,V)(24\,T)}{(254\,V - 2\,V)(.84)} = 3.55\,T$$

Make 4 turns.

The auxiliary winding should be two turns.

Transformer winding technique (refer to Section 3.5.9)

The transformer is going to be wound in an identical fashion as the PWM half-bridge design example (Section 3.15.4). That is, the secondary will be interleaved between two layers of primary windings. The auxiliary winding will be placed next to the core. Of course the necessary layers of Mylar tape and a 2 mm spacing from the side walls of the bobbin are necessary for VDE.

Considerations in the design of the L-C tank circuit

It is desired that the resonance frequency of the tank circuit be 1 MHz. In ZVS QR converters, the tank circuit is not responsible for storing and passing on energy as it is in ZCS QR converters. The tank circuit can be seen more as an off-time transition shaper similar to a snubber when used in PWM converters. Here a wide range of values for both the inductor and the capacitor will work as long as their combined resonant frequency is 1 MHz or

$$f_r = \frac{1}{2\pi\sqrt{L_r C_r}}$$

I am planning to use the leakage inductance of the primary as a part of the resonant inductor. For transformers of this nature, a leakage inductance of 0.5 to 1 uH is typical. I will specify the winding method to the transformer manufacturer so that the variation in the transformer's leakage inductance will be held to a minimum. I will have to add a small inductor external to the transformer. I will also use the C_{oss} (output capacitance) of the power MOSFET as a portion of the resonant capacitor. This is a highly variable and nonlinear capacitance and it is difficult to predict its value. The value of the C_{oss} is dependent upon the off-time VDSS of the MOSFET, which will be changing during the resonant off-time. Obviously, this requires some adjustment during the breadboard stage.

Just as an estimate, let us calculate the approximate value required for the resonant capacitor.

$$C_r = \frac{1}{L_r(2\pi f_r)^2} = \frac{1}{(1\,\mu\text{H})(2\pi(1\,\text{MHz}))^2}$$
$$= 0.025\,\mu\text{F}$$

This value is higher than that exhibited by the output capacitance of the MOSFETs so I will add an external resonant capacitor outside the MOSFET.

Selection of the power semiconductors

1. The power MOSFET (refer to Section 3.4):

$$V_{DSS} > V_{in} = 382\,\text{V make } 500\,\text{V}$$

$$I_D > I_{in(av)} = 2.75\,\text{Amps make } 4\,\text{Amps}$$

We can use the MTP4N50E, but the MTP8N50E will result in lower conduction losses.

2. Output rectifier:

$$V_R > 2V_{out} = 56\,\text{VDC make} > 70\,\text{VDC}$$

$$I_{FWD} > I_{out(max)} = 10\,\text{Amps, make } 20\,\text{Amps}$$

Select MBR20100CT

Design of the output filters

1. The minimum ac output filter inductance (refer to Section 3.5.5):

$$L_{o(\min)} = \frac{(40\,\text{V} - 28\,\text{V})(1\,\mu\text{S})}{1.4(1\,\text{Amp})} = 8.5\,\mu\text{H}$$

Using the LI2 method of determining the size of an MPP toroid, one determines the Magnetics P/N 55206A2 core part number.

The number of turns should be:

$$N_{\text{Lo}} = 1{,}000\sqrt{\frac{.0085}{68}} = 11.2 \text{ turns, make 12 turns}$$

The toroid should have an overall wire gauge of #12 AWG. We will use 100 strand Litz wire to minimize the skin effects.

2. The minimum output filter capacitance (refer to Section 3.6):

$$C_{o(\min)} = \frac{(10\,\text{A})(1\,\mu\text{S})}{.05\,\text{V}_{\text{p-p}}} = 200\,\mu\text{F}$$

I will use four capacitors at a value of 47 μF tantalum capacitors. This will make the RMS ripple current within the ratings of typical capacitors. I will also place a 0.5 μF ceramic capacitor in parallel with the other capacitors.

3. Design the output dc filter choke (refer to Section 3.5.7): Referring to the permeability versus dc bias graph (Figure 3–22) and selecting a permeability that does not excessively degrade at a reasonable dc bias level, I choose a permeability of 60 at an "H" level of 40 Oe.

Using the same core size as above one gets:

$$N = \frac{(300)(5.08\,\text{cm})}{0.4\pi(10\,\text{A})} = 12.12 \text{ turns, make 13 turns}$$

We still need #12 AWG wire. Litz wire is easier to wind around a toroid and reduces the skin effect so I will use it.

Design of controller IC related functions

The MC34067 data sheet contains the necessary equations and graphs in order to set the critical timer functions of a ZCS QR converter. Some of the times, such as the one-shot off-timer, will have to adjusted in the breadboard stage.

Setting the minimum operating frequency

The minimum operating frequency of the control IC is set by the combination of the R and C on the oscillator pin. First one select the oscillator capacitor from a graph. The discharge period to produce a 200 kHz period (5 μS) can be produced with a 200 to 300 pF capacitor. Make $C_{osc} = 220$ pF. The oscillator resistor is calculated from

$$R_{osc} = \frac{T_{\max} - 70\,\text{nS}}{0.348C_{osc}} = \frac{(2.5\,\mu\text{S} - .07\,\mu\text{S})}{0.348(220\,\text{pF})}$$
$$= 31.7\,\text{K, make } 33\,\text{K}$$

The maximum operating frequency is set by additional discharge current being drawn by the error amp through resistor $R_{VFO.}$ The equation to determine the additional discharge current (Imax) is

$$I_{\max} = 1.5 \cdot C_{osc} \cdot f_{\max} = 1.5(220\,\text{pF})(1\,\text{MHz})$$
$$= 330\,\mu\text{A}$$

The current flowing through the parallel oscillator resistor is

$$I_{R_{osc}} = \frac{1.5}{R_{osc}} e^{\left(\frac{-1}{f_{min} R_{osc} C_{osc}}\right)}$$

$$= \frac{1.5}{(33\,\text{K})} e^{\left(\frac{-1}{(200\,\text{kHz})(33\,\text{K})(220\,\text{pF})}\right)}$$

$$= 22.8\,\mu\text{A}$$

The value of the series discharge resistor (RVFO) from the output of the error amp to the oscillator is

$$R_{VFO} = \frac{2.5 - V_{ea(sat)}}{I_{max} - I_{R_{osc}}} = \frac{2.5 - .3}{330\,\mu\text{A} - 22.8\,\mu\text{A}}$$

$$= 7.16\,\text{K ohms, make } 6.8\,\text{K ohms}$$

Setting the one-shot timer

The one-shot timer can be set entirely from a graph and the approximate values from the graph are RT = 1.5 Kohms and the CT = 220 pF.

Design of the voltage feedback circuit

The voltage feedback circuit is to be an isolated variety using the error amplifier as a noninverting voltage-follower. This is to provide a buffer between the optoisolator and the VFO. Since this method of quasi-resonant control is a form of voltage-mode control a 2-pole–2-zero form of feedback compensation is necessary. The circuit will take the form shown in Figure 4–23. Refer to Figure 4–24 for part designations within these calculations.

I am designating the maximum voltage at the input to the error amplifier to be +4.5 V. The MOC8102 has a nominal C_{trr} of 100 percent and the TL431 requires a minimum of 1 mA passing through it to operate. This makes the value of the resistor R1:

$$R1 = (0.5\,\text{V})/(1\,\text{mA}) = 500\text{ ohms, make 470 ohms}$$

When the input to the error amp is at its lowest voltage, the input voltage shall be at the saturation of the output of the optoisolator or 0.3 V. The current that then must be sunk by the output of the optoisolator is

$$I_{max} = \frac{5.0\,\text{V} - 0.3\,\text{V}}{470} = 10\,\text{mA}$$

R2, then is then determined at the maximum current point of its operation, which is 10 mA. It becomes

$$R2 = \frac{28\,\text{V} - V_{fwd} - V_{TL431}}{I_{max}} = \frac{28\,\text{V} - 1.4\,\text{V} - 2.5\,\text{V}}{10\,\text{mA}}$$

$$= 2,410\text{ ohms, make 2.4 kohms}$$

The voltage-sense voltage divider is started by selecting a sense current through the divider. I will use 1 mA. The lower resistor (R3) becomes

$$R3 = 2.5\,\text{V}/1\,\text{mA} = 2.5\,\text{K or } 2.49\,\text{K } 1\%$$

The upper resistor of the divider (R4) is then

$$R4 = \frac{28\,\text{V} - 2.5\,\text{V}}{1\,\text{mA}} = 25.5\,\text{K } 1\%$$

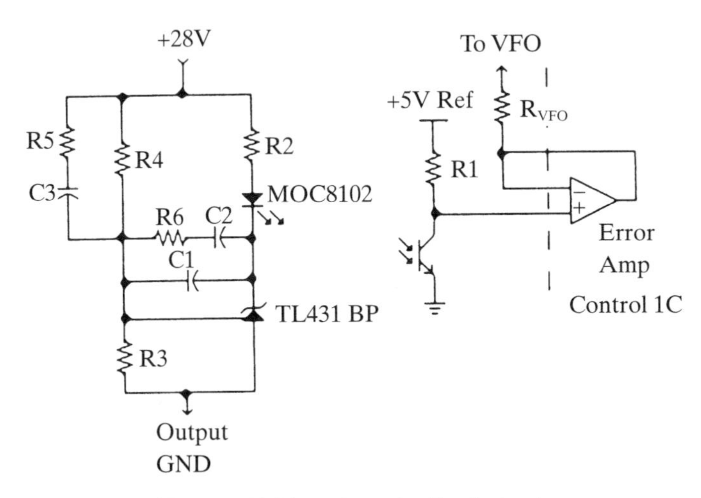

Figure 4–23 The voltage feedback circuit.

Design of the feedback loop compensation (refer to Appendix B)

The form of compensation is the 2-pole–2-zero method of compensation. This is to compensate for the effect of the double pole caused by the output filter inductor and capacitor. One starts by determining the control-to-output characteristic of the open-loop system.

The gain exhibited at dc of the system is

$$A_{DC} = \frac{V_{in} N_{sec}}{^2 V_e N_{pri}} = \frac{(340\,\text{V})(4\,\text{T})}{(1\,\text{V})(24\,\text{T})}$$

$$A_{DC} = 56.6$$

Expressing this in dB for the Bode plot, (GDC) is

$$G_{DC} = 20\,\text{Log}(A_{DC}) = +35\,\text{dB}$$

The frequency of the output filter pole is

$$f_{fp} = \frac{1}{2\pi\sqrt{L_o C_o}} = \frac{1}{2\pi\sqrt{(8.5\,\mu\text{H})(188\,\mu\text{F})}}$$

$$= 3.981\,\text{Hz}$$

The estimated frequency of the zero caused by the tantalum output filter capacitor and its ESR is around 20 kHz, so

$$f_{Z(ESR)} \approx 20\,\text{kHz}$$

Let us select the gain cross-over frequency to be 20 kHz. It could be selected as high as 40 KHz but the gain-bandwidth product of the TL431 is not extraordinary and we may have problems attaining that high a cross-over frequency.

Leaving the pole pair together in their frequency location, the location is then

$$f_{ez1} = f_{ez2} = 0.5 f_{fp} = 1991\,\text{Hz}$$

The location of the lower of the two error amplifier zeros (f_{ep1}) is placed at the location of the estimated zero of the output capacitor and its ESR. Hence,

$$f_{ep1} = 20\,\text{kHz}$$

The remaining pole (f_{ep2}) is placed higher than the gain cross-over frequency, so

$$f_{ep2} = 1.5 f_{xo} = 30\,\text{kHz}$$

The gain needed between the two compensating poles (G2) to achieve the desired gain cross-over frequency is

$$G_{xo} = G2 = 40\,\text{Log}\left(\frac{f_{xo}}{f_{fp}}\right)$$
$$= 40\,\text{Log}\left(\frac{20\,\text{kHz}}{3{,}981\,\text{Hz}}\right) - 36\,\text{dB}$$

A2 is 0.40

The gain exhibited at the location of the two compensating zeros (f_{ez1} & f_{ez2}) is

$$G1 = G2 + 20\,\text{Log}\left(\frac{f_{ez2}}{f_{ep1}}\right)$$
$$= -6.9\,\text{dB} + 20\,\text{Log}\left(\frac{1{,}990\,\text{Hz}}{20\,\text{kHz}}\right)$$
$$G1 = -26.9\,\text{dB}$$

A1 is 0.0451

The compensation values can now be calculated (refer to Figure 4–23).

$$C1 = \frac{1}{2\pi f_{xo}(A2)(R4)}$$
$$= \frac{1}{2\pi(20\,\text{kHz})(0.4)(25.5\,\text{K})}$$
$$= 780\,\text{pF, make } 750\,\text{pF}$$

$$R6 = A2\,R1 = (0.4)(25.5\text{k})$$
$$= 10.2\,\text{k ohms, make } 10\text{k ohms}$$

$$C3 = \frac{1}{2\pi f_{ez2}(R1)}$$
$$= \frac{1}{2\pi(1{,}991\,\text{Hz})(25.5\,\text{K})}$$
$$= 3{,}134\,\text{pF, make } 0.003\,\mu\text{F}$$

$$R3 = R2/A2 = (10\text{k ohms})/(0.0451) = 221\text{ kohms}$$

make 220 kohms

$$C2 = \frac{1}{2\pi f_{ep2}(R2)}$$
$$= \frac{1}{2\pi(30\,\text{kHz})(220\,\text{K})}$$
$$= 24\,\text{pF}$$

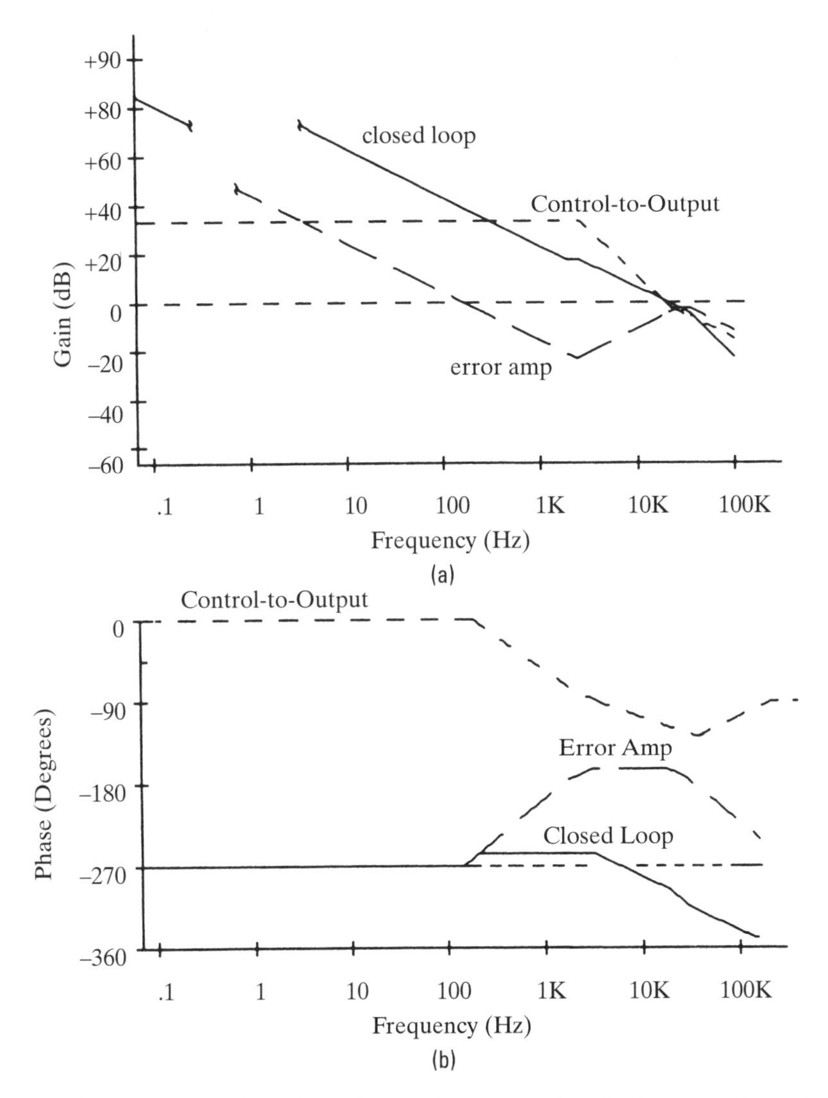

Figure 4–24 The gain and phase Bode plots for design example 4.7.2 (compensation design): (a) the gain plot for the power supply; (b) the phase plot for the power supply.

The resulting gain and phase Bode plots can be seen in Figure 4–24.

Design of the overcurrent protection circuit

The MC34067 has only a simple comparator with a 1.0 V threshold that then latches an R-S flip-flop. Either the input voltage must be removed and then reapplied, or the Vcc of the control IC must be interrupted momentarily to reset the failure latch.

I desire only a fairly rudimentary overcurrent protection that will be set at about 12 A. I will use a 100:1 current transformer followed by a rectifier bridge then an R-C integrator-smoother.

The peak current within the primary circuit is 3.1 A, which will occur at low-line and at a 200 to 400 kHz operating frequency. Then the output current from the secondary of the current transformer is

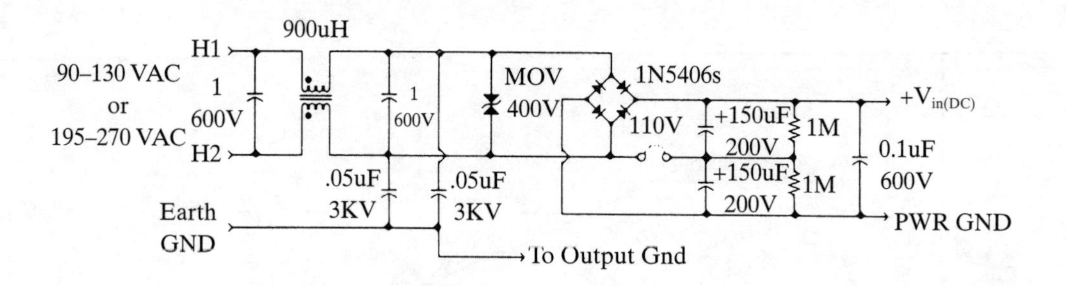

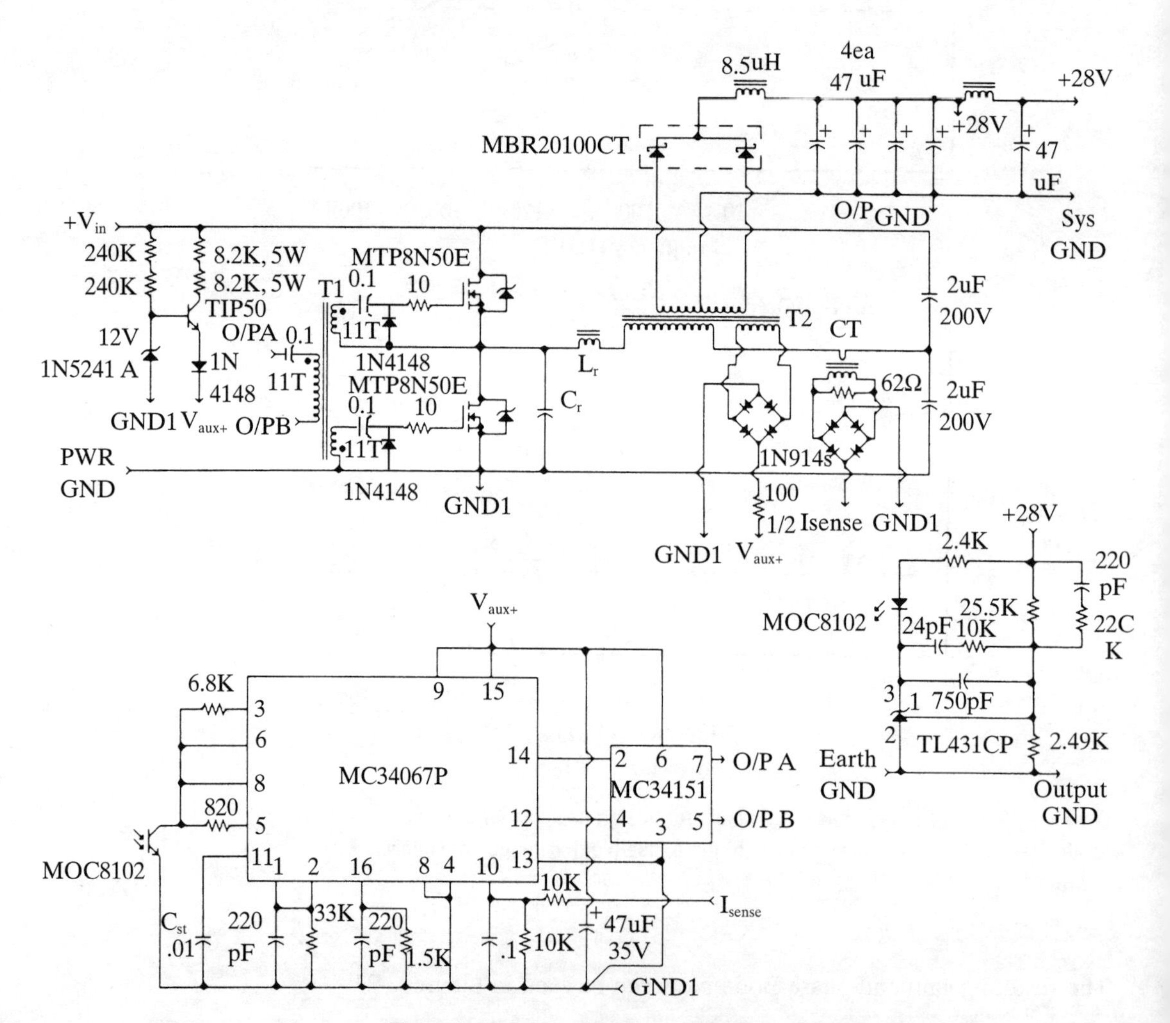

Figure 4–25 Schematic for the ZVS QR half-bridge converter.

$$I_{2CT} = (3.1\,\text{A})(1/100) = 31\,\text{mA}$$

The voltage needed from the output of the current transformer must overcome the forward drops of rectifiers and exceed 1.0 V threshold on the control IC, so

$$V_{2CT} = 1.0\,\text{V} + 2(0.6\,\text{V}) = 2.2\ \text{volts}$$

The resistor needed on the current transformer's secondary is

$$R_{2CT} = 2.2\,\text{V}/31\,\text{mA} = 70.9\,\text{K}$$

For a 20 percent overcurrent trip point, lower this resistance by 20 percent or a resistor value of 56 ohms.

I would like a time constant for the R-C filter to be about 1 mS. Selecting the value of the capacitor to be 0.1 uF, the charging resistor should be

$$R \approx T/C = 1\,\text{mS}/0.1\,\mu\text{F} = 10\,\text{kohms}$$

This completes the major design of the parts of the circuit that are different from design example 3.15.4. For the unchanging sections, please refer to that section. The finalized schematic is shown in Figure 4–25.

Appendix A. Thermal Analysis and Design

Proper thermal design is essential to the overall design of a power supply. Overdissipation failures account for probably the largest portion of the failures. Therefore, it is essential that the designer understands its basic principles.

Thermal analysis is really no more difficult than Ohm's Law. There are similar parameters to voltage, resistance, nodes, and branches. For the majority of electronic applications, the thermal "circuit" models are quite elementary and if enough is known of the thermal system, values can be calculated in a matter of minutes. If one has a temperature-measuring probe, the thermal components can also easily be measured and calculated.

There are two main goals in designing the thermal system: the first is to never allow any component to exceed its *maximum operating junction temperature* ($T_{J(max)}$); the second is to keep the components as cool as possible given the restrictions for space and weight. Failure to maintain the first condition will cause a component to fail within minutes. The second consideration affects the long-term life of the system. MIL-217, a reliability prediction tool for high-reliability applications, makes the following generalization, "*The life of a component will be halved every +10°C rise above room temperature.*" In most applications, the designer should be concerned if any component's case temperature exceeds +60°C.

A.1 Developing the Thermal Model

Thermal system analysis is actually a variation of Ohm's Law. There are equivalent circuit elements which map directly to the elements within the electrical domain (refer to Table A–1).

These elements always form a loop, with the power source providing the driving force for the entire model. Each circuit element and node corresponds to a physical structure or surface within the actual physical design. The power source corresponds to a heat-producing element within the circuit which creates a calculable or measurable power. The power semiconductors are typically the major heat-producing elements within a power supply. The power may be measured by graphically multiplying the terminal voltage and current waveforms from an oscilloscope and normalizing the energy product to one second (power = energy/second) or by measuring voltage and current directly using a digital volmeter (DVM), if it is a dc application. The result is expressed in watts.

The thermal resistance can represent two physical situations. The first is the resistance to heat flow across a surface boundary such as a power

Table A-1 Analogous Elements between the Thermal and Electrical Domains

Electrical element	Thermal equivalent
Voltage source	Power (heat) source
Resistance	Thermal resistance
Node voltage	Temperature of element
Current loop	Thermal loop
Circuit ground	Ambient air temperature

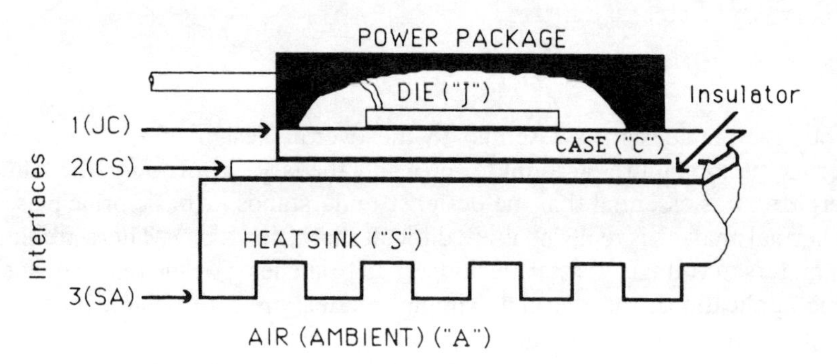

Figure A–1 Development of the thermal model for power packages.

transistor bolted to the surface of a heatsink. The second situation is how well the heat spreads through a body from the heat-emitting surface to a radiating surface. Both of these physical situations are simply represented by a single thermal element—the thermal resistor, which is represented by the Greek symbol theta. Its units are measured in degrees Celsius per watt (°C/W), which represents the temperature difference across a boundary given a certain power dissipation. Some of the thermal resistances related to the semiconductor are as follows.

Power packages

$R_{\theta JA}$ Thermal resistance from the junction to the air
$R_{\theta JC}$ Thermal resistance from the junction to the case
$R_{\theta CS}$ Thermal resistance from the case to the heatsink
$R_{\theta SA}$ Thermal resistance from the heatsink to the air

Diodes

$R_{\theta JL}$ Thermal resistance from the junction to the lead
$R_{\theta LA}$ Thermal resistance from the lead to the air

All of the semiconductor case-related parameters are published by the semiconductor manufacturers. The sink-to-air parameter is published by the heatsink manufacturers, if one buys a heatsink. If one makes his or her own, it is easy to measure these resistances from any model.

Every thermal model has as its ground, the ambient air temperature, unless the heat removing medium is water or a refrigerant, in which case the ambient temperature of that medium is used. This must be the case, since the power producing device can be no cooler than the coolest media around it and since heat flows from the warmer to the cooler body.

The nodes in the model are the respective surfaces of bodies along the path of flow of the heat. These can be transistor cases, heatsink surfaces, the semiconductor die, etc. The calculated temperatures of these surfaces can actually be measured using a temperature probe at their respective surfaces. If the power dissipation is not known but all the thermal resistances are known, one can extrapolate backwards within the model and determine the power being dissipated within the die by simply measuring the temperature difference across one of the thermal boundaries.

A.2 Power Packages on a Heatsink (TO-3, TO-220, TO-218, etc.)

This physical situation can be modeled as shown in Figure A–2. The thermal equation would look like

$$T_{j(max)} = P_D(R_{\theta JC} + R_{\theta CS} + R_{\theta SA}) + T_A \tag{A.1}$$

Since the heatsink performs the vast majority of the heat radiation, it is assumed that all the power flows through all the other thermal elements.

Temperature tests can be conducted at ambient room temperature, but the designer must remember that the typical product is enclosed in a case and its internal temperature rise must be added to the readings. Another consideration is the highest external ambient temperature the product may experience. In the desert, where this book was written, daytime temperatures may reach $^{+}43°C$ in the shade and exceed 55°C inside an automobile.

Some typical thermal resistances associated with the different power packages are given in Table A–2.

These thermal estimates are minimums and maximums for those types of packages. The thermal resistance values are highly dependent on the size of the die inside the package, so refer to the data sheet for the exact maximum value.

The insulating pad also adds to the thermal resistance of the case-to-heatsink. Choosing the proper insulating pad can minimize this thermal resistance. Two common technologies are mica and silicone. There are also some ceramic technologies but these are for highly specialized applications. In addition, some insulators require thermal grease to attain a good thermal contact, such as mica (see the references at the end of this Appendix).

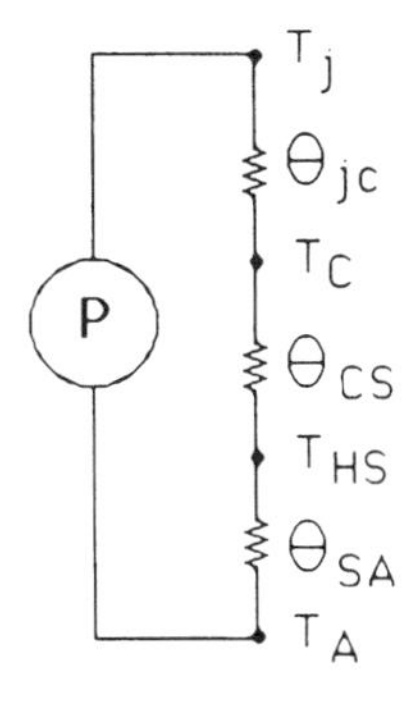

Figure A–2 The thermal model for a transistor on a heatsink.

Table A-2 Thermal Resistances of Common Thru-hole Power Packages

Package	Minimum	Maximum	Minimum	Maximum
TO-3	*	30.0	0.7	1.56
TO-3P	*	30.0	0.67	1.00
T0–218	*	30.0	0.7	1.00
TO-218FP	*	30.0	2.0	3.20
TO-220	*	62.5	1.25	4.10
TO-225	*	62.5	3.12	10.0
TO-247	*	30.0	0.67	1.00
DPACK	71.0	100.0	6.25	8.33

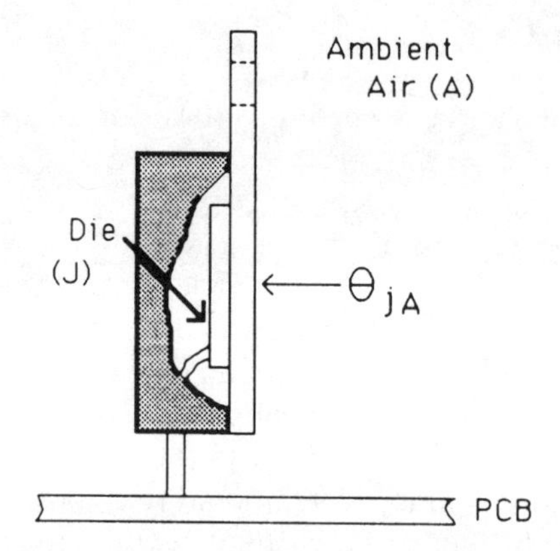

Figure A–3 A free-standing power package.

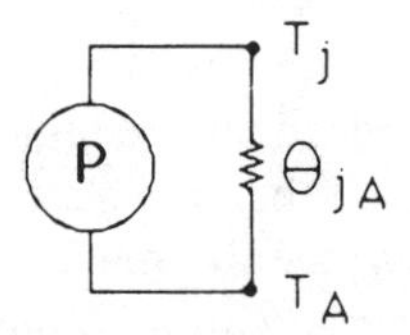

Figure A–4 Thermal model of a free-standing power package.

A.3 Power Packages Not on a Heatsink (Free Standing)

Power packages not mounted on a suitable heatsink can expect to dissipate less than five percent of the maximum specified power capability of the package. So 100 W devices will only dissipate 1 to 2 W when they are free standing. This also includes using the PC board copper plating as a heatsink. Thus, great discretion should be used when cost is the most important issue.

The thermal model for the case in Figure A–3 is shown in Figure A–4. The thermal equation becomes

$$T_{j(max)} = P_D \cdot R_{\theta JA} + T_A \tag{A.2}$$

As one can see by the typical values of the junction-to-air thermal resistance, it doesn't require much power to result in very high junction temperatures. If the designer can possible mount the power package on any metal surface to increase the radiating surface area, it will only improve the junction temperature.

A.4 Radial-leaded Diodes

The diodes within the power supply typically dissipate a large amount of power. These are the input rectifiers and the output rectifiers. In a bipolar centered switching power supply, the output rectifiers dissipate as much power as the bipolar power switches, so their contribution to the heat within the system is significant. The physical situation is shown in Figure A–5.

As one can see, the thermal parameters define a physically different situation. For a radial-leaded diode, the heat can only be conducted from the die, via the leads. The thermal resistance would then change as a function of lead-length and is published this way in data sheets. The thermal expression (see Figure A–6a) is

$$T_{j(max)} = P_D \cdot R_{\theta LA} + T_A \tag{A.3}$$

This is for the typical PC board-mounted application where only the PC board traces are used to conduct the heat away from the diode. The typical value range of the lead-to-air thermal resistance is between 30 to 40°C/W and is a variable which is dependent on the lead length.

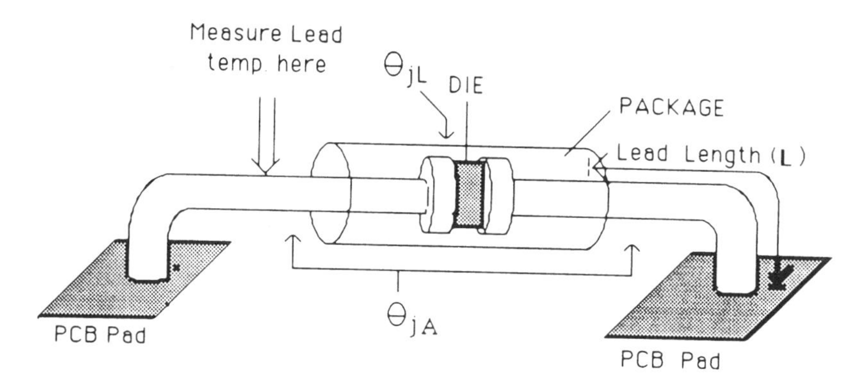

Figure A–5 Physical diagram of a mounted diode.

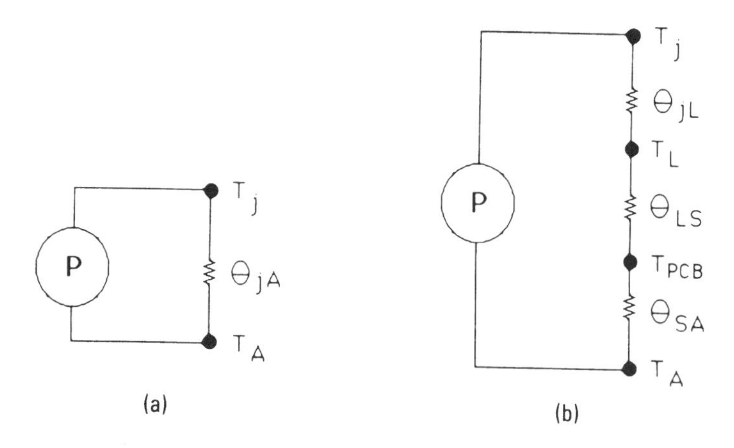

Figure A–6 The thermal model for an axial-leaded diode.

There are some heatsinks for radially-leaded diodes that solder to one of the leads. These are also available from the transistor heatsink manufacturers. In this situation the thermal equation (see Figure A–6b) becomes

$$T_{j(max)} = P_D(R_{\theta JL} + R_{\theta SA}) + T_A \quad (A.4)$$

These heatsinks will help a marginal heat situation. The alternative is to use a rectifier in a power transistor package such as a TO-220, TO-218, etc., and place it on a heatsink or to investigate a different technology of diode that exhibits a lower forward voltage drop such as a Schottky.

A.5 Surface Mount Parts

The use of surface mount parts is widespread. Surface mount parts can rid themselves of heat only through their leads which are soldered to a printed circuit board. The thickness and surface area of the copper island become the heatsink system. The thermal resistances in surface mount devices are much higher, therefore their designs have much less margin and room for error. Table A–3 has nominal values for thermal resistances of common surface mount packages. Please refer to the individual part data sheet for the exact value.

It is very important to select the package is appropriate for the function being performed. For switching signals, which are signal currents of less than 50 mA, the SOT23, SOD123, and other simple packages with gull-wing, J-leaded, and solder bump leaded packages are very compact and economical. For currents of 100 mA through amperes, the package must have a tab or multiple leads connected directly to the die. This is typically the drain, collector, or cathode. The common packages are the SOT223, DPAK, SMB, and SMC. These tabs offer a very low resistance channel to remove the heat from the die and get it onto the PC board for dissipation.

In surface mount printed circuit board applications, more than one issue usually must be considered. Heatsinking must be considered, along with signal and EMI/RFI considerations. The trace that must dissipate the greatest heat within a switching power supply is also the node that has the largest dv/dts which couple very easily to the surrounding traces.

Table A-3 Typical Surface Mount Package Thermal Resistances

Package	J-A[1]	J-C[2]
SOD123	340	150
SOT23	556	75
SOT223	159	7.5
SO-8	63	21
SMB		13
SMC		11
DPAK	80	6
D2PAK	50	2

[1] Theremal resistance for a reference pad size.

[2] Thermal resistance for a very large pad size.

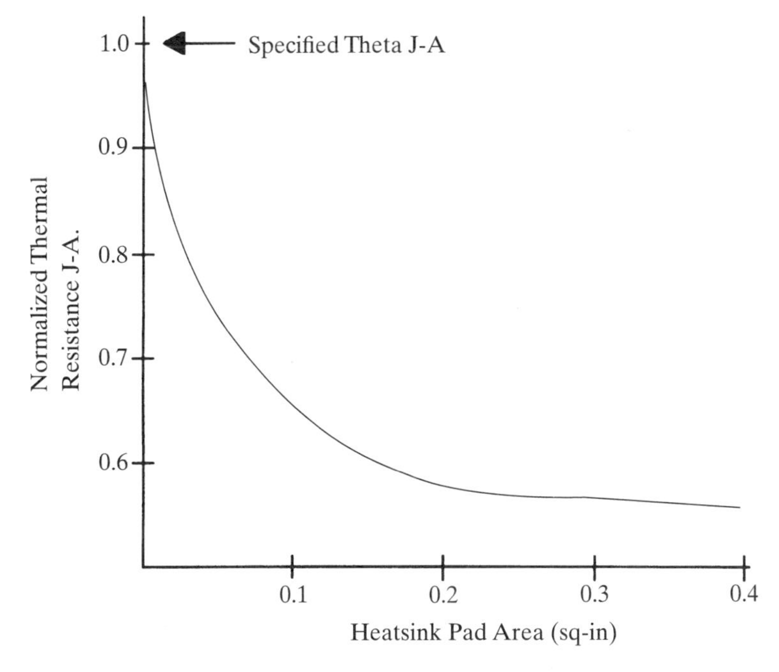

Figure A–7 Example of the effect of increasing pad area versus theta JA.

Laying out heatsinking systems for surface mount packaging technology systems is still an uncertain process. Semiconductor manufacturers still do not offer adequate information for each power package to feel confident about the adequacy of the heatsinking design. The graph in Figure A–7 is a normalized plot based upon a SOT223 package. The curve is 2 oz copper on the top of the PCB only. The curve, such as contained in Figure A–7, are needed to properly size the PC board heatsink island.

A.6 Examples of Some Thermal Applications

These examples will show the reader a typical application of thermal analysis but with common application variations. These variations are useful in defining thermal boundaries within a design.

A.6.1 Determine the Smallest Heatsink (or Maximum Allowed Thermal Resistance) for the Application

This approach is useful for determining the smallest possible heatsink that an application can use before the thermal limit of a power device is exceeded. This is an example of a consumer market approach to designing a heatsink system.

Specification
The device is an FDP6670 (Fairchild MOSFET) in a switching power supply. Convection cooling.

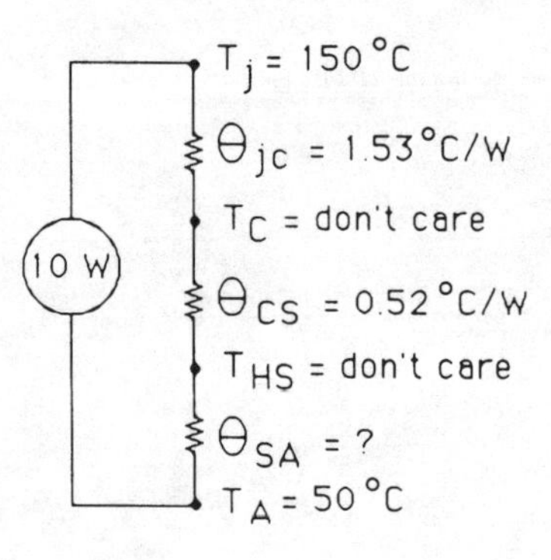

Figure A–8 Thermal model for design example A.6.1.

$$\text{PD} = 10 \text{ watts}$$
$$T_{\text{A(max)}} = +50°\text{C}$$
$$\theta_{\text{JC}} = 2.0°\text{C}/\text{W}$$
$$\theta_{\text{SA}} = 0.53°\text{C}/\text{W} \quad (\text{Thermalloy P}/\text{N 53-77-5})$$
$$T_{\text{J(max)}} = 175°\text{C}$$

The thermal model is shown in Figure A–8.

Rearranging Equation A.1 and solving for the thermal resistance of the heatsink,

$$\theta_{\text{SA(max)}} < (T_{\text{J}} - T_{\text{A}})/P_{\text{D}} - \theta_{\text{JC}} - \theta_{\text{CS}} \tag{A.5}$$

θ_{CS} is assumed at 1.0°C/W. Being conservative by not requiring the junction at its maximum temperature makes the maximum allowable junction temperature 150°C. The result is

$$\theta_{\text{SA(max)}} = 7.0\,°\text{C}/\text{W}$$

The PC board-mounted heatsink choices are: Thermalloy part numbers 7021B through 7025B for low-cost sheet-metal type heatsinks.

A.6.2 Determine the Maximum Power That Can Be Dissipated by a Three-terminal Regulator at the Maximum Specified Ambient Temperature without a Heatsink

A 3-terminal regulator's overcurrent protection is totally dependent upon the heatsinking system. When the die reaches approximately 165°C, the regulator shuts down. This example demonstrates the nonheatsink capabilities of a μA7805.

Specification

The desired 3-terminal regulator is a μA7805KC (TO220) (Texas Instruments).

$T_{\text{J(max)}}$	150°C
$T_{\text{A(max)}}$	+50°C
$V_{\text{in(max)}}$	10.0 VDC
$I_{\text{out(max)}}$	200 mA

$$\theta_{\text{JA}} = 22\,°\text{C}/\text{W}$$

The power dissipated by the regulator is

$$P_D = (V_{in(max)} - V_{out}) \cdot I_{out(max)} \quad (A.6)$$

or

$$P_D = 1.0\,W$$

The thermal model is that of Figure A–4, and the thermal equation is Equation A.2 rearranged to

$$T_{A(max)} = T_{J(MAX)} - P_D \cdot \theta_{JA} \quad (A.7)$$

$$T_{A(max)} = 150°C - (1.0\,W)(22\text{ deg }C/W)$$

$$T_{A(max)} = 128°C$$

So the μA7805KC will operate within its maximum junction temperature ratings for this application.

A.6.3 Determine the Junction Temperature of a Rectifier with a Known Lead Temperature

This is useful in verifying whether a diode's junction temperature is within its safe operating temperature.

Specification

This is a Zener diode, shunt regulator application. The diode is a 1N5240B ($10\,V_{(nom)}$, +/– 5%).

$I_{Z(max)}$	50 mA
$T_{A(max)}$	+50°C
T_L	+46°C (measured at TA = +25°C)
Lead length	3/8 inch (1.0 cm) each (175°C/W)

The worst-case power dissipation is

$$P_D = 1.05(10\,V)(50\,mA) = 525\,mW \text{ or } 0.525\,W$$

This situation would fit the thermal model as seen in Figure A–7b. It does not matter in this case not all the elements of the model are known since all the elements above the lead temperature node are known for this first step. The thermal expression for the temperature rise above the measured lead temperature is

$$T_{J(rise)} = P_D \cdot \theta_{JL} \quad (A.8)$$

or

$$T_j = (0.525\,W)(175°C/W) = 92°C\text{ rise}$$

The junction temperature at the specified maximum local ambient temperature is

$$T_{J(max)} = T_{J(rise)} + T_{A(max)} \quad (A.9)$$

$$T_{J(max)} = 142°C$$

The maximum junction temperature specified in the data sheet is +200°C, so the junction will be operating safely.

Appendix B. Feedback Loop Compensation

The heart of every linear and switching power supply is a negative feedback loop which maintains a constant value for the output voltage(s). To accomplish this, an error amplifier is used, which attempts to minimize the error between the output voltage and an ideal reference voltage. If the world were well behaved, a very high-gain inverting amplifier would be used and this job would be simple. The reality is that loads change, and the input voltage suddenly goes up or down. The error amplifier must respond to these changes quickly and without oscillating. This is complicated because the response in the power portion of the power supply is relatively "lethargic." If the error amplifier takes too long to respond to these changes, the supply behaves sluggishly. If the response is speeded up, the supply reaches a point where it may oscillate. So the problem becomes one of how fast and to what degree the response of the error amplifier should tailored to the power circuits.

Do not feel alone in your apprehensiveness about your knowledge in this area. There are few engineers that understand feedback loop compensation, because its been "cluttered" with too much fundamental mathematics that is not easily transferable to the actual circuit design. My approach is an easy step-by-step method which has always worked in my designs and can be completed in less than 20 minutes.

B.1 The Bode Response of Common Circuits Encountered in Switching Power Supplies

The Bode plot is a good method for working with feedback systems over a range of frequencies. It does employ logarithms, so a scientific calculator will be needed. The purpose of this section is not to teach the reader everything he or she needs to know about Bode plots, but instead should give a reasonable understanding of the behavior of the actual circuit elements and what their influences are on the responsiveness of the supply.

The Bode plot is actually composed of two graphs: a gain *vs*. frequency graph and a phase *vs*. frequency graph. It is a representation of the relative gain and phase shift of the output voltage signal referenced to the input voltage signal contributed by any 2-port circuit stage. When more than one circuit stage is cascaded together, their respective Bode responses can simply be added together to yield a combined Bode response.

Simple combinations of components produce responses which are called *poles* and *zeros*. A single pole (Figure B–1) produces a flat gain response from dc to

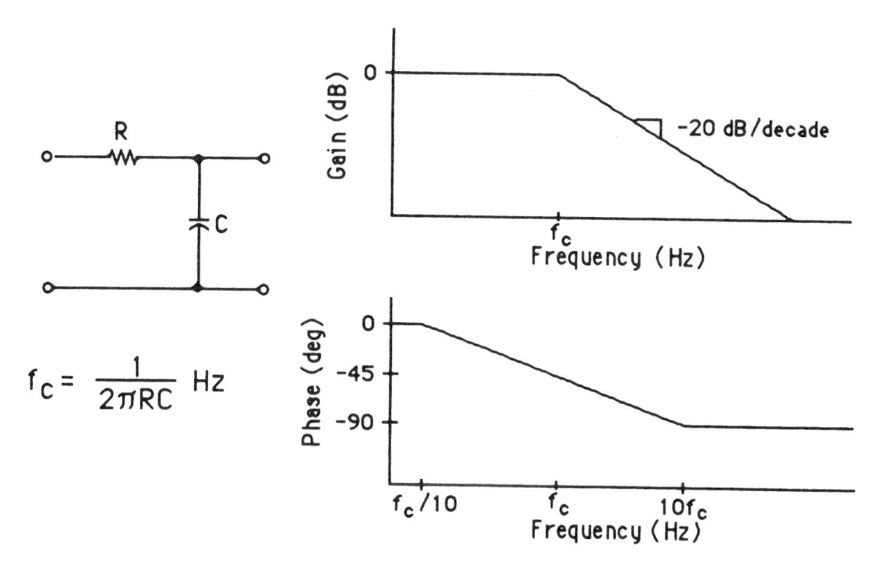

Figure B–1 A single-pole RC low-pass filter.

its *corner frequency*. It then produces a –20 dB/decade gain slope above its corner frequency. The corner frequency is the frequency at which the impedances of the two components equal one another. At least one of the components is *reactive*, which means the value of its impedance varies with the frequency. The impedance value of an inductor ($Z_L = j2\pi fL$) increases with frequency and its branch current always leads the branch voltage by 90 degrees. The impedance value of a capacitor ($Z_C = 1/j2\pi fC$) starts at infinity at dc and drops with frequency, and its current always lags the voltage by 90 degrees. In Figure B–1, a simple low-pass filter, the capacitor starts with an impedance at dc of infinity; then when the capacitor impedance equals the resistor value, an ac voltage divider is formed where the output amplitude is one-half the input amplitude. This is called the *6 dB point*. The output phase as compared to the input voltage is –45 degrees, which means that it is lagging the input signal. Eventually, its phase would reach 90 degrees as the impedance of the capacitor becomes magnitudes larger than that of the resistor. The rule-of-thumb for phase is that all phase influences from a pole or zero occur within a +/–1 decade about its corner frequency. A zero (Figure B–2) is just the opposite from a pole. It has a flat gain response from dc to its corner frequency, then proceeds at a +20 dB/decade gain response and a maximum phase lead of +90 degrees.

There are circuits within switching power supplies that exhibit double pole responses. This is where both elements in the stage are reactive, such as the L-C filter on the output stage of a forward-mode converter. This is best seen in Figure B–3. Here the response is flat from dc to its *resonance* frequency and then exhibits a –40 dB/decade gain response and a –180 degree lagging phase response at high frequencies. Lagging phase corresponds directly to a time delay through the forward-mode switching power supply output filter.

In switching power supplies, operational amplifiers (op amps) are used to alter the Bode functions (refer to Figure B–4). First, the op amp contributes an additional –180 degrees of lag (inverting amp) and any pole or zero adds or subtracts gain and phase from this –180 degree starting point. The generalized error amp is seen in Figure B–4. With op amps, a corner frequency of a simple pole or zero is defined as:

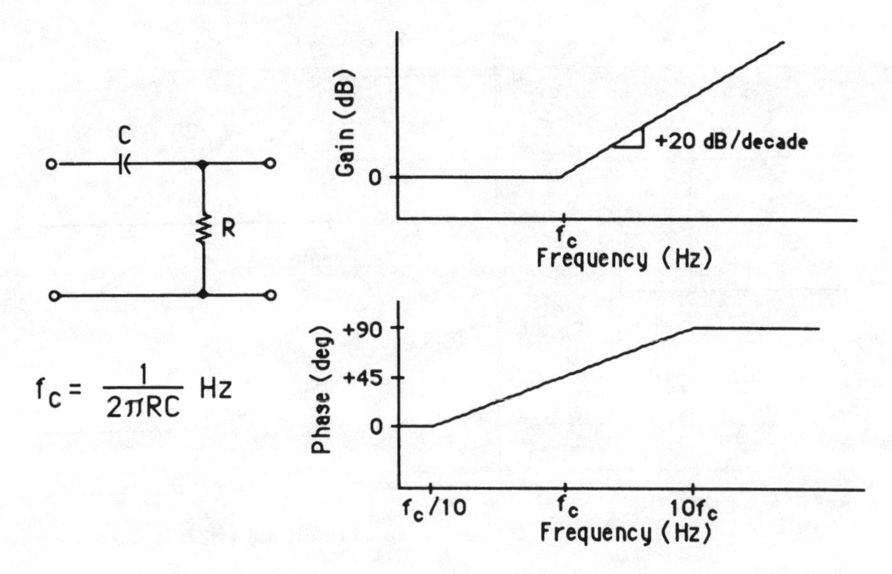

Figure B–2 A simple zero differentiator or high-pass filter.

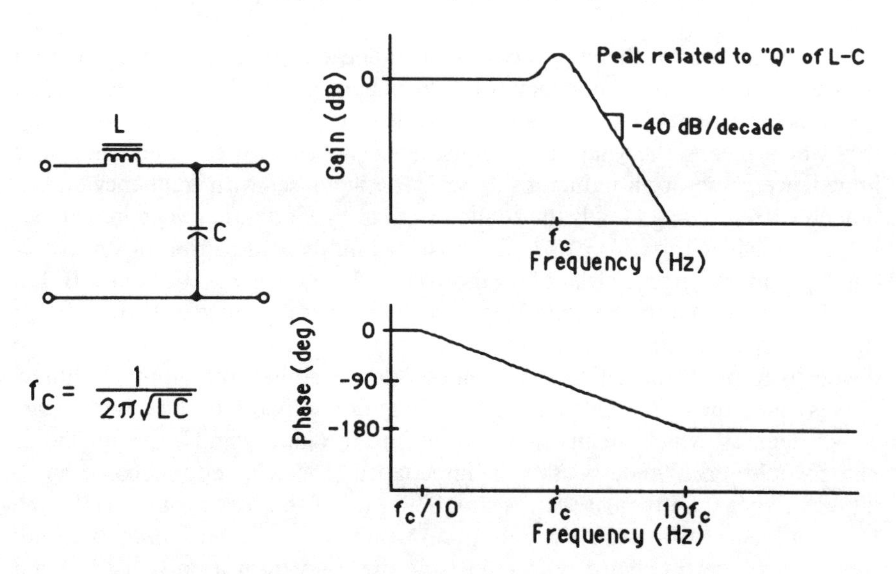

Figure B–3 A two-pole filter: choke input filter.

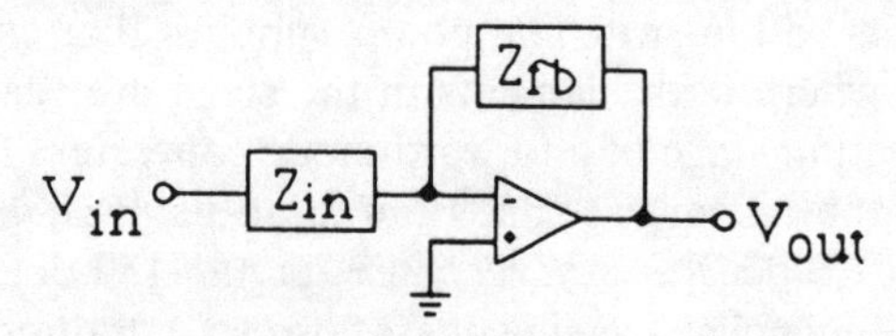

Figure B–4 A generalized error amplifier.

$$Z_{in} = Z_{fb} \tag{B.1}$$

Some implementations of these error amp circuits are shown in Figures B–5 through B–7. Some useful mathematical tools when working with Bode plots are given below.

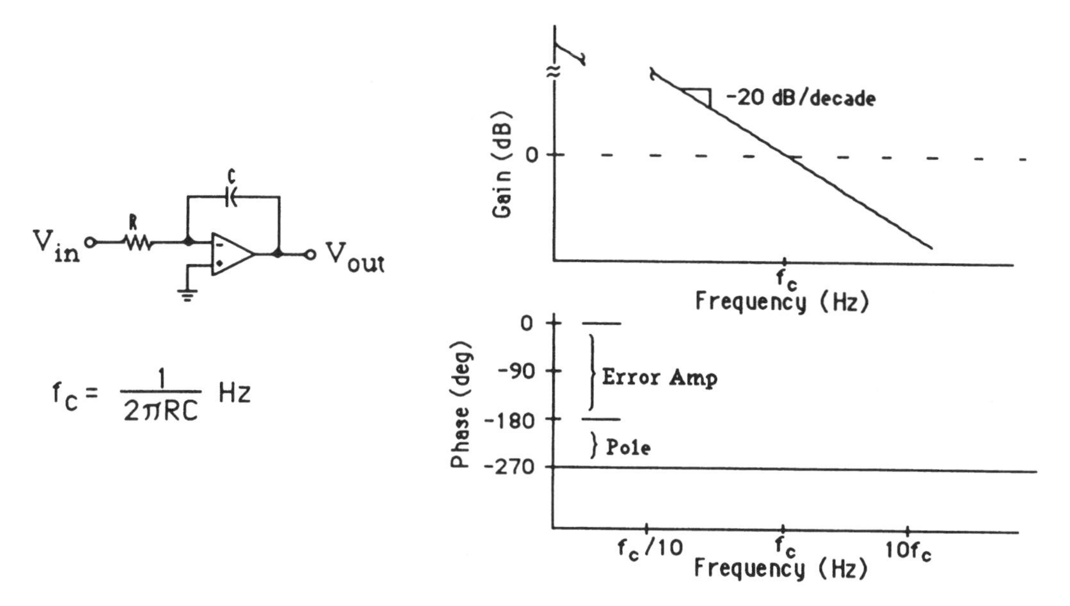

Figure B–5 An active single-pole filter.

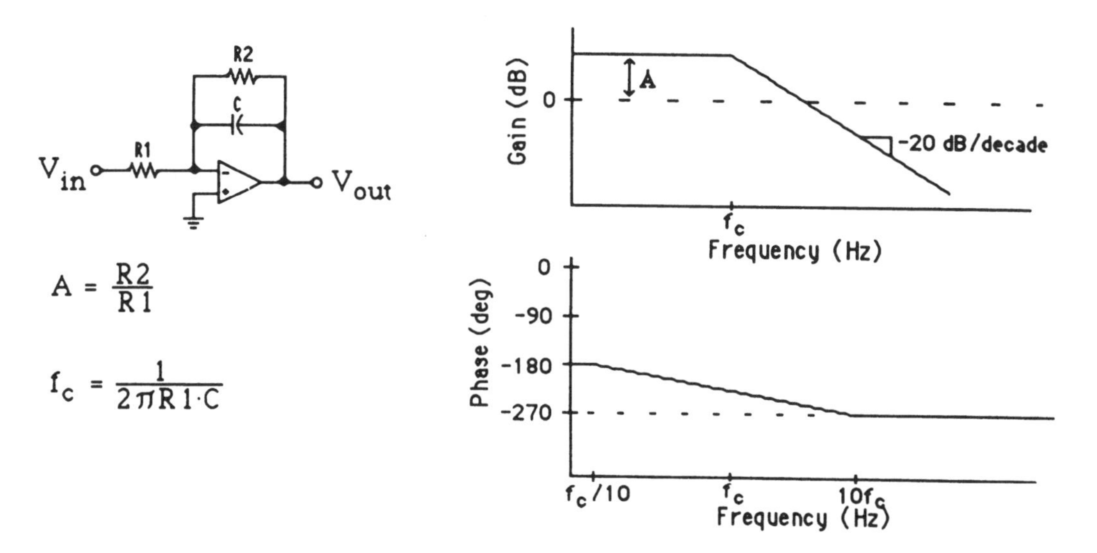

Figure B–6 An active single-pole filter with flat high- and low-frequency responses.

1. To find values of gain and phase at a different frequencies along a –20 dB/decade gain slope and its associated phase curve:

$$\Delta G(f_2 - f_1) = 20\log\left(\frac{f_2}{f_1}\right) \tag{B.2}$$

$$\phi(f_2 - f_1) = \tan^{-1}\left(\frac{f_2}{f_1}\right) \tag{B.3}$$

2. To find values of gain and phase at different frequencies along a –40 dB/decade gain slope and its associated phase curve:

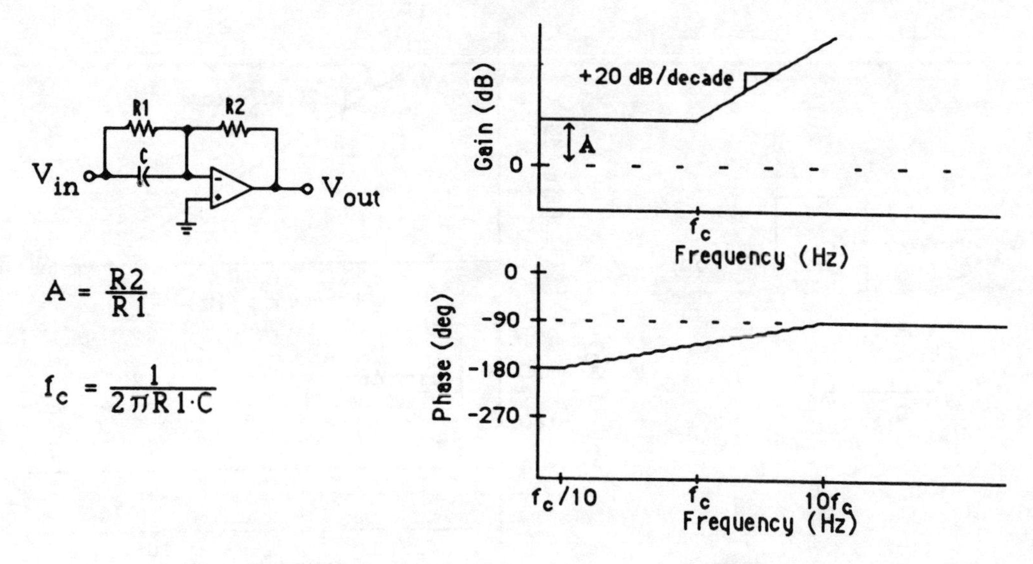

Figure B–7 An active high-pass filter (single zero) with flat high- and low-frequency responses.

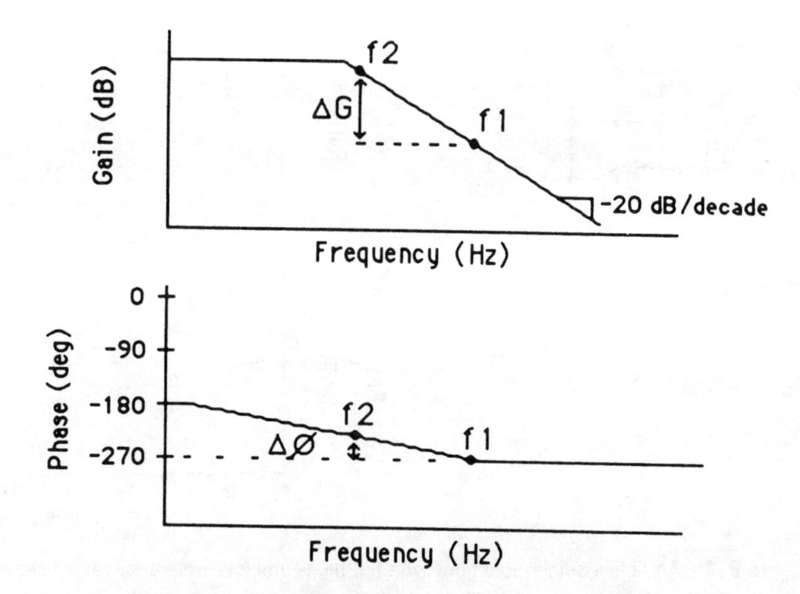

Figure B–8 Illustrating how the mathematical tools are used.

$$\Delta G(f_2 - f_1) = 40 \log\left(\frac{f_2}{f_1}\right) \tag{B.4}$$

$$\phi(f_2 - f_1) = 2 \tan^{-1}\left(\frac{f_2}{f_1}\right) \tag{B.5}$$

These tools and circuits form the basis of the design approach within the feedback loop compensation circuits in power supplies.

B.2 Defining the Open Loop Response of the Switching Power Supply—The Control-to-Output Characteristics

Before the designer can begin to develop a stable negative feedback circuit, the behavior of the system that is to be controlled must be defined. To do this, the designer should have a general understanding of what each of the major sections of the switching power supply contributes to the entire open-loop Bode response of the switching power supply. Fortunately, the commonly employed topologies, as described in this book, have previously been defined and fall into two major responses. The choice of which response to use is easily made from the design path that has been previously chosen by you. The response types are

1. Voltage-mode controlled forward-mode converters.
2. Voltage-mode controlled flyback-mode converters, and current-mode controlled forward- and flyback-mode converters.

The matter of whether or not the topology is transformer-isolated has an influence only on the dc characteristics of the model. The flyback-mode converters mentioned above are operating in the discontinuous mode of operation only.

The *control-to-output* characteristic is simply the behavior of the power supply when the error amplifier is removed from the system. The point where the error voltage enters the PWM-to-pulsewidth converter is viewed as the input port of the system. The point where the output voltage feedback enters the error amplifier's negative input is the output port of the system. This can be seen in Figure B–9. When the "input" is swept with a sweep generator, the resulting Bode plot is the control-to-output characteristic. This is the plot which is most meaningful for the purposes of stabilization. An approximate control-to-output plot may also be done by the following procedure with very satisfactory results.

B.2.1 The Voltage-mode Controlled Forward-mode Converter

The common topologies which are encompassed under this category are: the buck, half-forward, push-pull, half and full bridge, with only the traditional voltage-mode control method. Its representative circuit diagram is given in

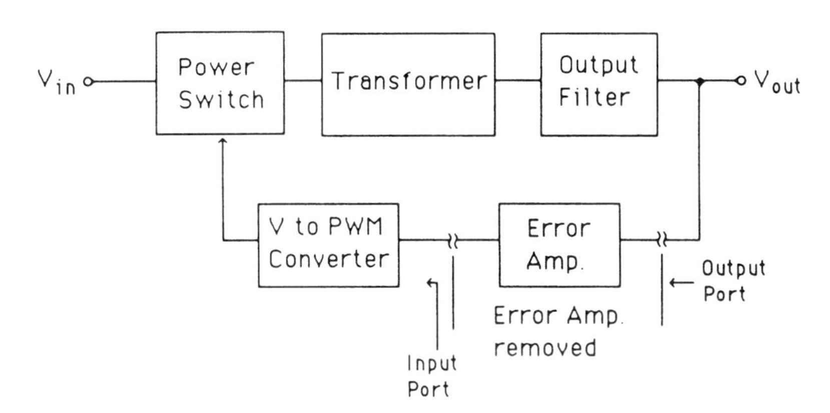

Figure B–9 The meaning of the control-to-output characteristic.

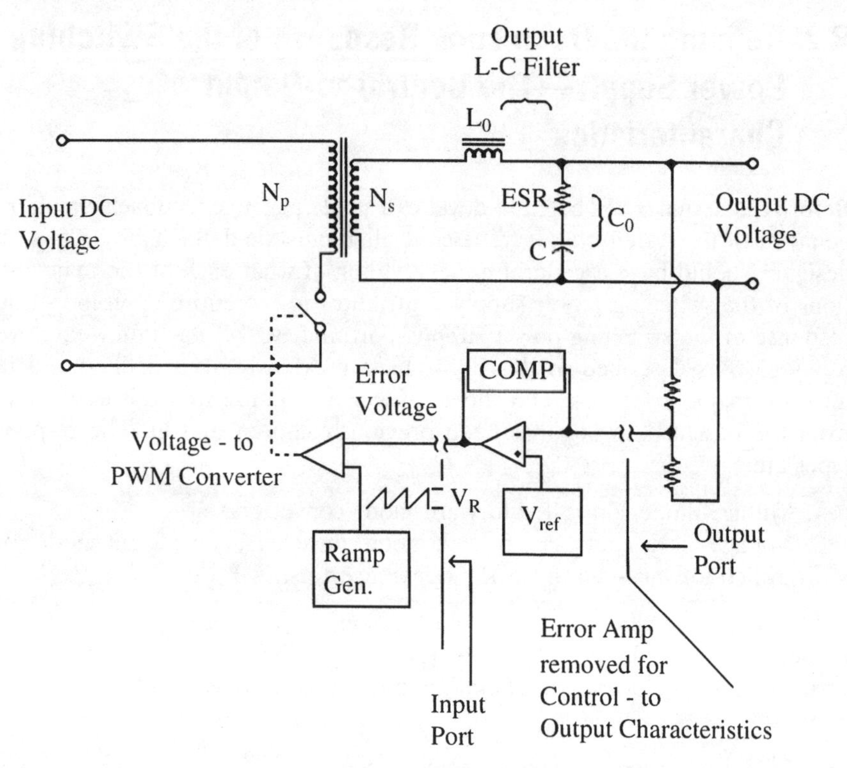

Figure B–10 The control-to-output model for a voltage-mode controlled forward converter.

Figure B–10. Figure B–10 shows the presence of a transformer. For the buck converter, the designer can assume that the turns ratio is 1:1.

The first step is to determine the gain of the system at dc. This is the starting point for developing the gain plot. The dc gain is found by

$$A_{DC} = \frac{V_{out}}{V_{in}} = \frac{N_{sec}}{N_{pri}} \cdot DC = \frac{V_{in}}{\Delta V_e} \cdot \frac{N_{sec}}{N_{pri}} \tag{B.6}$$

where DC is the maximum duty cycle (~95 percent).
ΔV_e is the peak-to-peak output voltage of the error amplifier.

To convert the dc gain into decibels,

$$G_{DC} = 20\,\mathrm{Log}(A_{DC}) \tag{B.7}$$

G_{DC} is then the starting point of the gain Bode plot at dc.

The first major pole is contributed by the output L-C filter. It represents a second order pole which exhibits a "Q" phenomenon, which is typically ignored, and a –40 dB/decade rolloff above its corner frequency. The phase plot will quickly begin to lag starting at a frequency of 1/10th the corner frequency, and will reach the full 180 degrees of lag at 10 times the corner frequency. The location of this double pole is found from

$$f_p = \frac{1}{2\pi\sqrt{L_o C_o}} \tag{B.8}$$

where L_o and C_o are the values of the inductor and capacitor in the output L-C filter (in henries and farads). If there are multiple outputs on the switching power supply, then use the filter values from the highest power output that is sensed.

The next item is the zero exhibited by the series combination of the ESR of the output filter capacitance and the actual value of the output filter capacitor itself. Its corner frequency is found from

$$f_{\text{ESR}} = \frac{1}{2\pi R_{\text{ESR}} \cdot C_{\text{o}}} \tag{B.9}$$

This causes the control-to-output characteristic to add gain and phase above the location of this zero. This can be a problem with respect to the stability of the supply. Unfortunately, many capacitor manufacturers do not present the value of the ESR for their capacitors. Typically, the zero caused by the output filter capacitor falls in the following range:

Aluminum electrolytic	1–5 kHz
Tantalum	10–25 kHz

As one can see, the choice of the type of output filter capacitor can influence the control-to-output characteristic, sometimes detrimentally.

The resulting control-to-output Bode plots for the voltage-mode controlled forward converter are given in Figure B–11.

B.2.2 Voltage-mode Controlled Flyback Converter and Current-mode Controlled Forward-mode Converter Control-to-Output Characteristics

The operation of a discontinuous-mode, flyback converter is quite different from that of a forward-mode converter, and likewise their control-to-output characteristics are very different. The topologies that fall into this category of control-to-output characteristics are the boost, buck/boost, and the flyback. The forward and flyback-mode converters operating under current-mode control also fall into this category. Only their dc value is determined differently. Their representative circuit diagram is given in Figure B–12.

The dc gain exhibited by the power section of the switching power supply of a current- or voltage controlled flyback converter is approximately

$$A_{\text{DC}} = \frac{(V_{\text{in}} - V_{\text{out}})^2}{V_{\text{in}} \cdot \Delta V_{\text{e}}} \cdot \frac{N_{\text{sec}}}{N_{\text{pri}}} \tag{B.10}$$

ΔVe in this case can be the peak-to-peak voltage of the oscillator ramp if the method of control is voltage-mode, or the maximum peak voltage representing the primary current within the current-mode method of control. The gain can be converted into decibels, which is shown in Equation B.7.

The current-mode controlled forward-mode converter exhibits the same dc gain as the voltage-mode controlled forward converter, as shown in Equation B.6.

The output filter pole in both voltage-mode controlled flyback converter and the current-mode controlled forward and flyback is highly dependent on the equivalent resistance of the load. This means that when the load current increases or decreases, the location of the output filter pole moves. The filter pole can be found from

$$f_{\text{p}} = \frac{1}{2\pi R_{\text{L}} C_{\text{o}}} \tag{B.11}$$

where R_{L} is V_{out}/I_{out}.

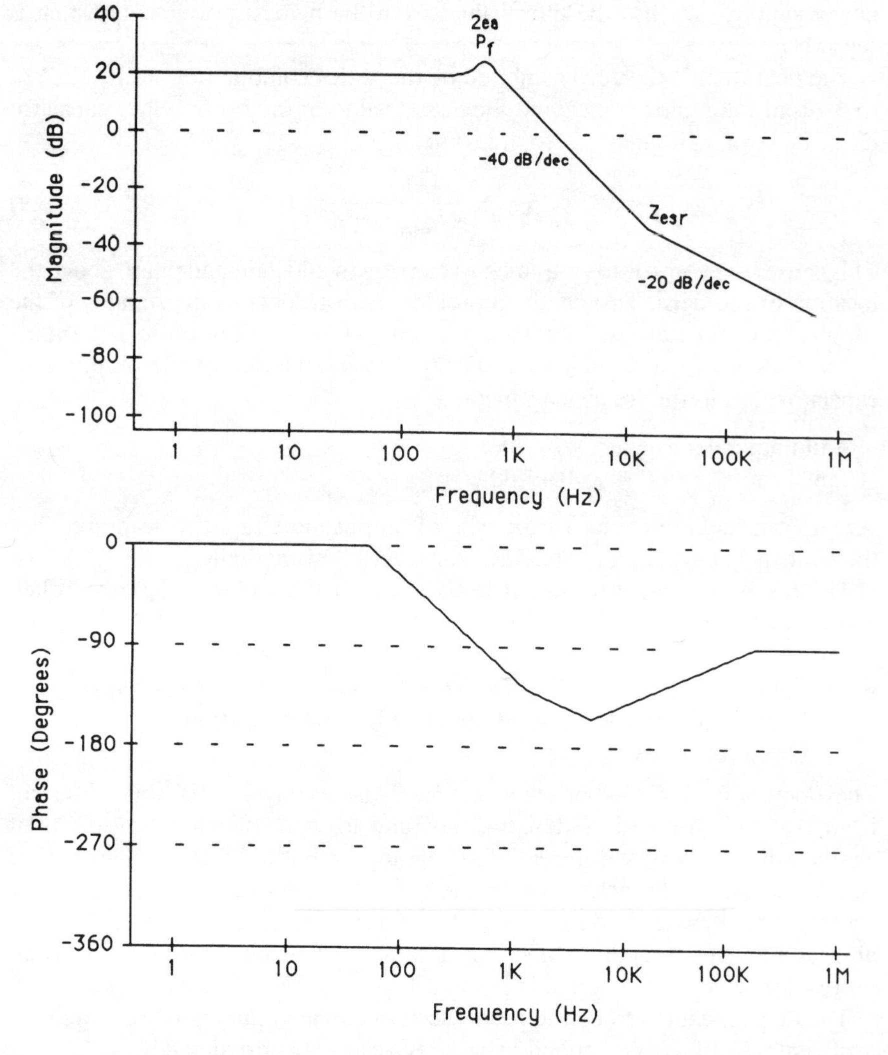

Figure B–11 The control-to-output curve for a forward-mode converter with voltage-mode control.

As one can see when the load current decreases, the pole frequency decreases. The considerations surrounding this fact for the error amplifier compensation will be discussed later.

The zero attributed to the output filter capacitor is still present in the control-to-output characteristics. Its location is found in Section B.2.1 and Equation B.9.

The resulting control-to-output characteristics are shown in Figure B–13. As one can see, both the input voltage and the equivalent load resistance have an influence on the gain and phase functions.

The current-mode controlled forward converter has one additional consideration: there is a double pole at one-half the operating switching frequency. The compensation bandwidth normally does not go this high, but it may cause problems if the closed-loop gain is not sufficiently low enough to attenuate its effects. Its influence on the control-to-output characteristic can be seen in Figure B–14.

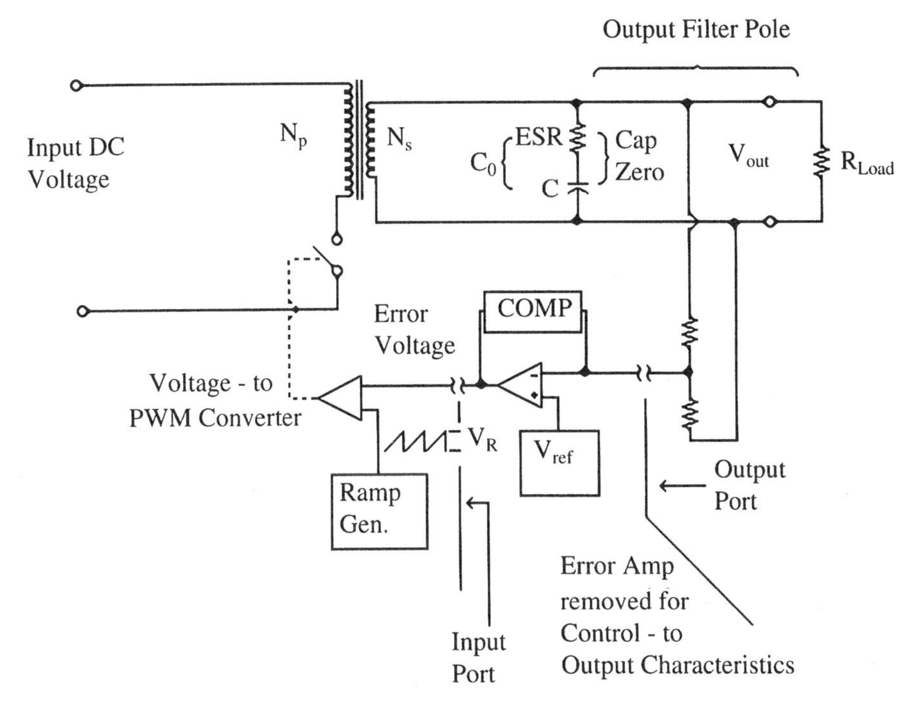

Figure B–12 The control-to-output model for a voltage-mode controlled flyback converter.

B.3 The Stability Criteria Applied to Switching Power Supplies

Before starting the design for the compensation of the error amplifier, it is desirable to know what constitutes a stable closed-loop system. The rule is elementary:

> The closed loop phase shall never exceed –330 degrees of lag whenever the gain of the closed-loop system is greater than 1 (or 0 dB).

Actually, a total closed-loop phase lag limit of 315 degrees is commonly used by designers; any closer to 360 degrees would constitute a *metastable* system. This could result in the power supply breaking out into periods of oscillation when large loads or line transients are experienced.

Figure B–15 shows some of the terms encountered in stability analysis.

1. *Phase margin*. This is the value of the phase of a closed-loop system at the gain cross-over frequency (G(s) = 0 dB).
2. *Gain margin*. This is the value of the gain when the phase crosses over –360 degrees.
3. *Excess phase*. This is the point of closest approach of the phase characteristic to –360 degrees anytime the gain is greater than 1 (0 dB).

The excess phase is the most important consideration of the three, since the gain cross-over frequency is usually much higher than the point of maximum phase lag caused by a filter pole.

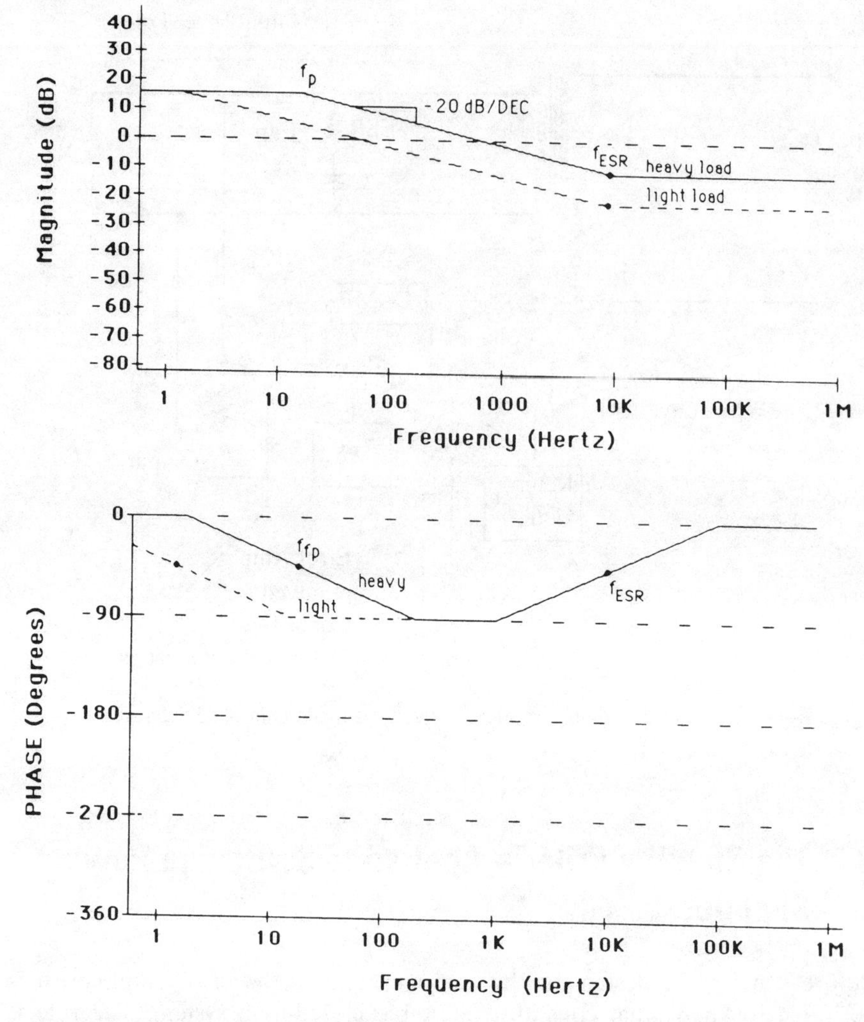

Figure B–13 The control-to-output characteristics for a voltage-mode controlled flyback converter.

B.4 Common Error Amplifier Compensation Designs

There are four guidelines that should be kept in mind when designing the error amplifier compensation. If reasonably followed, they will yield a good compensation design.

1. The closed-loop phase should be kept less than –300 degrees whenever the gain is greater than 0 dB.
2. The closed-loop gain cross-over frequency should be as high as practically possible. This quickens the transient response time of the supply.
3. The closed-loop gain at dc should be as high as possible. This has a direct bearing on the output load regulation of the supply.
4. The resulting average slope of the closed-loop gain curve should be an average of –20 dB/decade.

An additional consideration is the gain-bandwidth product specification of the operational amplifier used as the error amplifier. If the Bode character-

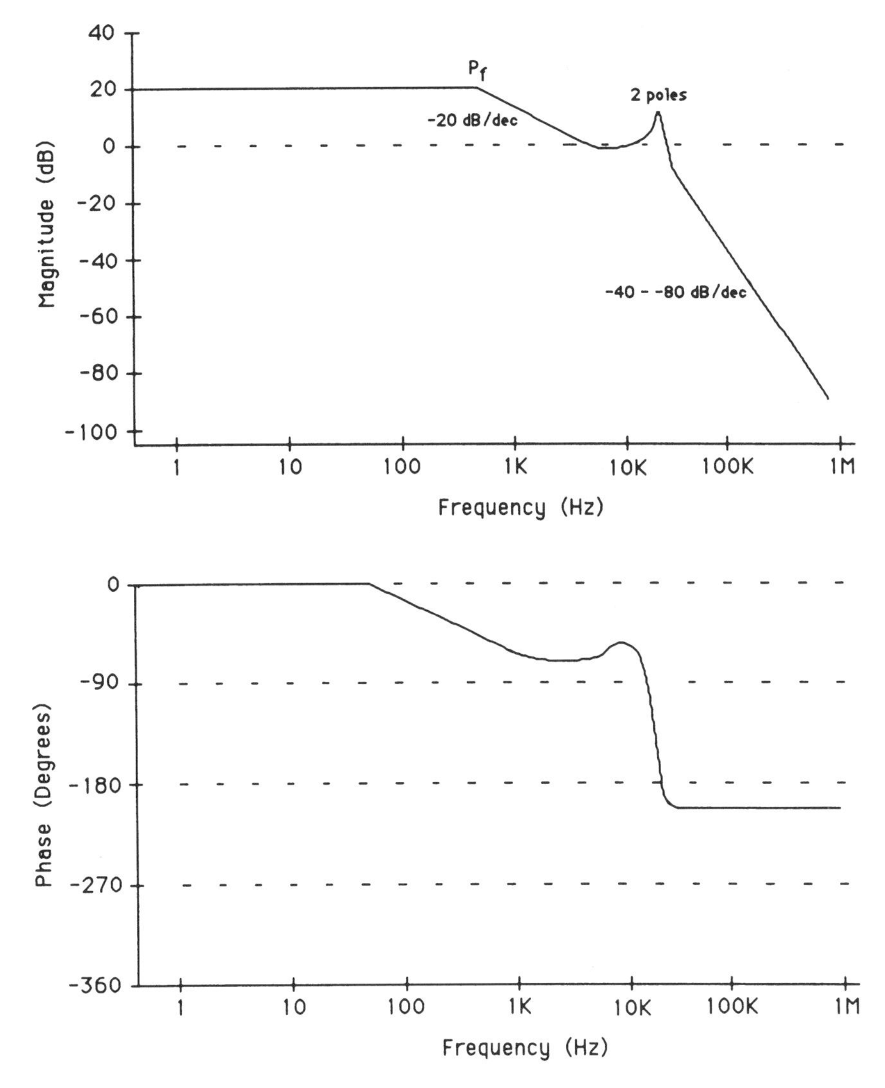

Figure B–14 The control-to-output characteristics for a current-mode controlled forward converter.

istics of the op amp are too low in frequency, the desired compensation scheme may not work completely.

The compensation schemes shown in this book are the most common methods. Several of these will work in each applications, but only one for each is considered the optimum method insofar as the dc gain and closed-loop bandwidth aspects are concerned. Tables B–1 and B–2 will help match a possible compensation method with your application.

B.4.1 Single-pole Compensation

This type of compensation is used for those converter topologies that exhibit minimal phase shift prior to the anticipated closed-loop gain cross-over point. This topology would be that of the forward-mode converters using voltage-mode control. This compensation method, though requiring the minimum amount of parts and excellent load regulation, yields a very poor closed-loop bandwidth that produces a sluggish transient response time.

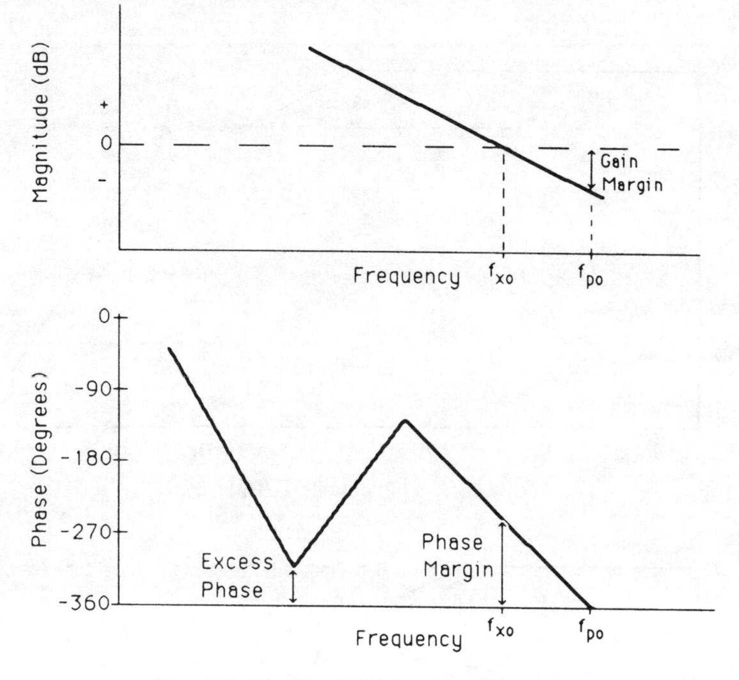

Figure B–15 The definition of stability terms.

Table B–1 Which Compensation Schemes Work for Which Case

Supply Type	Single Pole	Single Pole with Gain Limitation	1-Pole–1-Zero	2-Pole–2-Zero
Forward-mode with voltage-mode control	×			×
Boost-mode with voltage- or current-mode control		×	×	
Forward-mode with current-mode control		×	×	

Table B–2 The Relative Merits of the Various Common Methods of Compensation

Compensation Type	Load Regulation	Transient Response
Single-pole	Good	Poor
Single-pole with in-band gain limiting	Fair	Good
Pole–zero	Good	Good
2-pole–2-zero	Good	Good

The schematic and Bode plot for the single-pole method of compensation are given in Figure B–16. At dc it exhibits the full open-loop gain of the op amp, and its gain drops at −20 dB/decade from dc. It also has a constant −270 degree phase shift. Any phase shift contributed by the control-to-output characteristic

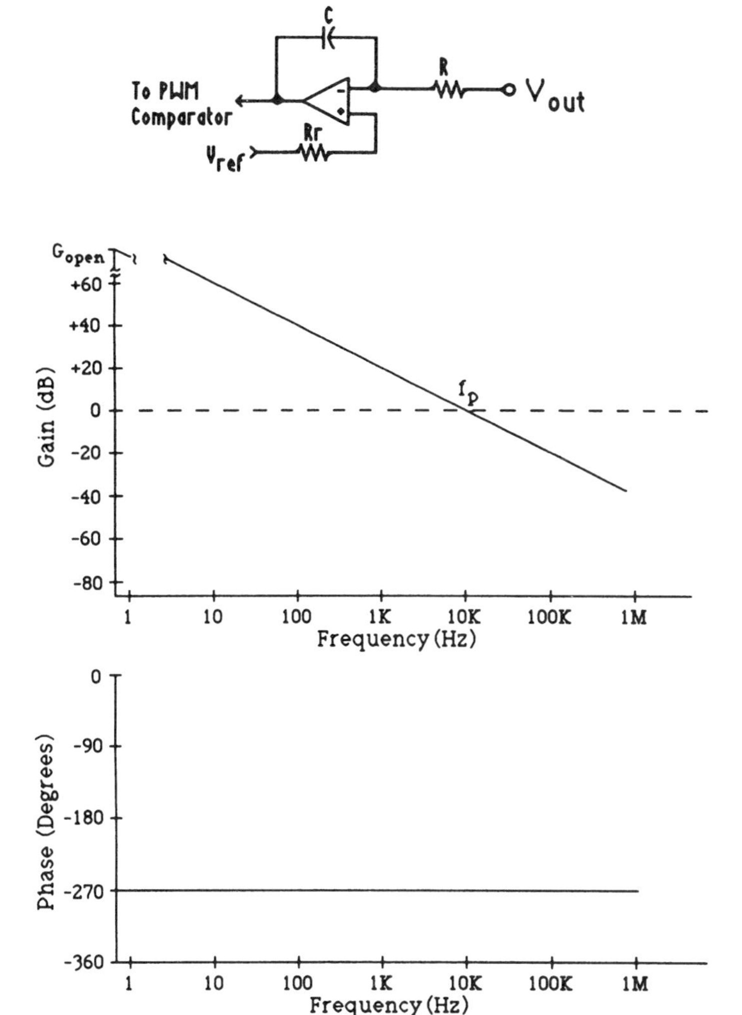

Figure B–16 The single-pole method of compensation.

cannot then add more than the desired excess phase (–315 to –330 degrees) or an additional 30 to 45 degrees.

The first step in determining the component values for this compensation method is to determine the inherent dc gain of the power stages. The calculation should be performed with the maximum value of the input voltage, since this is where the system will have its widest exhibited bandwidth. This is done using Equation B.12.

$$A_{DC} = \frac{V_{in}}{\Delta V_e} \cdot \frac{N_{sec}}{N_{pri}} \tag{B.12}$$

Refer to Section B.2.1 for any additional help. Do not yet convert this value to dB.

Next find the closed-loop gain cross-over frequency by deciding how much phase margin you desire in your system. A good value is 45 degrees. Ignoring any effect of the Q of the L-C filter, the gain cross-over point is found from

$$f_{xo} = f_{fp} \operatorname{Tan}(\phi_{pm}/2) \tag{B.13}$$

Next find the cross-over frequency of the error amplifier from

$$f_{xe} = f_{xo} \cdot 10^{-(A_{xo}/20)} \tag{B.14}$$

Then find the value of the feedback capacitor C. The designer knows the value of the input resistor (R). It is the upper resistor in the voltage divider responsible for the voltage feedback to the error amplifier. One then performs Equation B.15.

$$C = \frac{A_{DC}}{2\pi R f_{xe}} \tag{B.15}$$

This compensation method, although easy to design, exhibits very sluggish transient load response. This is because the gain cross-over point is always much lower than that of the output filter pole. This usually results in loop bandwidths of 50 to 500 Hz. This may cause a problem to the load circuitry by having the supply voltages exhibit momentary excursions outside their operating specifications. Figure B–17 shows an implementation of this compensation method. As one may be able to see, a high Q characteristic of the L-C filter can cause

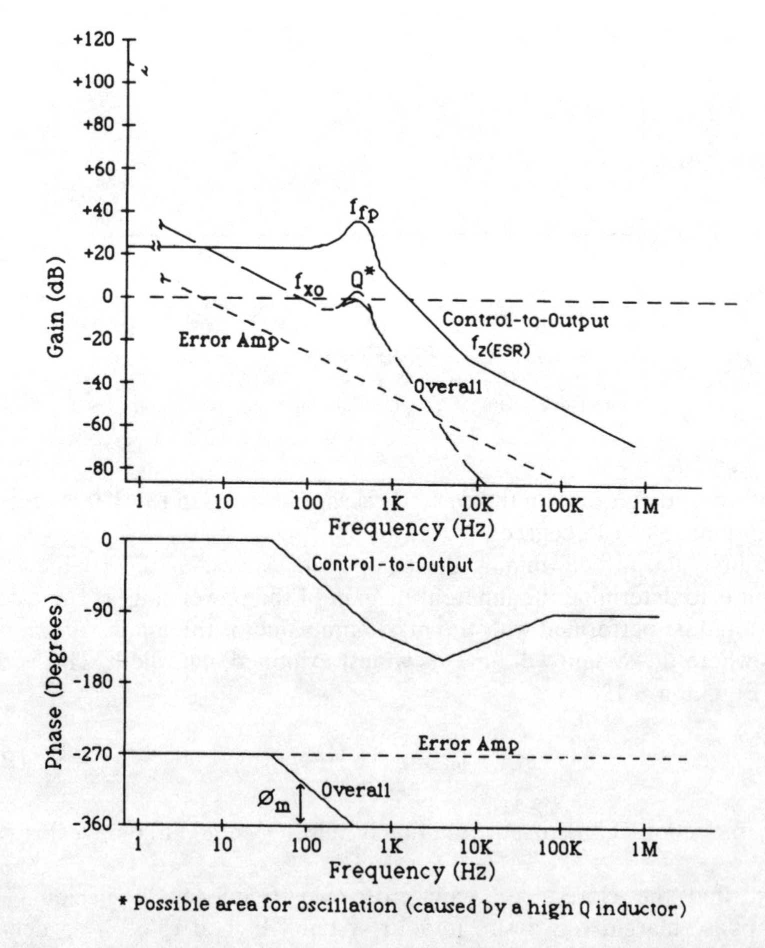

Figure B–17 Single-pole compensation used with a voltage-mode controlled forward regulator.

oscillation at the resonance frequency of the output filter. The designer would then have to further lower the closed-loop gain cross-over frequency to avoid this problem. This yields even worse transient performance.

B.4.2 Single-pole Compensation with In-band Gain Limiting

This method of compensation is recommended for only those topologies that exhibit a single-pole filter response. These would be the current-mode controlled forward converters and current- or voltage-mode controlled flyback mode converters. This method's bandwidth can go beyond the frequency of the single output filter pole and the method's only drawback is it has lower dc gain than the other compensation methods, which would then make the switching power supply exhibit worse load regulation. Its circuit diagram and Bode plots are shown in Figure B–18.

This compensation method now exhibits a –180 degree phase lag at low frequencies, then beginning at one-tenth the error amplifier's filter pole (f_{EP}) the phase lag increases to its high frequency limit of –270 degrees.

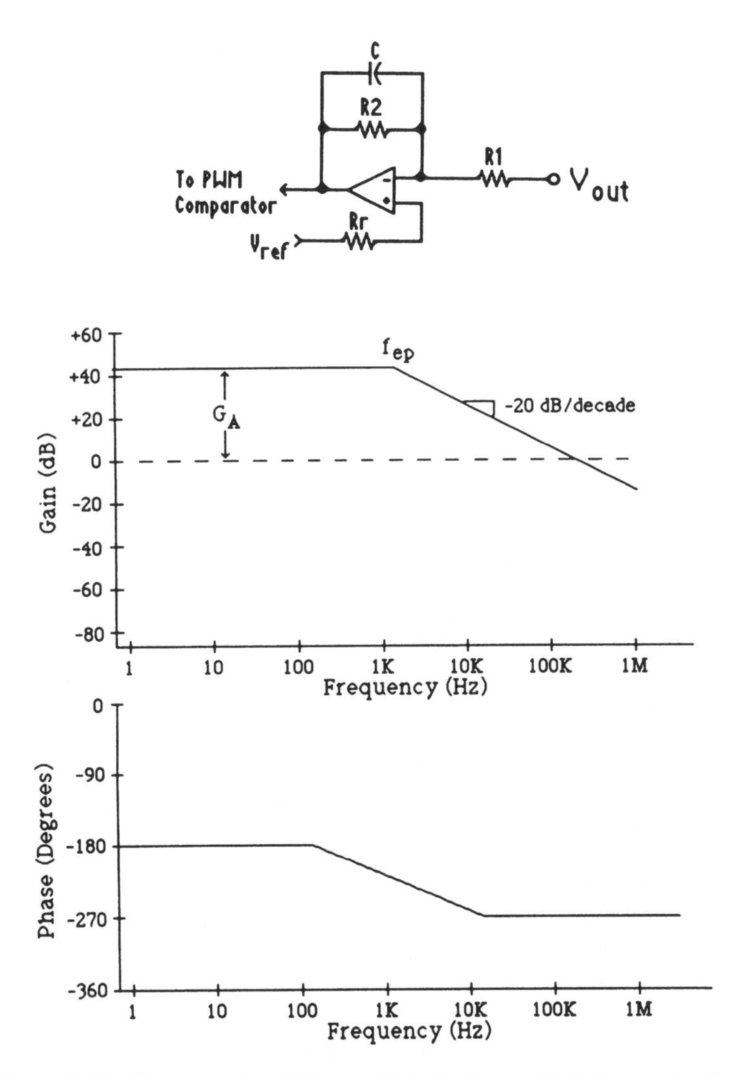

Figure B–18 The one-pole with in-band gain limiting method of compensation.

To begin the design process, one of course needs the inherent maximum dc gain exhibited by the power circuit's control-to-output characteristic (A_{DC}). The worst-case condition is considered at the highest input voltage. The closed-loop gain cross-over frequency is limited to one-fifth the switching frequency. This is because if the cross-over frequency were higher, too much of the switching frequency would be amplified by the error amplifier which would bbe detrimental. So the closed-loop gain cross-over frequency is

$$f_{xo} < 0.2 f_{sw} \quad \text{(B.16)}$$

where f_{sw} is the switching frequency of the switching power supply.

The location of the error amplifier's pole is used to counteract the effects of the output filter capacitor's zero caused by the ESR. The error amplifier's pole should be located by

$$f_{EP} \approx f_{Z(ESR)} \quad \text{(lowest frequency)} \quad \text{(B.17)}$$

Next, one must determine the amount of gain needed by the error amplifier to bring the control-to-output function up to 0 dB at the closed-loop cross-over frequency:

$$G_{xo} = 20 \log\left(\frac{f_{xo}}{f_{fp}}\right) - G_{DC} \quad \text{(in dB)} \quad \text{(B.18)}$$

To find the amount of gain needed below the error amplifier pole;

$$G_A = G_{xo} + 20 \log\left(\frac{f_{xo}}{f_{ep}}\right) \quad \text{(B.19)}$$

To find the value of the feedback capacitor, one evaluates the following:

$$A_{xo} = 10^{(G_{xo}/20)} \quad \text{(absolute gain of } G_{xo}) \quad \text{(B.20)}$$

$$C = \frac{1}{2\pi R_1 A_{xo} f_{ep}} \quad \text{(B.21)}$$

To find the feedback resistor value,

$$A_A = 10^{(G_A/20)} \quad \text{(absolute gain of } G_a) \quad \text{(B.22)}$$

$$R_2 = A_A \cdot R_1 \quad \text{(B.23)}$$

This completes the design of the single-pole with in-band gain limiting method of compensation. Its closed-loop Bode plot can be seen in Figure B–19. If you attempt this method, there is one drawback. Due to the high-gain and wide bandwidth nature of this method, it is very easy to encounter the maximum gain-bandwidth characteristics of the op amp itself. That is, the op amp may not have enough gain at the frequencies required. There are two alternatives: lower the crossover point, or use an external op amp with a higher G_{BW} specification.

B.4.3 Pole-Zero Compensation

This compensation method is intended for those topologies that exhibit a single filter pole response. That would include: the voltage-mode controlled, discontinuous, flyback converter, and the current-mode controlled forward and flyback converters. It possesses a high dc gain, and a phase lead characteristic. This method gives the designer the opportunity to tailor the compensation to the power supply. Its circuit and Bode response can be seen in Figure B–20.

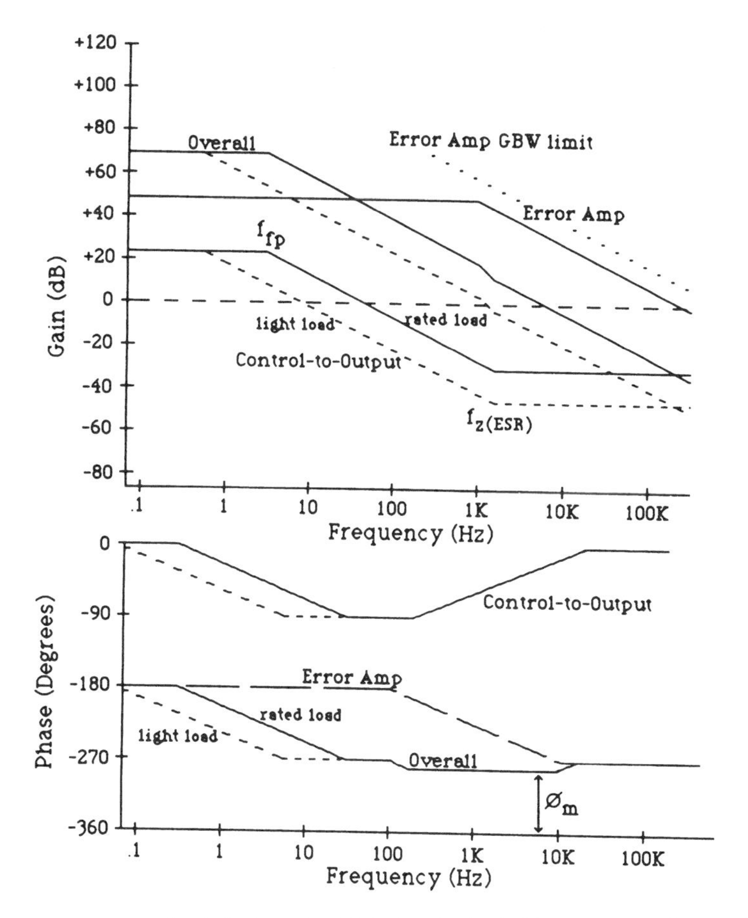

Figure B–19 An example of single-pole with in-band gain limiting compensation used with a voltage/current-mode controlled flyback converter.

This compensation method has a pole at dc that provides the open-loop op amp gain for good output regulation. A zero is introduced at or below the supply's lowest manifestation of the output filter pole in order to compensate for the phase lag by that pole. The phase lag of the error amplifier actually decreases between the error amplifier's zero and pole. Its theoretical limit is –180 degrees (or +90 degree "phase bump"). This phase bump should be placed where the greatest phase lag from the output filter pole will exist. In that way, good control over the excess phase of the system can be maintained. The last pole is put into the compensation to roll off the gain at high frequencies and to counteract the capacitor's zero caused by the ESR. Its closed-loop Bode plot can be seen in Figure B–21.

To begin the design, the inherent dc gain of the control-to-output characteristic should be found. For voltage and current-mode controlled flyback converters, refer to Equation B.10. For current controlled forward-converters refer to Equation B.6. These should be calculated at the highest input voltage since that will yield the highest dc gain (worst case).

Next determine the maximum closed-loop gain cross-over frequency. Less than one-fifth of the switching power supply's operating frequency is a good rule.

$$f_{xo} \leq .2 \cdot f_{SW} \tag{B.24}$$

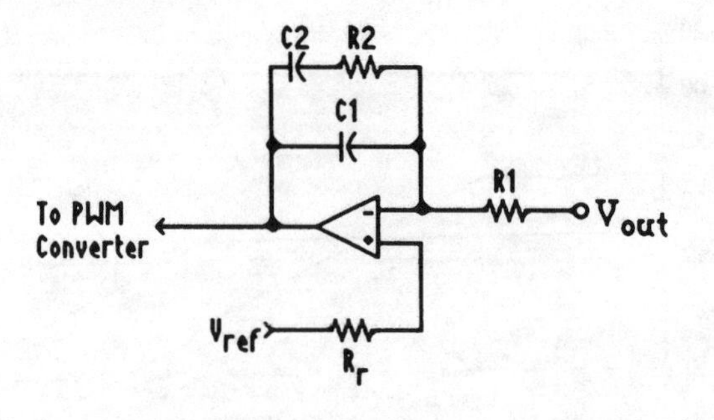

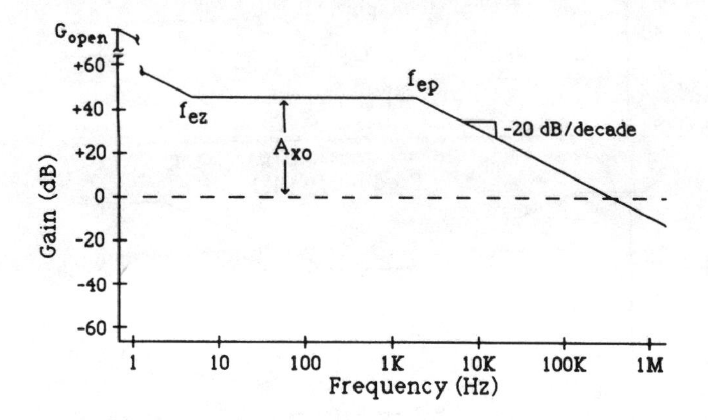

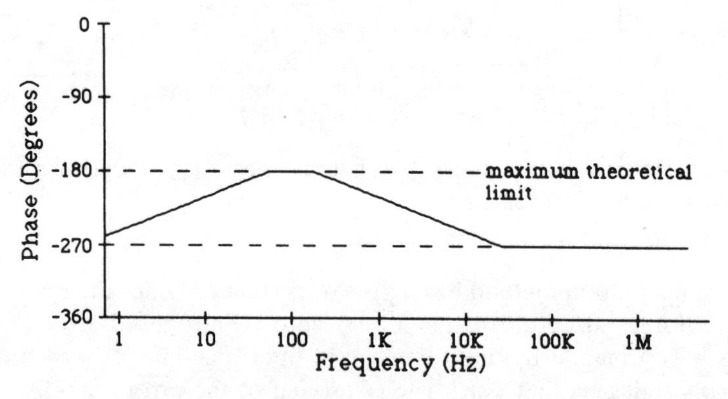

Figure B–20 The pole-zero compensation method.

Determine how much gain is required to elevate the control to output gain curve to 0 dB at the gain cross-over frequency.

$$G_{xo} = 20\log\left(\frac{f_{xo}}{f_{fp}}\right) - G_{DC} \tag{B.25}$$

The next task is to determine the placement of the compensating zero and pole within the error amplifier. The zero is placed at the lowest frequency manifestation of the filter pole. Since for the voltage-mode controlled flyback converter, and the current-mode controlled flyback and forward converters, this pole's frequency changes in response to the equivalent load resistance. The lightest expected load results in the lowest output filter pole frequency. The error amplifier's high frequency compensating pole is placed at the lowest anticipated zero frequency in the control-to-output curve cause by the ESR of the capacitor. In short:

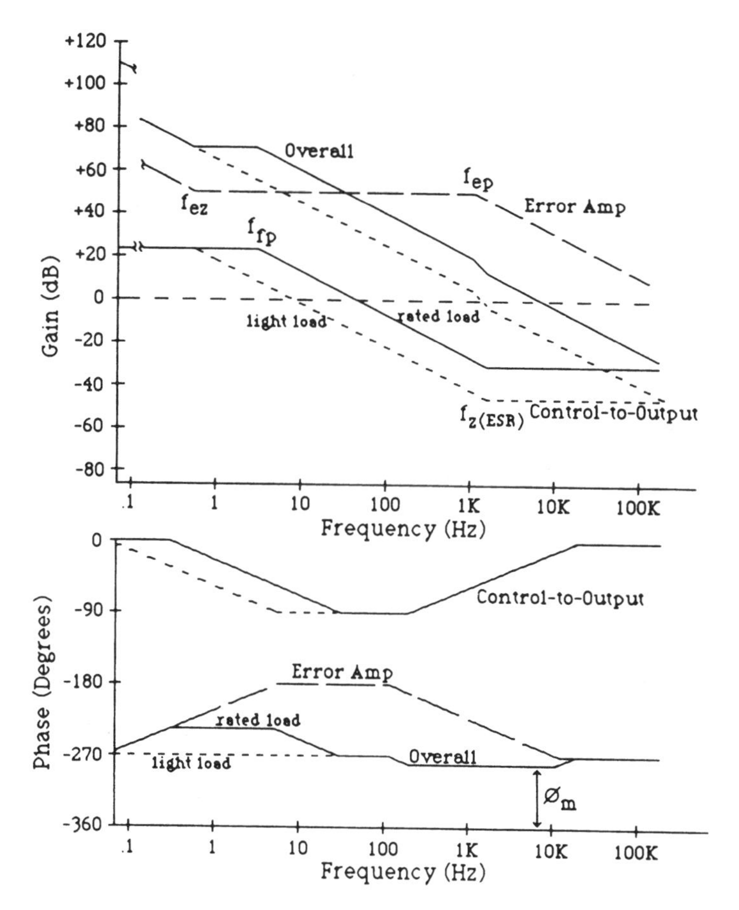

Figure B–21 An example of pole-zero compensation used with a voltage/current-mode controlled flyback converter.

$$f_{ez}: \quad f_{ez} = f_{fp(\text{light load})} \tag{B.26}$$

and

$$f_{ep}: \quad f_{ep} = f_{Z(\text{ESR-min})} \tag{B.27}$$

Now the component values can be calculated. The value of the input resistor (R1) is known since it is the upper resistor in the feedback voltage divider. The value of the feedback components are found as follows:

$$C1 = \frac{1}{2\pi f_{xo} \cdot A_{xo} \cdot R1} \tag{B.28}$$

$$R2 = A_{xo} \cdot \mathrm{R1} \tag{B.29}$$

where A_{xo} is the absolute value for the gain needed at the cross-over frequency (non dB).

$$C2 = \frac{1}{2\pi f_{ez} R2} \tag{B.30}$$

The amount of phase boost exhibited by the error amplifier is

$$\phi_{\text{Boost}} = 2\tan^{-1}\left(\frac{\sqrt{f_{\text{ep}}}}{\sqrt{f_{\text{ez}}}}\right) - 90° \tag{B.31}$$

The phase boost is proportional to the separation of the zero-pole pair in the error amplifier, but that is of secondary nature since the error amplifier's pole and zero were placed to compensate the worst case zero and pole in the control-to-output characteristics. The actual location of the zero caused by the ESR, which amounts to the choice of a vendor and part number, will affect the amount of excess phase the supply will exhibit. So the designer may have to relocate the compensating pole if there is any possibility that the excess phase will fall to less that 30 degrees (–330 degree lag).

B.4.4 2-Pole–2-Zero Compensation

This compensation is intended for voltage-mode controlled forward converters which exhibit a second order output filter pole characteristic. This would also include a quasi-resonant forward-mode converter that uses variable frequency, voltage-mode control. The L-C filter has a severe 180 degree phase lag and a –40 dB/decade gain rolloff. To get any sort of wide bandwidth from the supply at all, this type of compensation must be used.

This method has a pair of zeroes that counteract the gain and especially the phase of the double filter pole (see Figure B–22). This results in a closed-loop slope of –20 dB/decade above the filter pole. It also has a high-frequency pole to counteract the capacitor induced zero caused by the ESR. Lastly, it has a very high frequency pole to guarantee that the phase and gain margins at the closed-loop gain and phase cross-over frequencies are well behaved.

The more complicated methods of compensation, such as this, allow the designer much more control over the final closed-loop bode response of the system. The poles and zeros can be located independently of one another. Once their frequencies are chosen, the corresponding component values can be easily determined by the step-by-step procedure below. The zero and pole pairs can be kept together in pairs, or can be separated. The high-frequency pole pair appear to yield better results if they are separated and placed as below. The zero pair are usually kept together, but can be separated and placed either side of the output filter pole's corner frequency to help minimize the gain effects of the "Q" of the L-C filter (refer to Figure B–23).

To begin the design process, the inherent dc gain of the control-to-output gain function (G_{dc} and A_{dc}) is found using equations B.6 and B.7.

Next, the maximum closed-loop gain cross-over frequency (f_{xo}) is determined and is, once again, no more than one-fifth the minimum switching of the switching power supply:

$$f_{\text{xo}} < .2(f_{\text{sw}}) \tag{B.32}$$

Determine the amount of gain that is required to bring the control-to-output gain function up to 0 dB at the gain cross-over frequency. This can be approximated from

$$G_{\text{xo}} = 40\log\left(\frac{f_{\text{xo}}}{f_{\text{fp}}}\right) - G_{\text{DC}} \tag{B.33}$$

Determine the placement of the compensating zeroes (f_{ez1} and f_{ez2}). If the zeroes are to be placed at the same frequency, then

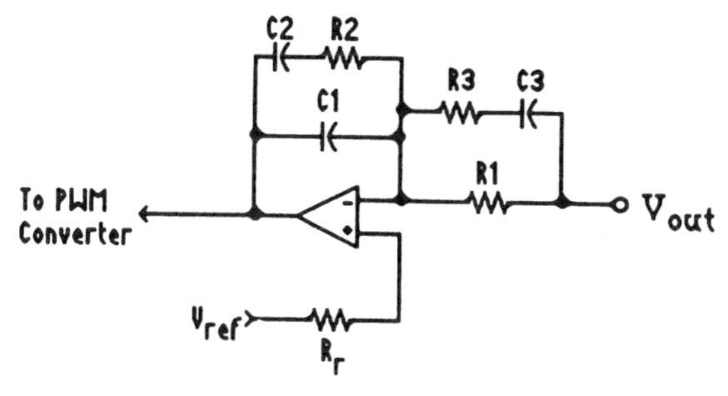

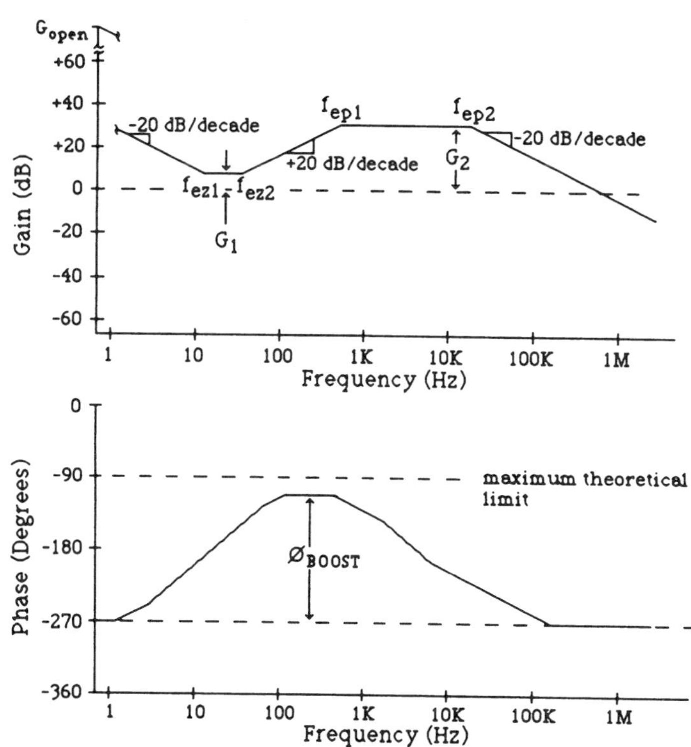

Figure B–22 2-Pole–2-zero compensation.

$$f_{ez1} = f_{ez2} = f_{fp}/2 \tag{B.34}$$

If the zeroes are to be placed on either side of the filter poles, then

$$f_{ez1} = f_{fp}/5 \tag{B.35}$$

and

$$f_{fp} \leq f_{ez2} \leq 1.2 f_{fp} \tag{B.36}$$

Next, place the lowest frequency error amplifier pole (f_{ep1}) at the lowest anticipated capacitor ESR zero frequency:

$$f_{ep1} = f_{z(ESR)} \tag{B.37}$$

The highest frequency compensating pole (f_{ep2}) is slightly higher in frequency than the gain cross-over frequency:

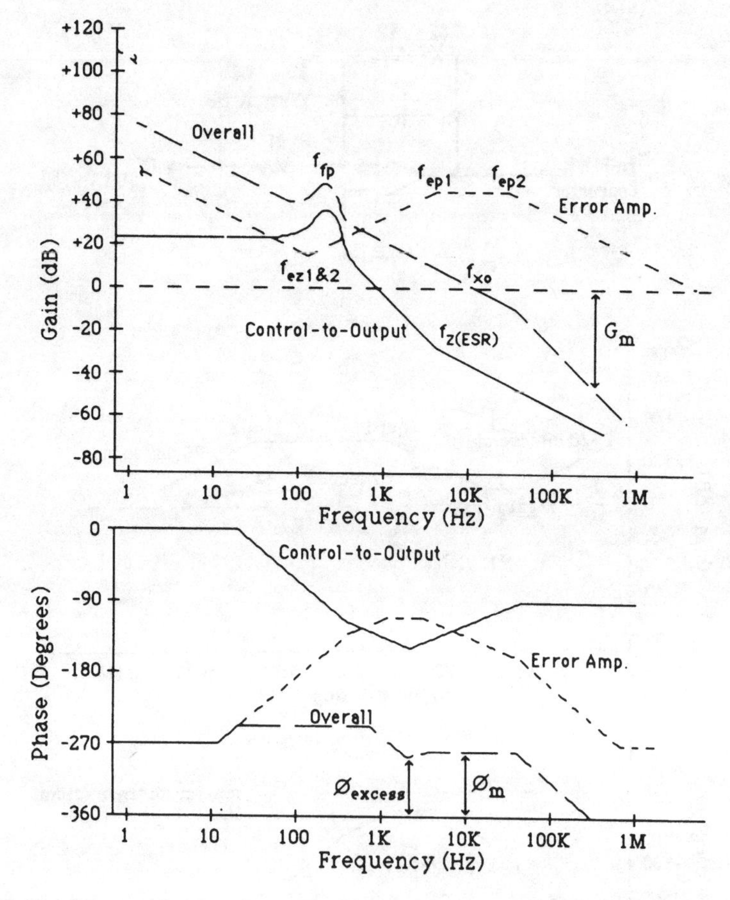

Figure B–23 An example of 2-pole–2-zero compensation used with a voltage-controlled forward converter.

$$f_{ep2} \geq 1.5 f_{xo} \tag{B.38}$$

Now one can determine the component values. One first calculates the gain at the location of the two compensating zeros (A_1). (See Figure B–22 for the meanings of the variable.)

$$G1 = G2 + 20\log\left(\frac{f_{ez2}}{f_{ep1}}\right)$$

$$A1 = 10^{(G1/20)} \tag{B.39}$$

Then

$$C1 = \frac{1}{2\pi f_{xo} \cdot A1 \cdot R1} \tag{B.40}$$

$$R2 = A1 \cdot R1 \tag{B.41}$$

$$C3 = \frac{1}{2\pi f_{ez2} R1} \tag{B.42}$$

$$R3 = \frac{R2}{A2} \tag{B.43}$$

$$C2 = \frac{1}{2\pi f_{ep2} R2} \tag{B.44}$$

where A_1 and A_2 are the absolute values of the gain (not dB).

The method performed above with the placement of the poles and zeroes will yield a minimum value for the excess phase of 45 degrees, which is satisfactory. If other pole and zero locations are attempted, then locate the maximum phase lag point of the L-C filter at the geometric mean frequency between f_{ez2} and f_{ep1}. This will guarantee the best phase performance. The amount of phase boost of the compensation design will be

$$\phi_{Boost} = 4\tan^{-1}\left(\frac{\sqrt{f_{ep1}}}{\sqrt{f_{ez2}}}\right) - 180° \tag{B.45}$$

Appendix C. Power Factor Correction

Power factor correction is becoming a very important area in the power world. Adding more generating capacity to the world's electrical pool is very costly and would consume additional resources. One method of creating about 30 percent excess generating capacity is to use the ac power more efficiently through the broad use of power factor correction. Motors, electronic power supplies, and fluorescent lighting consume about 40 percent of the power in the world and each of these would benefit from power factor correction. From the mid-1990s to 2002, many of the countries of the world are adopting requirements for power factor correction for the new products marketed within their borders. The added circuitry will add about 20 to 30 percent to the cost of power supplies, but the near-term energy savings will greatly outweigh the initial costs.

The term "power factor" in the field of power supplies is a slight departure from the traditional usage of the term, which applied to reactive ac loads, such as motors, powered from the ac power line. Here, the current drawn by the motor would be displaced in phase with respect to the voltage. The resulting power being drawn would have a very large reactive component and little power is actually used for producing work. Since power meters do not measure phase, the power measured is the scaler voltage times current. Capacitor banks are typically used to bring the phase more towards zero degrees if motors are the primary load.

In switching power supplies, the problem lies in the input rectification and filter network. The typical input circuit and its associated waveforms are shown in Figure C–1. As one can see, the input rectifiers can only conduct current when the ac line voltage exceeds the voltage on the input filter capacitor. This typically occurs within 15 degrees of the crest of the ac voltage waveform. The result is that current pulses are 5 to 10 times higher than the expected average current draw. This also can lead to distortion of the ac voltage waveform and an imbalance of the three-phase power lines feeding the circuits. This produces a neutral line current where no current flow is expected. Another drawback is that no current is drawn when the rectifiers are not conducting, thus throwing away a significant portion of the power system's energy capability.

Power factor correction circuits are intended to increase the conduction angle of the rectifiers and to make the ac input current waveform sinusoidal and in phase with the voltage waveform. The input waveforms can be seen in Figure C–2. This means that all the power drawn from the power line is real power and not reactive. The net result is that the peak and RMS current drawn from the line is much lower than that drawn by the capacitive input filter circuit traditionally used.

Active power factor correction circuits can take the form of nontransformer isolated switching power supply topologies, such as buck, boost, and buck/boost. The buck topology in Figure C–3 produces an output dc voltage lower than found at its input, whenever the PFC stage is operating ($V_{in} > V_{out}$). In other

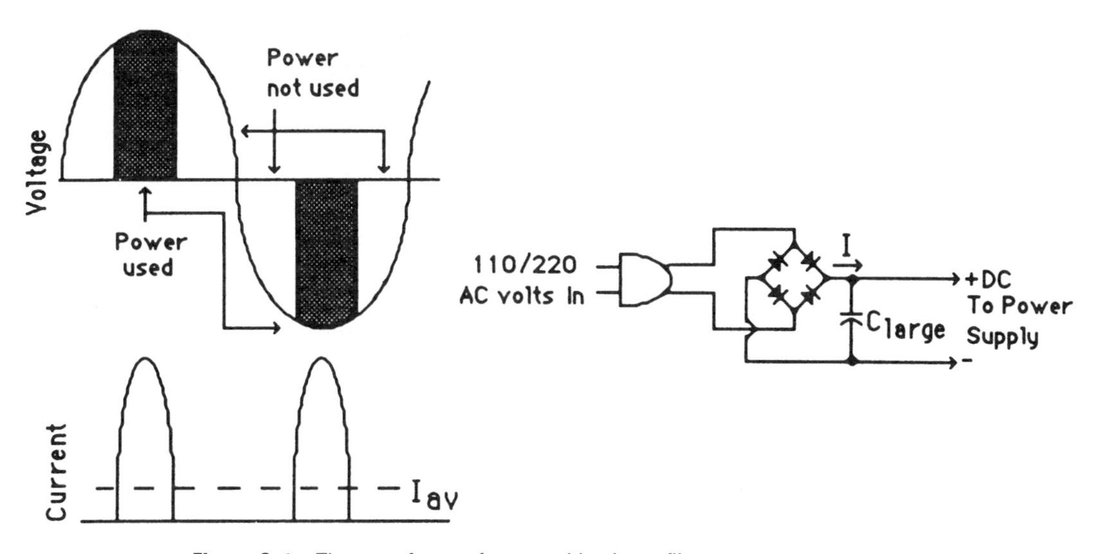

Figure C–1 The waveforms of a capacitive input filter.

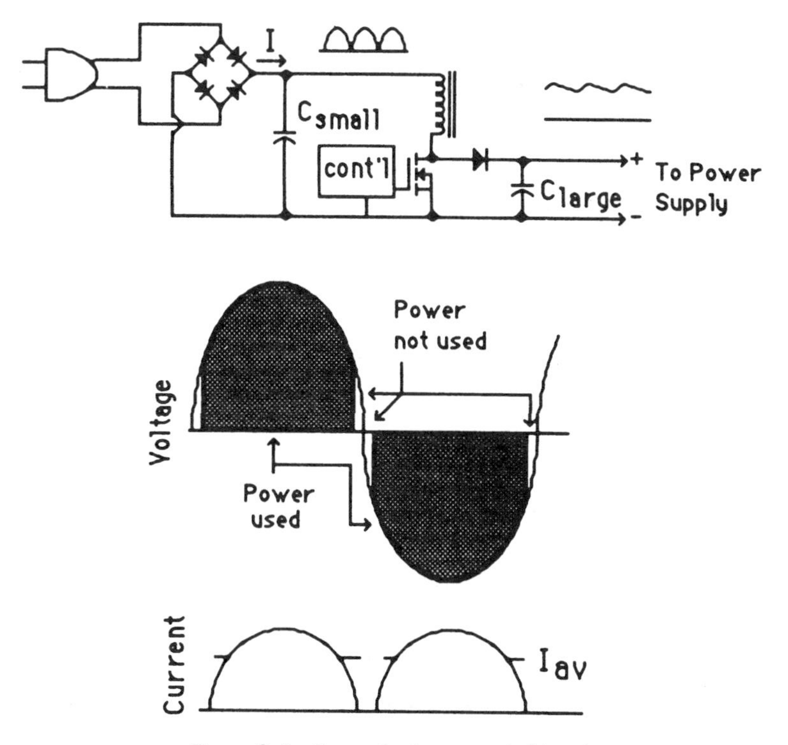

Figure C–2 Power factor corrected input.

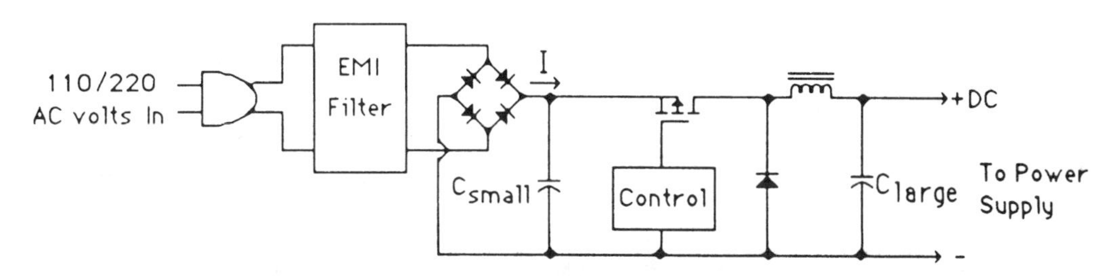

Figure C–3 A buck power factor correction circuit.

words, the output voltage is typically in the 30 to 50 VDC range. This can present a problem for higher powered loads which would then draw a large amount of current from the PFC circuit. The boost and the buck/boost topologies are popular within the field, since they produce a higher dc output voltage than the peak input voltage, which means lower average output currents. These are seen in Figures C–4 and C–5.

The buck/boost develops an output voltage that is negative with respect to the input ground following the rectifiers. The cascaded power supply and the PFC voltage sense networks must work with a negative voltage, but the dc output voltage can be independent from the values of the rectified input ac waveform. The major disadvantage is the need for a high-side power switch and high breakdown voltage requirements for the semiconductors. The boost topology has become the most popular topology. It has a low-side power switch that is easy to drive. Its only restriction is that the dc output voltage must be higher than the highest expected ac crest voltage. This means that for a PFC circuit to be useful in all power grids in the world, the output voltage must be greater than 390 VDC, and will pass voltage surges onto the load. Otherwise, it requires the least parts and hence costs the least.

Control of the power factor correction stage is a point of debate and battling patents. There are three general methods of control: fixed on-time, critical conduction-mode (just discontinuous), and continuous-mode. Fixed on-time has the minimum amount of circuitry, but limits the instantaneous current that can be drawn from the input line. Critical conduction-mode has no output rectifier reverse recovery loss, but is limited in output to about 300 to 600 W. Continuous-mode boost PFC circuits can go much higher in output power, but suffer in severe rectifier reverse recovery losses, unless some zero-transition loss circuits are added, which can add cost.

Added cost is very important in this area since power factor correction is an unseen benefit to the customer who does not want to pay for anything he or she cannot directly see. An example is PFC corrected, electronic fluorescent light-

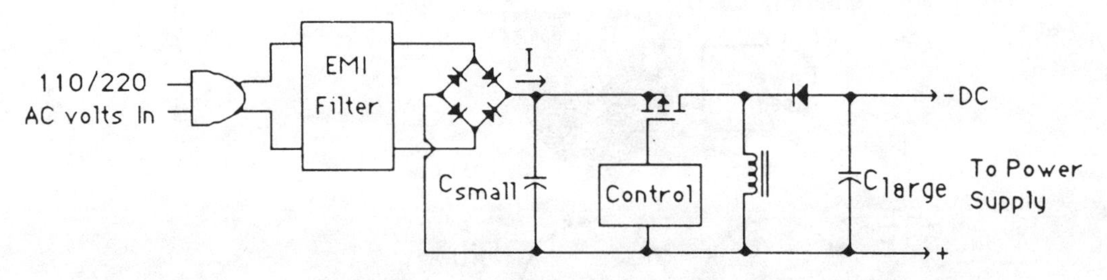

Figure C–4 A buck/boost power factor correction circuit.

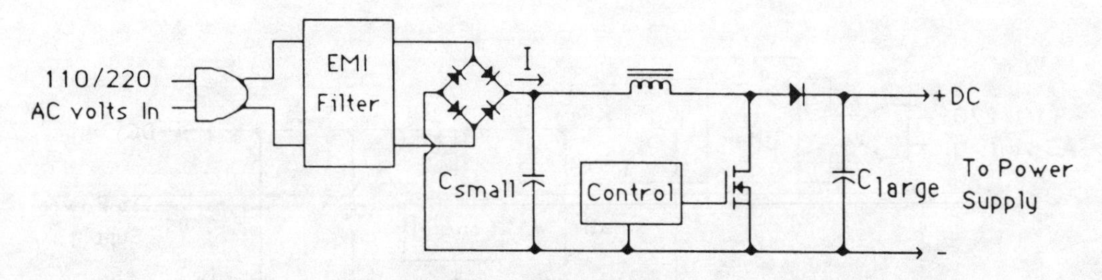

Figure C–5 A boost power factor correction circuit.

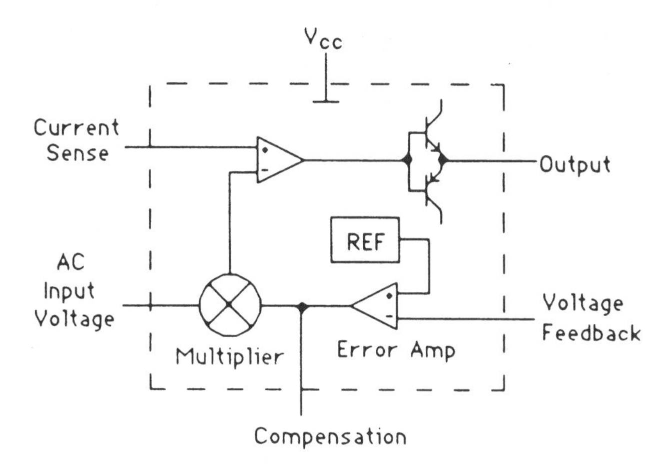

Figure C–6 A generalized typical power factor control IC.

ing ballasts at two times the purchase price versus magnetic fluorescent ballasts. Only industrial customers are thoughtful enough to see the vast difference in their yearly electric bill 12 month later.

The basic PFC controller takes the form as shown in Figure C–6, which would include the critical and discontinuous-mode circuits and the continuous-mode circuits. There is a multiplier subcircuit inside the control IC which multiplies the instantaneous value of the input full-wave rectified voltage waveform with the output of the error amplifier. This produces a current limit signal which makes the input current follow the voltage sinusoidal waveshape. The ac input is filtered by the input EMI filter to produce a 50 to 60 Hz input current waveform that is free of switching artifacts.

The inductor operating mode is a major consideration in designing a PFC circuit. The discontinuous-mode of operation is typically used for power levels less than 300 to 600 W. It has high peak currents that limit its use at the higher input power levels. For powers greater than 300 W, the continuous-mode of operation is typically used. This lowers the peak currents seen by the power switch and output rectifier and is much easier to filter in the input EMI filter since there are no rapid transitions in the input switched current waveforms. The only disadvantage is that the diode-related switching losses rise significantly since the power switch must force the output rectifier to turn off at the beginning of each on-time period. The choice of output rectifier (low Trr) becomes critical to the operation of the PFC stage.

How power factor and harmonics are specified

I strongly recommend that your company engage a third-party EMI testing house to test your products. The minimum-level of test equipment required to test for the discussed factors is very expensive and there is a long learning period involved.

The following discussion is primarily based upon EN61000-3-2, which is more industrial-based products. EN60555 is the newer version of IEC-555 and covers the emitted harmonics of household products. Knowing the product class is very important for designing the power factor interface for the ac line. Products that draw less than 75 W today and 50 W in the near future, do not have to comply with the relative limits, only to absolute maximums laid out by the specifications. The limits presented in Table C–1 may change, so please refer to the latest released version. This is a developing field so be aware of the most recent specifications at the time of your product's release.

The real power delivered to a load is given by

$$P_{in} = V_{in} \cdot I_{in} \cdot (\text{Power Factor}) \tag{C.1}$$

where

$$\text{Power Factor} = \frac{\text{Real Power}}{\text{Real Power} + \text{Reactive Power}} \tag{C.2}$$

In terms of strictly passive reactive loads, the power factor is the resulting phase between the voltage and the current waveforms. In power supplies though, it is the distortion to the voltage waveform resulting from the time which input rectifiers conduct. Power factor is measured from 0 to 1 where 1 is where all the power is used by the load (purely resistive). The typical capacitive input filter found in power supplies has an average power factor of 0.5 to 0.7.

In running the tests, a power analyzer must be used such as the Voltech PM1000, PM1200, or PM3000. Also an audio spectrum analyzer is needed to measure the amplitude of the harmonic components of the ac current. The total input voltage and currents are given by

$$V_{RMS(total)} = \sqrt{V^2_{fund(RMS)} + V^2_{1(RMS)} + V^2_{2(RMS)} \cdots} \tag{C.3}$$

and

$$I_{RMS(total)} = \sqrt{I^2_{fund(RMS)} + I^2_{1(RMS)} + I^2_{2(RMS)} \cdots} \tag{C.4}$$

where the subscript of 1, 2, . . . are the harmonics of 50 or 60 Hz. In power supplies the third harmonic is by far the next largest amplitude and therefore the largest problem. Harmonics cause problems because, in a pure sense, only the fundamental current frequency produces real power. So the reduction of harmonics produces a better power factor.

A term used in PFC is *total harmonic distortion*. This is defined as

$$T.H.D. = \frac{I_{1(RMS)} + I_{2(RMS)} \cdots}{I_{RMS(total)}} \tag{C.5}$$

and it is an indication as to the performance of a PFC circuit.

From the power analyzer or the spectrum analyzer, one can measure the amplitude values needed to verify compliance to the PFC specifications. EN61000-3-2 has the limits shown in Table C–1. Class A and D are shown because they are common product categories.

These limits must be measured with a LISN (line impedance stabilization network) as specified by the regulatory agencies. This makes the input power line a 50 ohm impedance and serves the basis of all of these tests. The test results are highly dependent upon the ac line impedance.

Some comments on the design of PFC circuits. First, the EMI filter is an integral part of any PFC circuit. It filters out the switching harmonics from the input current waveform. Without an EMI filter, your product will fail the EMI/RFI tests which are in addition to the power factor tests. Please refer to Appendix E for information on how to design an EMI filter. Secondly, using a variac during the measurements will affect the input line impedance and thus affects

Table C-1 IEC555-2 Harmonic Current Limits

Harmonic	Class A RMS-Amps	Class D RMS-Amps
2	1.08	2.30
3	2.30	—
4	0.43	—
5	1.44	1.14
6	0.30	—
7	0.77	0.77
9	0.40	0.40
11	0.33	0.20
13	0.21	0.33
$8 < n < 40$	$0.23 \times 8/n$	
$11 < n < 39$	$0.15 \times 15/n$	$0.15 \times 15/n$

the validity of the data you are trying to measure. Many units will pass testing without the use of a LISN, but fail when the LISN is used. The added impedance of a LISN distorts the waveforms more than the impedance of the typical raw ac line in use at that moment. Thirdly, all voltage measurements must be differential and use the specified current measuring apparatus.

C.1 A Universal Input, 180W, Active Power Factor Correction Circuit

This design example demonstrates the design process of a 180 W discontinuous-mode boost PFC circuit. It can be scaled to provide output powers up to 200 W. The PFC stage is designed to work from every residential ac power system within the world; that is, from 85 to 270 VRMS at 50 and 60 Hz without the need for a jumper.

Design specification

AC input voltage range: 85–270 VRMS
AC line frequencies 50–60 Hz
Output voltage: 400 VDC +/– 10 V
Input power factor at rated load: >98%
Total harmonic distortion (THD) under EN1000-3-2 limits

Predesign considerations

Having a rating less than a 200 W has many benefits for a power factor correction stage. The major benefit is that it can operate in the discontinuous-mode. Within higher power PFC designs the continuous-mode must be employed which presents a significant loss within the circuit due to the reverse recovery time of the output rectifier. In fixed frequency discontinuous-mode PFC controllers, there is still a period when the circuit operates in the continuous-mode (Vin < 50 V (approx.)). By employing a *critical conduction-mode* controller, the designer can guarantee that the continuous-mode is never entered.

The first consideration is to determine the peak ac input voltages.

110 V input:

$$V_{\text{in(nom)}} = 1.414(110\,\text{V}) = 155.5\,\text{V}$$

$$V_{\text{in(hi)}} = 1.414(130\,\text{V}) = 183.8\,\text{V}$$

240 V input (Britain—worst case):

$$V_{\text{in(nom)}} = 1.414(240\,\text{V}) = 339.4\,\text{V}$$

$$V_{\text{in(hi)}} = 1.414(270\,\text{V}) = 381.8\,\text{V}$$

The output voltage should be higher than the highest anticipated input peak crest voltage. The output voltage of the PFC stage is now chosen to be 400 VDC.

The maximum value for the peak inductor current will occur at the crest voltage of the minimum expected ac input voltage. This is

$$\begin{aligned} I_{\text{pk(max)}} &= 1.414(2)(P_{\text{out(rated)}})/(eff_{\text{est}})(V_{\text{in(min)RMS}}) \\ &= 1.414(2)(180\,\text{W})/(0.9)(85V_{\text{RMS}}) \\ &= 6.6\,\text{A} \end{aligned}$$

Inductor design

In designing the boost inductor, one would designate the point of reference as the crest voltage of the minimum expected ac input voltage. For any set of operating conditions with this method of PFC control (i.e., fixed load and ac input voltage, the on time pulsewidth remains constant over the entire half-sinusoid waveform). To determine the on time at the minimum peak ac input voltage one would do the following operations:

$$R = \frac{V_{\text{out(DC)}}}{\sqrt{2}V_{\text{in-AC(min)}}} = \frac{400\,\text{V}}{1.414(85V_{\text{RMS}})}$$

$$R = 3.3$$

The maximum on time which occurs at this point is

$$T_{\text{on(max)}} = \frac{R}{f(1+R)} = \frac{3.3}{(50\,\text{KHz})(1+3.3)}$$
$$= 15.3\,\mu\text{S}$$

The approximate maximum value of the boost inductor is

$$L \approx \frac{T_{\text{on(max)}}(\sqrt{2}V_{\text{in-AC(min)}})^2(\text{eff})}{2P_{\text{out(max)}}}$$
$$\approx \frac{(15.3\,\mu\text{S})(1.414(85V_{\text{RMS}})(0.9)}{2(180\,\text{W})} \approx 552\,\mu\text{H}$$

The power winding of the inductor (transformer) not only must support the maximum average input current but the output current as well. So, the wire gauge of the winding should be

$$V_{\text{w(max-av)}} = \frac{P_{\text{out}}}{\text{eff}(V_{\text{in(RMS)}})} + \frac{P_{\text{out}}}{V_{\text{out}}}$$
$$= \frac{180\,\text{W}}{(0.9)(8.5V_{\text{RMS}})} + \frac{180\,\text{W}}{400\,\text{V}} = 2.8\,A_{\text{mps}}$$

The wire gauge to accommodate this average current would then be #17 AWG. I will use three strands of #22 AWG (which adds up to the same wire cross-sectional area), which is more flexible during the winding process and will help reduce the ac resistance of the winding due to the skin effect. Also, due to the high voltages present within the same winding, I will be using quad-thickness insulation to reduce the threat of interturn arc-overs.

I am selecting a PQ core style. A major concern is the length of air-gap required for various core styles in unipolar applications. The larger air-gaps (>50 mils) cause excessive electromagnetic radiation into the immediate environment thus making it harder to RFI filter. To reduce the air-gap, one needs to find a ferrite core with the largest core cross-sectional area for a given core size. The PQ core has this characteristic. Referring to the WaAc *vs*. power charts provided by Magnetics, Inc., the resulting PQ core part number is P-43220-XX. (XX is the gap length in mils).

The approximate air-gap needed in the core is

$$l_{gap} \approx \frac{0.4\pi L \cdot I_{pk} 10^8}{A_c B_{max}^2}$$

$$\approx \frac{0.4\pi(552\,\mu H)(6.6\,A)10^8}{(1.70\,cm^2)(2{,}000\,G)^2} \approx 66\,mils$$

Let us make the air-gap 50 mils, which is a custom air-gap. Magnetics has no problem with this practice and usually adds only a couple of percent to the core cost. The inductance factor (AL) for this core with this gap is estimated at 160 mH/1000 T (using a linear extrapolation of AL reduction versus air-gap length).

The number of turns needed for this inductance is

$$N = 1000\sqrt{\frac{0.55\,mH}{160\,mH}} = 59\ turns$$

Checking to see if the core will support this many turns (neglecting the auxiliary winding area):

$$\frac{A_W}{W_A} = \frac{(59\,T)(.471\,mm^2)}{47\,mm^2} = 59\% \quad OK$$

Designing the auxiliary winding. The auxiliary winding will have the low frequency (100 to 120 Hz) variation on its output peak rectified voltage, so the controller filter capacitor needs to be large to minimize the droop in the Vcc of the controller. The highest flyback-mode rectified voltage will occur at low input voltages and will be of the form

$$v_{aux} \approx \frac{N_{aux}(V_{out} - V_{in})}{N_{pri}}$$

This ac waveform is seen in Figure C–7.

The MC34262 has a high-side driver clamp of 16 VDC, so in order to keep the high-side driver dissipation to a minimum, the peak voltage of the rectified auxiliary voltage should be around 16 V. Determine the turns ratio needed for this from

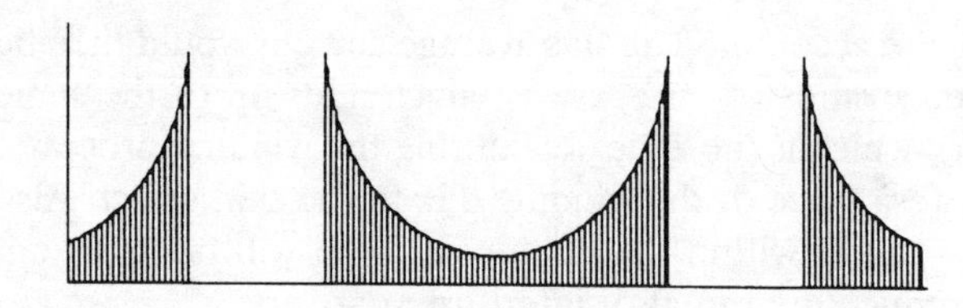

Figure C–7 The rectified ac waveform present on the auxiliary winding.

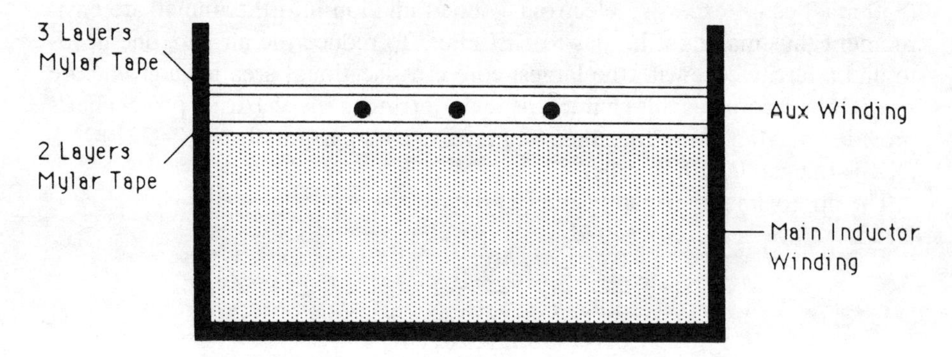

Figure C–8 Construction of the PFC boost inductor.

$$N_{\mathrm{aux}} = \frac{(59\,\mathrm{T})(16\,\mathrm{V})}{(400\,\mathrm{V} - 30\,\mathrm{V})} = 2.5\ \mathrm{turns}$$

I will make this winding three turns because of concern about low ac line operation. I will use one strand of #28 AWG heavy insulated magnet wire.

The capacitor needed to filter this voltage with approximately 2 V of voltage ripple is

$$C_{\mathrm{aux}} \approx \frac{I_{\mathrm{dd}} T_{\mathrm{off}}}{V_{\mathrm{ripple}}} = \frac{(25\,\mathrm{mA})(6\,\mathrm{mS})}{2.0\,\mathrm{V}}$$
$$= 75\,\mu\mathrm{F} \quad \text{make } 100\,\mu\mathrm{F} \ @\ 20\ \mathrm{VDC}$$

Transformer construction

The two-winding transformer will be constructed by first winding the 59 turns of the three strands of #22 AWG quad-thickness magnet wire onto the bobbin. Then place two layers of Mylar tape. Then the three turns for the auxiliary winding, and lastly three layers of Mylar tape. The internal layers of tape are to discourage any arcing that may occur due to the high voltages between the primary winding and the auxiliary winding.

Designing the start-up circuit

I will use a passive resistor for starting up the control IC and to provide current to the gate drive of the MOSFET. For the resistor I need to use two resistors placed in series, since the 370 V peak on the rectified input is comparable to the breakdown voltages of the resistors themselves. The start-up resistors will charge the 100 μF bypass capacitor and the subsequent energy stored in the capacitor must be sufficient to operate the control IC for the 6 mS before the worst-case rectified peak voltage from the auxiliary winding is available to operate the IC. The start-up voltage threshold hysteresis is 1.75 V minimum. Checking whether the bypass capacitor is large enough to start the circuit before the turn-off threshold is reached:

$$V_{drop} = \frac{I_{dd}T_{off}}{C} = \frac{(25\,mA)(6\,mS)}{100\,\mu F}$$
$$= 1.5\,V_{olts} \text{—OK}$$

I would like to keep the dissipation less than 1 W at the high input voltage line. To do this one needs to determine the maximum current that that should pass through the start-up resistors.

$$I_{start} < \frac{1.0\,W}{270\,V_{RMS}} = 3.7\,mA$$

The total resistance is then:

$$R_{start} = \frac{270\,V - 16\,V}{3.7\,mA} = 68\,K(min)$$

Make the total resistance about 100 K or two 47 Kohm, 1/2 W resistors.

Designing the voltage multiplier input circuit

The minimum specified maximum linear limit of the input to the multiplier (pin3) is 2.5 V. This level should be the peak value of the divided rectified input waveform at the highest expected ac input voltage at the crest of the sinusoid (370 V). If a sense current of 200 μA is selected at this point the resistor divider becomes

$$R_{bottom} = \frac{2.5\,V}{200\,uA} = 12.5K \quad \text{make it } 12\,K.$$

The true sense current is 2.5 V/12 K = 208 μA.

The top resistor becomes

$$R_{top} = \frac{370\,V - 2.5\,V}{208\,\mu A} = 1.76\,Mohms$$

Make this two resistors in series each with a value of 910 Kohms.

The power rating of these resistors are P = (370 V)2/1.76 MΩ or 0.8 W. Each resistor should have 1/2 W power rating.

Design of the current sensing circuit

The current sense resistor should be sized in order to reach the 1.1 V current sense threshold voltage at the low ac input voltage. The value then becomes

$$R_{CS} = \frac{1.1\,V}{6.6\,A} = 0.3\,ohms$$

A leading edge spike filter of 1 K and 470 pF will also be added before inputting the current signal to pin 4.

Designing the voltage feedback circuit

For the output voltage sense resistor divider, selecting the sense current as 200 μA, the lower resistor becomes

$$R_{bottom} = \frac{V_{ref}}{I_{sense}} = \frac{2.5\,V}{200\,uA} = 12.5K \quad \text{Make } 12.0K.$$

This makes the true sense current 2.5 V/12 K = 208 μA. The upper resistor is

$$R_{upper} = \frac{(400\,V - 2.5\,V)}{208\,uA} = 1.91\,\text{Mohms}$$

Make this resistor a 1 Mohm and a 910 Kohm resistor in series, each with a 1/2 W rating.

The compensation of the voltage error amplifier should be a single-pole rolloff with a unity gain frequency of 38 Hz. This is required to reject the fundamental line frequencies of 50 and 60 Hz. The feedback capacitor around the voltage error amplifier becomes

$$C_{fb} = \frac{1}{2\pi f R_{upper}} = \frac{1}{2\pi(38\,Hz)(1.82\,M)}$$
$$= 0.043\,\mu F \text{ or } .05\,\mu F$$

Designing the input EMI filter section

I will be using a second order, common-mode filter. The difficulty in considering an input conducted EMI for this power factor correction circuit is its variable frequency of operation. The lowest instantaneous frequency of operation occurs at the crests of the sinusoid voltage waveform. This is where the core requires the longest time to completely discharge the core. The estimated frequency of operation has been 50 kHz, so I will use this as an assumed minimum frequency.

A good starting point is to assume that I will need 24 dB of attenuation at 50 kHz. This makes the corner frequency of the common-mode filter

$$f_C = f_{SW} \cdot 10^{\left(\frac{Att}{40}\right)}$$

where Att is the attenuation needed at the switching frequency in negative dB.

$$f_C = (50\,kHz)10^{\left(\frac{-24}{40}\right)} = 12.5\,kHz$$

Assuming that a damping factor of 0.707 or greater is good and provides a − 3 dB attenuation at the corner frequency and does not produce noise due to ringing. Also assume that the input line impedance is 50 ohms since the regulatory agencies use an LISN test which make the line impedance equal this value. Calculating the values needed in the common-mode inductor and "Y" capacitors:

$$L = \frac{R_L \cdot \zeta}{\pi \cdot f_C} = \frac{(50)(0.707)}{\pi(12.5\,kHz)} = 900\,\mu H$$

$$C = \frac{1}{(2\pi f_C)^2 L} = \frac{1}{[2\pi(12.5\,kHz)]^2 (900\,\mu H)}$$
$$= 0.18\,\mu F$$

Real-world values do not allow a capacitor of this large a value. The largest value capacitor that will pass the ac leakage current test is 0.05 μF. This is 27

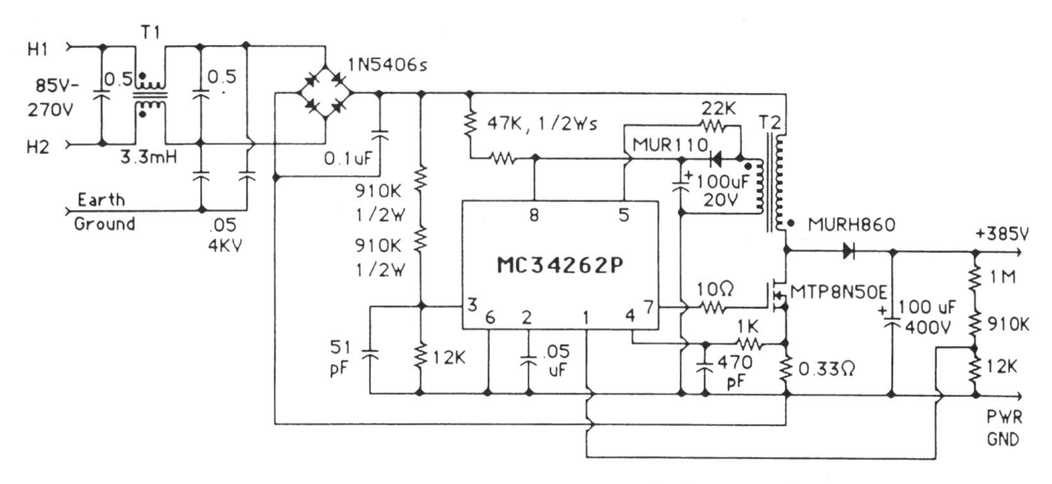

Figure C–9 The schematic for the 180 W power factor circuit (with EMI filter).

percent of the calculated capacitor value, so the inductor must be increased 360 percent in order to maintain the same corner frequency. The inductance then becomes 3.24 mH and the resultant damping factor is 2.5 which is acceptable.

Coilcraft offers off-the-shelf common-mode filter chokes (transformers) and the part number closest to this value is E3493. With this filter design I can expect a minimum of –40 dB between the frequencies of 500 kHz and 10 MHz. If later during the EMI testing stage, I find I need additional filtering, I will add a third order to the filter design by using a differential-mode filter.

The resulting schematic of the power factor correction circuit is given in Figure C–9.

Printed circuit board considerations

The unit in which this power factor correction circuit resides is going to be marketed everywhere in the world. The toughest safety requirements are issued by VDE in Germany. Here the creepage distance, or the distance that an arc must travel over a surface, is 3.2 mm for those signals that are opposite phases of an ac power line up to 300 VRMS. This means that there must be 3.2 mm spacing between traces of H1 and H2 (Hot and Neutral), and their rectified dc signals. Also there must be a 3.2 mm (minimum) surface distance between the windings on the input common-mode filter transformer and between high and low pins of the flyback inductor. The spacing of the 400 V output must be more than 4.0 mm from all other traces carrying less voltage. The creepage between any earth ground trace and the other traces must be more than 8.0 mm.

All current-carrying traces should be as wide and as short as possible. One-point grounding practices between the input, output, and low-level grounds should be done at the ground side of the current sense resistor.

Appendix D. Magnetism and Magnetic Components

The magnetic elements form the backbone of the switching power supply's operation. Understanding their basic operation, and the practical tradeoffs during their design is crucial to the entire power supply's proper operation. Factors such as efficiency and reliability are highly dependent on the magnetic component's design.

Unfortunately, the typical engineer's college curriculum only includes about a course and one-half of magnetism-related theory, which is easily forgotten. The purpose of this appendix is to refresh your memory with some magnetic theory highly slanted to switching power supply applications.

D.1 Basic Magnetic Theory Applied to Switching Power Supplies

Magnetic fields are the invisible companion to the readily visible electric signals within electronics. Every time there is an electron current flowing, there are associated electric and magnetic fields. Their orientations are easily remembered by the use of the *right-hand rule*, as seen in Figure D–1a. As one can see, the electric field emanates radially from the wire carrying the current. If the wire is oriented as in Figure D–1a, and the current is flowing towards the reader, the magnetic field flows in a counterclockwise direction around the wire. When a wire is "coiled" upon itself, as in inductors, the magnetic fields flow around the entire coil as in Figure D–1b. The *lines of flux* shown in Figure D–1 are a graphical representation *of flux density*. Within the coil, the magnetic field is compressed within a small area in coil's center; therefore, the flux density is higher. Outside the coil, the area is boundless, and the magnetic field is permitted to spread out over a much larger area, therefore the flux density is lower.

When a coil of wire is wrapped around a ring of magnetic material such as a ferrite toroid, the magnetic field travels almost exclusively within the ferrite core material. This is because the magnetic resistance, called *reluctance*, is much lower than that of air and it forms a complete magnetic loop. When a second, identical winding is placed on the toroid, and the test set-up is assembled as shown in Figure D–2, a familiar curve is traced out. This is called the *B–H curve*. This curve is the unique "fingerprint" for magnetic materials and their alloys. If the ac excitation voltage is driven high enough, soon the curve becomes "flattened" at both the top and the bottom. This condition is called *saturation* when

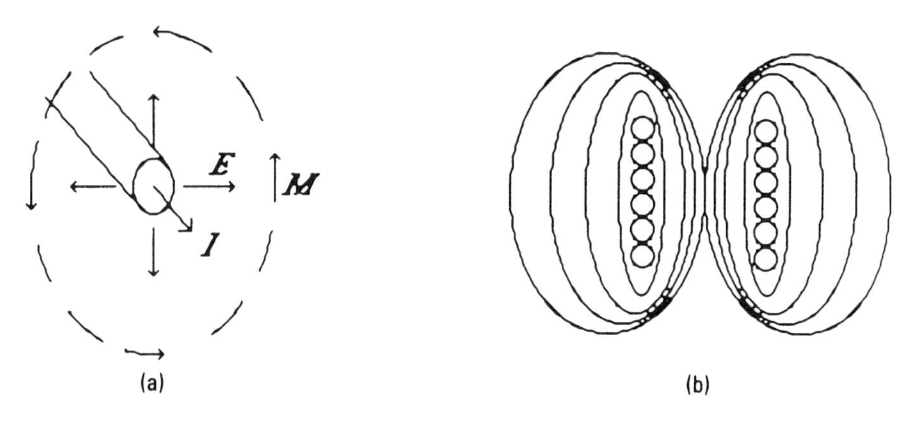

Figure D–1 The fields around a conductor in free air: (a) field around a free wire; (b) field around a coil in free air.

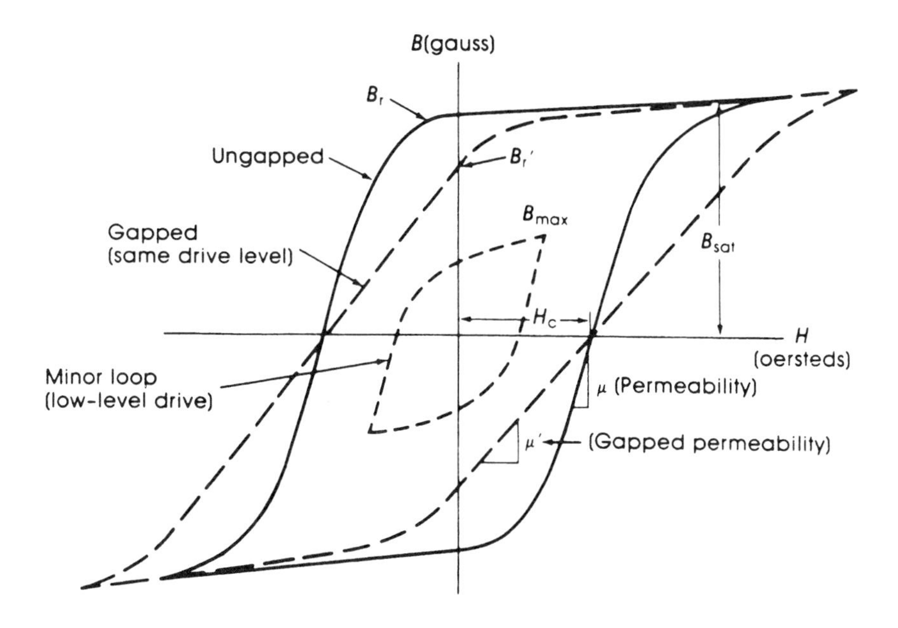

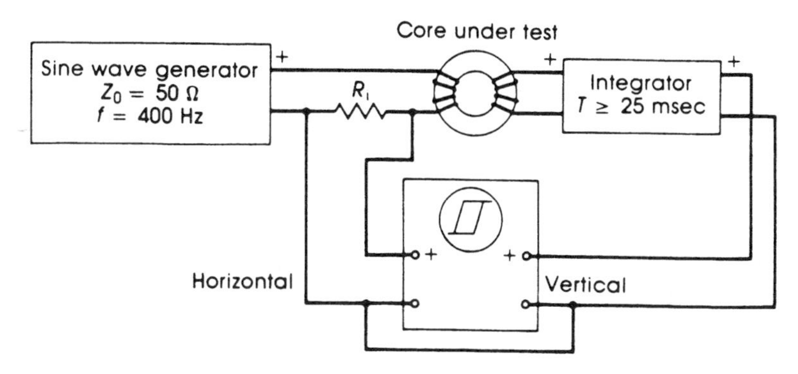

Figure D–2 The B-H curve and how to display it.

nearly all the magnetic domains within the magnetic material become aligned with the applied magnetic field. The significance of the curve is that it represents the amount of work required to reorient the magnetic domains within the magnetic material in the direction of the magnetic field created by the winding and the driving voltage. This work is lost and ends up as heat within the core. This is called the *hysteresis loss* of the core material. This loss is necessary to perform the work needed by the power supply during each cycle of operation. It can be seen as a delivery truck. The energy expended in moving the weight of the truck from its starting point to all its delivery locations and back again is waste, but it does accomplish additional work in the process. The X-axis is the *coercive force* or *magnetic field strength* (H) and has the units of ampere-turns per meter or oersteds (Oe). This provides the driving force to develop a magnetic field. Its closest electrical equivalent is voltage.

$$H = \frac{4\pi N 1}{1_m} \tag{D.1}$$

where N is the number of turns on the drive winding.
I is the peak current through the drive winding (amps).
l_m is the length of the magnetic path (cm).

The Y-axis is the *flux density* (B) measured in gauss (G) or webers per square centimeter in the U.S., and teslas or webers per square meter within the metric system. Its behavior can be seen by a useful relationship given by Faraday's law:

$$B_{(max)} = \frac{E \cdot 10^8}{k \cdot N \cdot A_c \cdot f} \quad \text{Gauss (U.S.)} \tag{D.2a}$$

where k is 4.0 for rectangular waves and 4.4 for sinewaves.
A_c is the core cross-sectional area (cm^2).
E is the voltage applied to the drive winding (V).
f is the frequency of operation (Hz).

In the MKS system used outside the U.S., the equation becomes

$$B_{(max)} = \frac{E}{k \cdot N \cdot A_c \cdot f} \quad \text{Teslas (metric)} \tag{D.2b}$$

where k is 4.0 for rectangular waves and 4.4 for sinewaves.
A_c is the core cross-sectional area ($meters^2$).
E is the voltage applied to the drive winding (V).
f is the frequency of operation (Hz).

This equation is useful for determining how close to saturation an inductor or transformer is operating, which could avoid a catastrophe.

The slope of the sides of the B-H curve is referred to as the *permeability* of the material. It can be seen as the degree of ease it takes to reorient the magnetic domains within the material. The greater the slope, the less magnetic field strength and current it requires to create a given flux density. Its value has a great bearing on how much inductance one gets per turn of a winding. The higher the value of permeability (or steeper slope), the more inductance one gets per turn added:

$$\mu = \Delta B / \Delta H \tag{D.3}$$

The relationship between them is given by

$$B = \mu H \tag{D.4}$$

When using an inductor or transformer within a switching power supply, the core is never operated to the point of saturation. Instead it is operated in what is called a minor loop. These are B-H curves that are wholly contained within the boundary of the saturated B-H curve. In 20 to 50 kHz PWM switching power supplies, the peak excursion of the flux density (B_{max}) is usually half of the saturation flux density (B_{sat}). This results in a core loss of two percent in overall converter efficiency, which is considered acceptable. For higher frequencies of operation, B_{max} should be lowered to keep the core losses at or below two percent loss of efficiency. The common minor-loop curves are seen in Figure D–3. Curve A is the B-H curve within the transformer of a "push-pull" style forward converter such as the push-pull, half-bridge, and full-bridge converters. Curve B is the B-H curve of a discontinuous-mode flyback converter. Curve C is the B-H operation of a forward-mode filter choke and a flyback transformer operating in the continuous-mode. For dc and unipolar flux applications, it is desirable to place a small air-gap within the magnetic path of the core. Its effect on the B-H curve can be seen in Figure D–3. As one can notice, the permeability of the overall inductor drops. This drop is in proportion to the length of the air-gap introduced. This offers an advantage to the inductor's or transformer's operation in that it requires a greater current through the drive winding to drive the core's flux density to enter a state of saturation. Most of the energy placed within the core is now stored in the air-gap and the result is the flux density within the magnetic core material drops. For these applications, additional turns will have to be added to the core to maintain the same

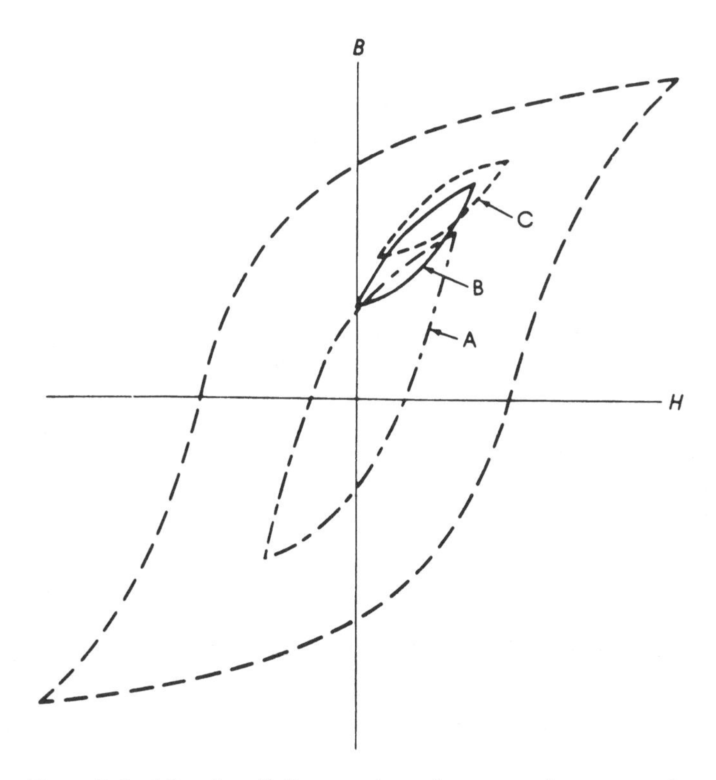

Figure D–3 Minor loop B-H curves for various magnetic components.

inductance value. This will make the volume of the overall inductor or transformer grow, but it is necessary for a greater margin of safe operation for the power supply.

The major losses within any core material are the *hysteresis loss* and *eddy current loss*. These losses are typically lumped together by the core manufacturer and given in a graph of watts lost per unit volume *vs*. the peak operational flux density (B_{max}) and frequency of operation. Hysteresis loss is given as

$$P_H = k_h \cdot v \cdot f \cdot (B_{max})^2 \qquad \text{(D.5)}$$

where k is the hysteresis loss constant for the material.
v is the volume of the core (cm^3 (CGS), m^2 (MKS)).
f is the frequency of operation (Hz).
B_{max} is maximum flux density (Gauss(CGS), Teslas (MKS)).

The eddy current loss is caused by electron currents that are induced within the core material by the high local magnetic fields. These usually flow in circular paths and are encouraged to flow when large unbroken volumes of core material are present. They also are common in cores with corners. They can be discouraged either by using a core with high reluctance that offers resistance to the flow of current or by using a laminated core that breaks the core into small cross-sections, thus discouraging a circular current path. The eddy current losses are described by

$$P_e = k_e \cdot v \cdot f^2 \cdot (B_{max})^2 \qquad \text{(D.6)}$$

As one can see, both losses increase dramatically with the increasing levels of B_{max} and the eddy current loss increases drastically with the frequency of operation. These losses cause an increase in the size of the inductor or transformer for an increased frequency of operation. Increasing the frequency of operation of a switching power supply does not necessarily reduce the size of the core.

In designing the magnetic components within the power supply, most of the problems arise from the very noticeable difference in nomenclature between the normal electronic design literature and the core manufacturer's literature. The difference in units between U.S. and non-U.S. manufacturers also can be confusing, especially when they do not clearly present the exact form of the equations and units they expect you to use. For core loss information, each manufacturer seems to use its own units. Some use watts per unit volume (cm^3 or m^3) or per unit weight (lb). The designer can only use a point of reference within each manufacturer's graph which is typically one-half of B_{sat} at 50 kHz, which should give two percent loss in overall supply efficiency. The designer then can readjust the B_{max} to maintain that two percent loss figure. You should review the utilized units carefully, and use the appropriate magnetic equation. Usually the core manufacturer will have applications literature that presents the equations that work with their respective units.

D.2 Selecting the Core Material and Style

Selecting the core material and style for a switching power supply application is often viewed as a "dart board" type of selection process by a designer starting his or her first transformer design. Although almost every core material and style will work in all applications, their behavior within the application dictates which is best. There really is some sense to the selection process.

Selecting the core material is the first issue to be addressed. All core materials are alloys based on ferrite. The major factor in a material's worthiness is its loss at the frequency of operation and the flux density of the application. A good place to start is with the materials the core manufacturer's themselves recommend for PWM switching power supplies and those that are commonly used by the designers in the field (see Table D–1).

Using one of the core materials listed in Table D–1, the designer can feel reasonably confident that he or she has made the best choice for a ferrite. Mopermalloy is a ferrite alloy that has nonmagnetic molybdenum mixed with it. The molybdenum acts as a distributed air-gap within the material, which makes the material excellent for dc biased or unipolar applications. Unfortunately, it is only available in toroid core styles, and it typically used for output filter chokes.

What if a new material emerges onto the scene and you are asked to review it? The primary points of interest are the core loss (W/cm^3), the amount of B-H degradation at elevated temperatures, and whether it is offered in the desired core style (with air-gaps). The primary issue is the core loss. This is composed of both the hysteresis and eddy current losses combined. Manufacturers utilize graphs that plot loss versus frequency of operation versus maximum operational flux density, which makes it easy to compare materials (refer to Figure D–4). Be careful though, the manufacturers use differing units of measurements such as teslas or gauss, or different bases such as volume or weight. The conversion factors are given in Appendix F. To use these graphs, the designer should already have a good idea as to what frequency of operation he or she is going to use. The second factor needed is the maximum flux density (B_{sat}). The industry's rule-of-thumb for the amount of allowable loss within the magnetic elements should be no more than a two percent loss in overall power supply efficiency. For instance, at 50 kHz the nominal B_{max} should be half that of the B_{sat}. B_{max} should follow the guidelines presented in Table D–2 to maintain the same amount of

Table D–1 Common Core Materials Used within the Industry

	Core Material (Ferrites)	
Manufacturer	<100 kHz	<1 MHz
Magnetics, Inc.	F, T, P	F, K, N
TDK	P7, C4	P7, C40
Philips	3C8	3C85
Siemens	N27	N67

Table D–2 Recommended Flux Density Limits vs. Frequency

Frequency	Maximum Operational Flux Density (B_{max})
<50 kHz	$0.5B_{sat}$
<100 kHz	$0.4B_{sat}$
<500 kHz	$0.25B_{sat}$
<1 MHz	$0.1B_{sat}$

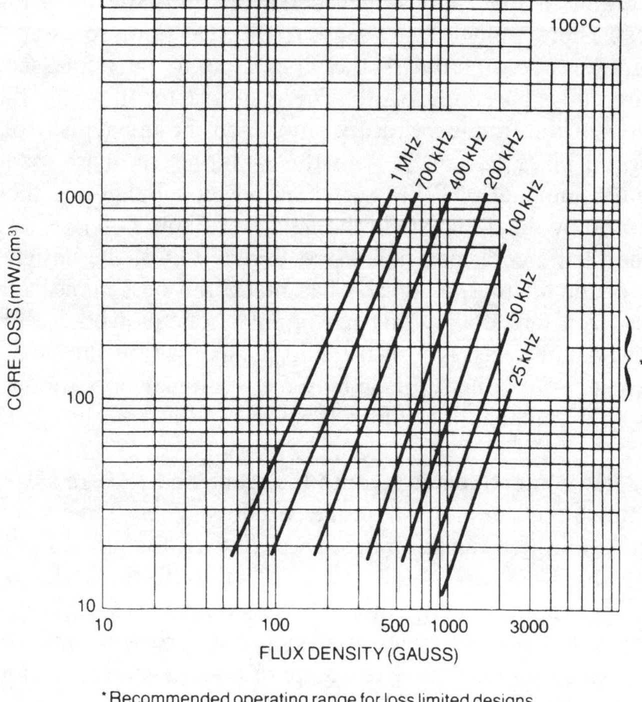

Figure D–4 Curves showing volumetric core loss vs. frequency and B_{max} (3C8 material shown). (Courtesy of Philips Components.)

core loss within the power supply design. To use a chart similar to Figure D–4, locate your B_{max} on the x-axis; go vertically until the desired frequency curve is intersected and read the volumetric core loss from the y-axis.

The second consideration is how much the B_{sat} of the material degrades with increased operating temperature (see Figure D–5). Some materials degrade more than others. In general for the commonly used materials, a drop of 30 percent in B_{sat} is expected at 100°C. This tells the designer to never have flux excursions above 70 percent of the rated saturation flux density of the material. Lastly, some materials exhibit lower core losses at elevated core temperatures. Cores will always be hotter than the ambient temperature. Core temperature rises of 10–40°C are not uncommon. Sometimes a graph is provided by the core manufacturers illustrating this point at a given drive level. If the material reaches a minimum loss at 50°C, this will be an advantage to the designer.

Once a core material has been selected, the core style must be considered. Many different core styles are offered by the core manufacturers; they fall basically into the categories shown in Figure D–6. Each has an advantage of size, cost, or shielding, and these factors should be considered in the light of the application. They fall basically into two styles: toroid and bobbin style cores. Toroidal transformers are more expensive to build because of the special machinery needed to wind the turns onto the core, but they are superior in the amount of radiated flux escaping from the transformer. Bobbin cores are typi-

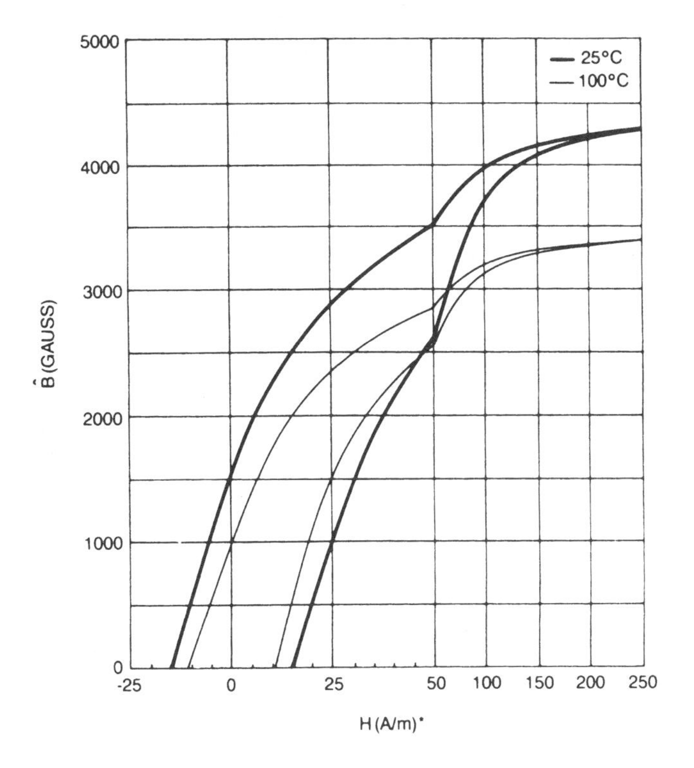

Figure D–5 Curve illustrating the degradation of B_{sat} with core temperature (3C8 material shown). (Courtesy of Philips Components.)

Table D–3 Core Styles and Their Merits

Basic Core Style	Core Material: Permalloy	Core Material: Ferrite	Gapped	Winding Shielding	Core Cost	Manufacturing Cost
Toroid	×	×	Yes/No	No	Low	High
E-Core		×	Yes	No	Low	Low
U-Core		×	Yes	No	Low	Low
Pot-Core		×	Yes	Yes	High	Medium
EP-Core		×	Yes	No	Low	Medium

cally less expensive to build than toroids, which is a distinct advantage, but cost more than toroids for the basic core parts. Some of these considerations are outline in Table D–3. Pot cores and their derivatives (PQ, RS, etc.) are expensive to buy, but are inexpensive to have a transformer built. Pot cores offer good magnetic shielding of the windings and the gap. Unfortunately, the lack of airflow around the windings causes them to operate at a higher temperature. E-E and E-I cores are less expensive than pot cores and have a generally larger winding area. Since, in the winding of transformers, the winding area is what typically determines a core size, this makes E-E and E-I cores the most preva-

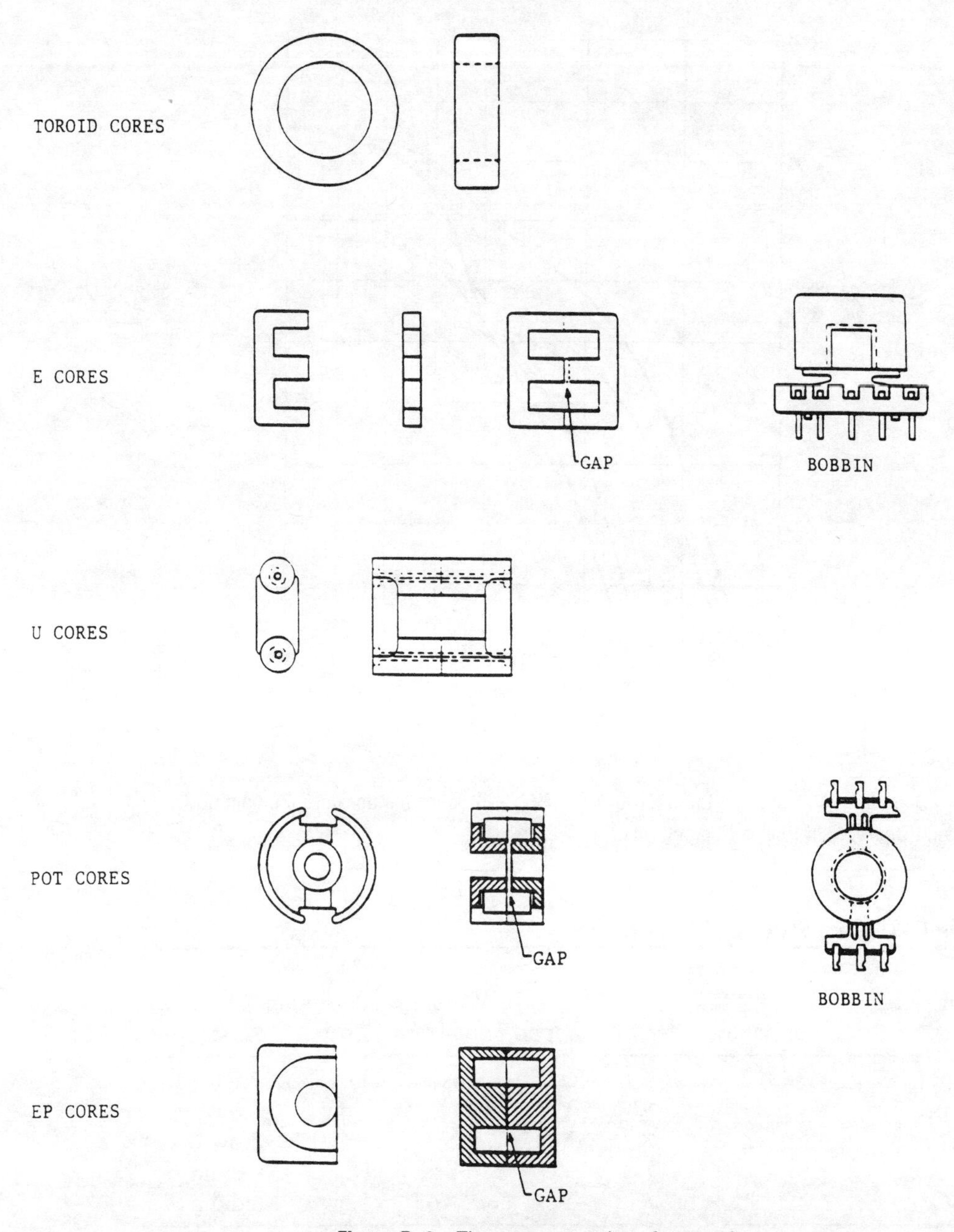

Figure D–6 The common styles of magnetic cores.

lent choice among designers. The windings are exposed to the air so that the windings operate at cooler temperatures, but radiate more to the environment because of the exposed air-gap.

Trade-offs in basic core cost, final transformer cost, and RFI are the primary considerations.

Appendix E. Noise Control and Electromagnetic Interference

Controlling high frequency noise generation and radiation is the blackest of the "black box" art in switching power supply and product-system design. It is a subject that warrants a book all to itself and it is the final area that will interfere with the release of your product into the market. This appendix cannot adequately cover the subject, but will overview the major considerations involved with product design.

Most companies cannot afford the expense of setting up a noise testing laboratory suitable for regulatory agency testing. The equipment is expensive and the operators must have special training. It is recommended that a company employ a regulatory testing consultant company to help it through this phase of the program. The majority of products initially submitted for RFI/EMI approval fail one or both of the radiated or conducted EMI tests. Almost always the design needs last-minute changes in order to pass the tests. The consultant engineers have been through this exercise many times before and are familiar with the problem areas and their solutions.

With help from this appendix, and Sections 3.12 and 3.14, it is hoped that your design will at least have an acceptable PC board layout, input EMI filter, and enclosure design that can serve as a basis for minor modifications at the time of testing. The PC board layout is the first major thing that the designer can do to minimize the effects of noise. The use of waveshaping techniques are the second, and the enclosure design is the third most important thing. One general rule is that if you design for the most stringent of your regulatory requirements, you will be better off when it comes time to test the product. Most of the countries in the world are "harmonizing" on their testing limits within their specifications.

If one passes one country's EMI/EMC specification, it is likely that the product will have no trouble in passing another country's requirements.

E.1 The Nature and Sources of Electrical Noise

Noise is created whenever there are rapid transitions in voltage and/or current waveforms. Many waveforms, especially in switching power supplies, are periodic. That is, the signal that contains pulses with high frequency edges repeats itself at predictable pulse repetition frequencies (PRF). For rectangular pulse-trains, the inverse of the period dictates the fundamental frequency of the waveform itself. The fourier conversion of a rectangular waveform generates a wealth of harmonics of this fundamental frequency. The inverse of twice the edge risetime or falltime of these pulses are estimates of the spectral fundamental frequency of the edges. This is typically in the megahertz range and their harmonics can go much higher in frequency.

In PWM switching power supplies, the pulsewidth of the rectangular waveshape is continuously changing in response to the supply's operating conditions. The result is typically an almost white noise energy distribution that exhibits some peaks and the amplitude rolls off with higher frequencies. Figure E–1 is a near-field radiated spectrum of an off-line PWM flyback switching power supply with no snubbing. As one can see, the spectral components extend well over 100 MHz (far right) and would interfere with consumer electronics equipment, if not filtered and shielded.

Quasi-resonant and resonant transition switching power supplies have a much more attractive radiated spectral shape. This is because the transitions are forced to be at a lower frequency by the resonant elements, hence only the low frequency spectral components are exhibited (below 30 MHz). The lower rate of change during the transitions are responsible for behavior. The higher frequency spectral components are almost non existent. The near-field radiated spectrum of a quasi-resonant, flyback converter are shown in Figure E–2. The quasi-resonant and soft switching families of converters are much "quieter" and easier to filter.

Conducted noise, that is, noise currents that exit the product enclosure via the power lines and any input or output lines, can manifest itself in two forms: common-mode and differential-mode. Common-mode noise is noise that exits the case only on the power lines and not the earth ground and can be measured with respect to the power lines (refer to Figure E–3a). Differential-mode noise is noise that can only be measured from the earth ground to one of the power leads. Noise currents are actually exiting via the earth ground lead. Its model can be seen in Figure E–3b. Each mode of noise can only be controlled by specific filter topologies and in each power supply design may require two types of input filtering. These filters have inductors and capacitors which are called "X" and "Y" elements. The X elements go across the power lines filtering the

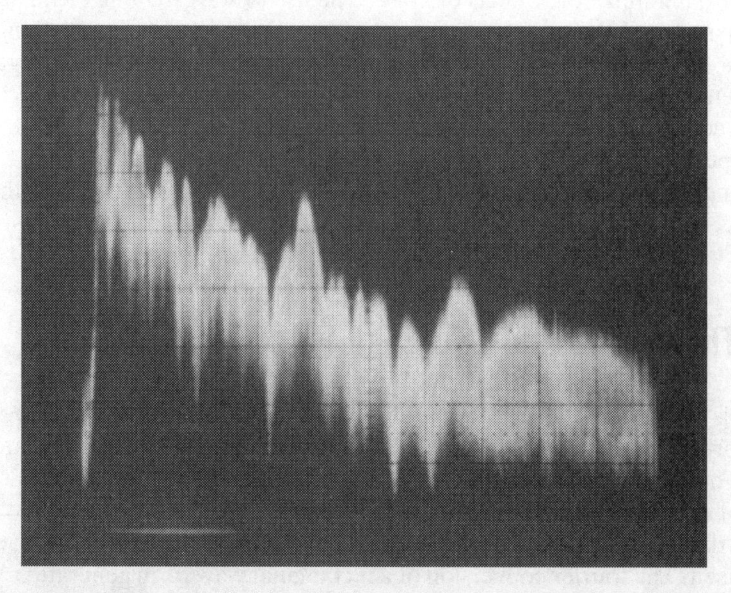

Figure E–1 The radiated spectrum of a typical off-line PWM flyback converter.

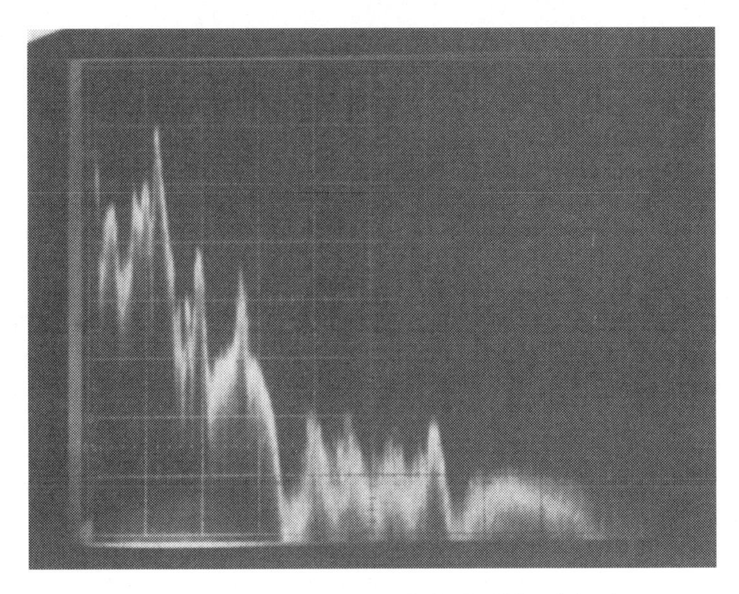

Figure E–2 The radiated spectrum of a ZVS QR off-line flyback converter.

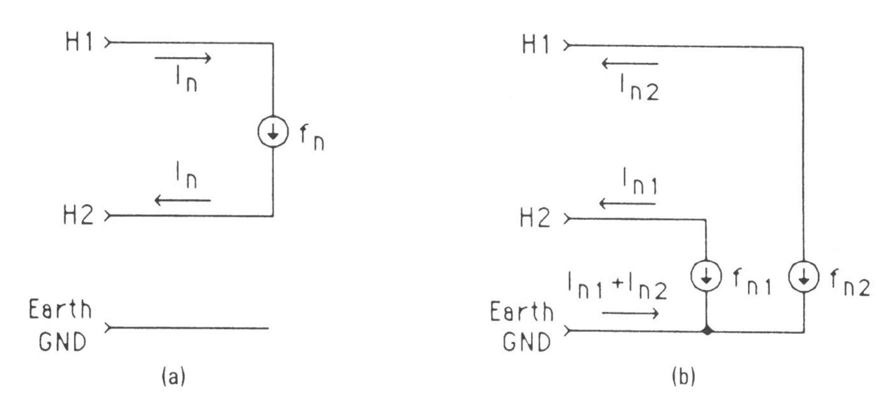

Figure E–3 Common-mode and differential noise models: (a) common mode; (b) differential mode.

common-mode noise artifacts and the Y elements go between the power lines and earth ground filtering the differential noise artifacts.

Regulatory approval bodies check for both radiated and conducted noise during their certification testing. Radiated noise is checked by locating a calibrated antenna and receiver at a specified distance (1 meter) from the product and plotting the resulting spectrum well into the GHz region. Radiated noise causes interference with other equipment, but conducted noise uses the power and I/O lines to radiate its noise and therefore is also checked. Conducted noise is checked by coupling into the input power lines via a high-frequency current transformer and the resultant spectrum is checked beyond 1 GHz.

E.2 Typical Sources of Noise

Noise, especially radiated noise, can be reduced by understanding its sources and what design techniques can reduce its effects. There are several major sources of noise within a PWM switching power supply that create the majority of radiated and conducted noise. These sources can be easily located and their design can be modified to reduce the noise generation of the power supply.

Noise sources are part of noise loops which are printed circuit board connections between high-frequency current sinks and current sources. Following the PC board design practices in Section 3.14 will help greatly in reducing the radiated RFI. Appreciation of the high-frequency characteristics of the common components and PC boards is needed.

The first major source of noise is the input power circuit, which includes the power switch, the primary winding of the transformer, and the input filter capacitor. The input filter capacitor provides the entire trapezoidal current waveforms needed by the power supply, since the input line is always heavily filtered with a bandwidth much less than the operating frequency of the power supply. The PCB traces must be as physically short and as fat as possible. Fat traces have lower inductance than thin traces. The trace length dictates the frequencies above which noise will be easily radiated into the environment. Shorter traces radiate less energy at the higher frequencies. The input filter capacitor and the power switch should be next to the transformer to minimize the trace lengths. A high-frequency ceramic or film capacitor also should be placed in parallel with any aluminum electrolytic or tantalum input capacitor since they have poor high-frequency characteristics. The worse the ESR and ESL characteristics of the input filter capacitor, the more high frequency noise energy the power supply will draw directly from the power line, thus promoting poor common-mode conducted EMI behavior.

Another major source of noise is the loop consisting of the output rectifiers, the output filter capacitor, and the transformer secondary windings. Once again, high-peak valued trapezoidal current waveforms flow between these components. The output filter capacitor and rectifier also want to be located as physically close to the transformer as possible to minimize the radiated noise. This source also generates common-mode conducted noise mainly on the output lines of the power supply.

One subtle, but major noise source is the output rectifier. The shape of the reverse recovery characteristic of the rectifiers has a direct affect on the noise generated within the supply. The *abruptness* or sharpness of the reverse recovery current waveform is often a major source of high-frequency noise. An abrupt recovery diode may need a snubber placed in parallel with it in order to lower its high-frequency spectral characteristics. A snubber will cost the designer in efficiency. Finding a soft recovery rectifier will definitely be an advantage in the design.

One structure that encourages the conduction of differential-mode noise is the heatsink. Heatsinks are typically connected to earth ground as a protection to the operator or service person. Any power switch or rectifier that is bolted to a heatsink allows capacitively coupled noise into the heatsink through the insulating pad. This noise then exits the case via the green, earth ground wire. One way of reducing the injection of this noise onto the ground is to use a power device insulator pad with an embedded foil pad. This reduces the mounting capacitance by pacing two capacitor in series, or the designer can connect the internal foil layer to the internal power supply common.

E.3 Enclosure Design

Product enclosure should act as an electromagnetic shield for the noise radiated by the circuitry within the package. A metal-based, magnetic material should be used in the enclosure construction. The material should be iron, steel, nickel, or Mu metal. For plastic enclosures, there are an assortment of conductive paints that can be used to add EMI/RFI shielding to the case. Also, any vent openings may need magnetic screening covering the openings.

The philosophy of any EMI shield is to encourage eddy currents to flow within the surfaces, thus dissipating the noise energy. Also, the assembled enclosure should act as a gaussian enclosure where there is good electrical conduction totally around the enclosure. So removable hatches and enclosure members need very good electrical connections around their peripheries. RF gasketing is sometimes used in particularly troublesome cases.

Leads that enter or exit the enclosure ideally should have their associated EMI filters at the point of entry or exit from the enclosure. Any unfiltered leadlengths that run within the enclosure will inductively pick-up noise within the case and allow it to exit the case, thus making any EMI filtering less effective. Likewise, any unfiltered leads within the case will radiate any transients from outside the case into the case, which may affect the static discharge behavior of the contained circuits to external static events.

E.4 Conducted EMI Filters

There are two types of input power buses. DC power buses are single-wire power connections such as found in automobiles and aircraft. The ground connection forms the other leg of the power system. The other form of input connection is the ac, or two or three-wire feed systems as found in ac power systems. The design of the EMI filter for dc systems is covered in Section 3.12 and takes the form of a simple *L-C* filter. All the noise is common-mode between the single power wire and the ground return. The dc filter is much more complicated, because of the parasitic behavior of the components involved.

To design a filter for the input of a switching power supply, the designer first needs to know which of the regulatory specifications is appropriate for the product. The specifications dictate the conducted and radiated EMI/RFI limits the product must meet to be sold into the particular market. A company's marketing department should know which areas of the world the product will be sold and hence the designers can determine the requirements that are appropriate. It is always a good idea to design for the most stringent specification that is applicable to your market.

The purpose of an input conducted EMI filter is to keep the high-frequency conducted noise inside the case. The main noise source is the switching power supply. Filtering on any of the input/output (I/O) lines is also important to keep noise from any internal circuit, like microprocessors, inside the case.

Design of the common-mode filter

The common-mode filter essentially filters out noise that is generated between the two power lines (Hot and Neutral or H1 and H2). The common-mode filter schematic is shown as part of Figure E–4.

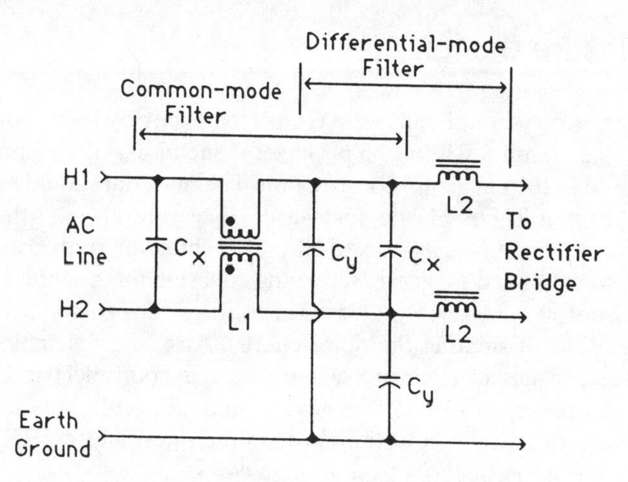

Figure E–4 A complete third-order, input EMI filter (common-mode and differential-mode).

In the common-mode filter the windings of the "transformer" are in phase, but the ac currents flowing through the windings are out of phase. The result is that the common-mode ac flux within the core for those signals that are equal and opposing phases on the two power lines cancel out.

The problem with designing the common-mode filter is that at high frequencies, where one wants and needs the filtering, the ideal characteristics of the components are compromised by their parasitic behavior. The major parasitic element is the interturn capacitance of the transformer itself. This is the small capacitance that exists between all windings, where the voltage difference (volts/turn) between turns behaves like a capacitor. This capacitor, at high frequency, effectively acts as a shunt around the winding and allows more high-frequency ac current to go around the windings. The frequency at which this becomes a problem is above what is called the *self-resonance* of the winding. A tank circuit is formed between the winding inductance itself and this distributed interturn capacitance. Above the self-resonance point the effects of the capacitance become larger than the inductance which then reduces the level of attenuation at high frequencies. The effect of this within the common-mode filter can be seen in Figure E–5. Its affect can be reduced by purposely using a larger X capacitor. The self-resonance frequency is the point where the greatest possible attenuation for the filter is exhibited. So by choosing the winding method of the transformer, one can locate this point on top of a frequency that needs the greatest filtering, such as a harmonic peak in the unfiltered system noise spectrum.

Another area of concern is the "Q" of the filter at self-resonance. If the Q is too high, or in other words, the damping factor is too low, the filter will actually generate noise in the form of narrow-band ringing. This can be dealt with during the design.

Some major transformer manufacturers build standard off-the-shelf components used in the design of common-mode filter transformers such as Coilcraft (Cary, IL). These transformers have various inductance values, and current ratings and also provide the needed creepage dimensions. This can make the designer's job a lot easier.

The initial common-mode filter component values can be determined in a step-by-step process (like everything else in this book). To begin this process, either a baseline measurement of the unfiltered conducted noise spectrum is

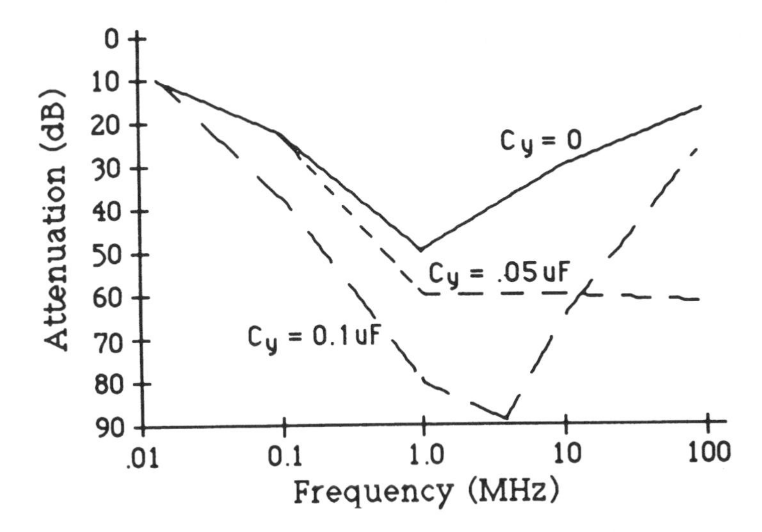

Figure E–5 Frequency response of a second-order common-mode filter ($L = 1$ mH).

needed or some assumptions need to be made. This is in order to know how much attenuation is needed and at what frequencies. Obviously, making the measurement will yield a good result (with minor tweeks) the first time. Assuming that one needs a particular filter response on paper may lead to additional "fudging" of the circuit on the test table.

A reasonable beginning is that one needs about 24 dB of attenuation at the switching frequency of the switching power supply. This, of course, should be modified in response to the actual conducted noise spectral shape. One determines the corner frequency of the filter by

$$\text{Attenuation}\,(-\text{dB}) = 40\,\text{Log}\left(\frac{F_C}{f_{SW}}\right)$$

or

$$f_C = f_{SW} \cdot 10^{\left(\frac{Att}{40}\right)}$$

where: f_c is the desired corner frequency of the filter.
f_{sw} is the operating frequency of the power supply.

For this example, the switching frequency is assumed to be 50 kHz. The corner frequency to produce −24 dB of attenuation at this point is

$$f_C = (50\,\text{kHz})10^{\left(\frac{-24}{40}\right)} = 12.5\,\text{kHz}$$

One assumes that the line impedance is 50 ohms (because that is what the LISN test's impedance is). This impedance is then the damping element within the reactive filter circuit.

Choosing the damping factor

The minimum damping factor (ζ) should be no less than 0.707. Less than that would allow ringing to occur and produce less than 3 dB of attenuation at the corner frequency.

Calculating the initial component values

$$L = \frac{R_L \cdot \zeta}{\pi \cdot f_C} = \frac{(50)(0.707)}{\pi(12.5\,\text{kHz})} = 900\,\mu\text{H}$$

$$C = \frac{1}{(2\pi f_C)^2 L} = \frac{1}{[2\pi(12.5\,\text{kHz})]^2 (900\,\mu\text{H})}$$

Choosing "real world" available components

The largest value of capacitor that is available in the 4 KV voltage rating is 0.05 μF. This is 27 percent of the calculated value. In order for the corner frequency to remain the same, the inductor value should be increased by a factor of 3.6. This would make the value 3.24 mH. The damping factor is directly proportional to the value of the inductance so the resultant damping factor is 2.5 which is acceptable.

The closest Coilcraft common-mode inductor part number is E3493 and its self-resonance frequency is 1 MHz. The calculated capacitors are what are typically called "Y" capacitors. These are placed between each phase and the earth ground and must meet the full HIPOT test voltage of 2500 VRMS. "X" capacitors are the ones that are placed between the power lines and need only meet the 250 VRMS rating of the power line and be able to withstand any surge that may be anticipated. Choosing the value of the X capacitors is mainly arbitrary and usually they fall in the 0.001 to 0.5 mF range.

One can reasonably expect this filter to provide a minimum of 60 dB of attenuation between the frequencies of 500 kHz and 10 MHz.

Once the component values have been calculated, the physical construction of the transformer and the PCB layout become critical for the effectiveness of the filter stage. Magnetic coupling due to stray inductive pick-up of high-frequency noise by the traces and components can circumvent the filter all together. Added to this is the fact that the common-mode filter choke looks more and more capacitive above its self-resonance frequency. The net result is the designer needs to be concerned about the high-frequency behavior of the filter typically above 20 to 40 MHz.

Physical layout of the PCB is important. The filter should be laid out in a linear fashion so that the input portion of the filter is physically distant from the output portion. Large, low-inductance traces should be also used, but keep in mind the creepage requirements of the regulatory specifications.

Sometimes the high-frequency attenuation is insufficient to meet the specifications and a third pole needs to be added to the EMI filter. This filter is typically a differential-mode filter and will share the Y capacitors from the common-mode filter. Its corner frequency is typically the same as the common-mode filter. This filter is made up of a separate choke on each power line, and is placed between the input rectifiers and the common-mode filter.

The differential-mode filter should have a lower damping factor than the common-mode because the combined damping response of the entire filter section would be too sluggish if higher damping factors were used. A damping factor of a minimum of 0.5 is acceptable.

Calculating the differential-mode choke value

$$L_d = \frac{R_L \cdot \zeta}{2\pi \cdot f_C} = \frac{(50)(0.5)}{2\pi(12.5\,\text{kHz})}$$
$$= 318\,\mu H$$

The addition of this stage of filtering will bring the very high-frequency attenuation under control and further attenuate any differential-mode noise on the earth ground lead. It will also produce a combined attenuation of –36 dB at the switching frequency of the power supply.

Real-world considerations

If one was to build the inductive elements instead of buying off-the-shelf parts from a manufacturer, the following guidelines are common industry practices.

Common-mode chokes (transformers)

1. A toroid is best for this application because it produces very little stray magnetic fields.
2. A high permeability ferrite is used such as the W material from Magnetics, Inc. which has a permeability of 10,000.
3. If an E-E core is used (which is a common choice), there should be no air-gap and the mating surfaces of the cores must be polished. Any surface imperfections would lower the permeability.
4. The bobbin should be a two-section bobbin and not be completely filled with windings. A 2 mm space from the outside surface of the bobbin is required for the 4 mm creepage requirement of VDE.

Differential-mode chokes

1. These are wound on separate cores (not mutually coupled).
2. Use a powdered iron material such as available from MicroMetals (Evanston, IL).
3. Bar cores are typically used because of cost.

Appendix F. Miscellaneous Information

This appendix contains a potpourri of miscellaneous information that may be needed occasionally.

F.1 Measurement Unit Conversions

This is by far the most confusing area of global cooperation. There are those countries which have completely converted to metric and use the MKS (meters-kilogram-second) system, there are countries that use a hybrid metric system with a mixture of metric units, such as Japan, and then there is the U.S. which mixes metric CGS (centimeter-gram-second) system with old English units such as inches, mils, and circular mils. All three systems are used by major core manufacturers around the world. The designer, depending upon where he or she wants to buy magnetics, must be extremely cautious as to which equations he or she is using and the units of measurement of the variables each particular equation. Core manufacturers rarely elaborate on their units of measurement. As an aide to the designer, the conversion constants between the systems are given below.

Flux Density
1 Tesla (Webers/m^2) = 10^4 Gauss (Webers/cm^2) (Europe)
1 Gauss (Webers/cm^2) = 10^{-4} Tesla (Webers/m^2) (USA)
1 milliTesla = 10^{-3} Tesla (Japan)
1 milliTesla = 10 Gauss (Japan)

Linear Measurements
1 centimeter = 0.394 inches
1 millimeter = 0.0394 inches
1 inch = 2.54 centimeters
1 inch = 25.4 millimeters

Area Measurements
1 square inch (in^2) = 6.45 square centimeters (cm^2)
1 in^2 = 645 square millimeters (mm^2)
1 cm^2 = 0.155 in^2
1 mm^2 = 0.00155 in^2
1 circular mil = 7.854×10^{-7} in^2
1 circular mil = 5.07×10^{-6} cm^2
1 in^2 = 1.273 × 106 circular mils
1 cm^2 = 1.974 × 105 circular mils

F.2 Wires

The specification of wires can be confusing. All wires diameters are based upon the American Wire Gauge (AWG) table, published in the early 20th century. The metric countries directly converted these dimension (inches) to millimeters and created what is now the IEC R20 wire table. This is shown below in both measurement systems in Table F–1.

The R20 chart is being eventually replaced with the IEC R40 standard as shown in Table F–2. The wire diameters are still very close to the AWG table.

Table F–1 Wire Table (USA and IEC R20)

AWG	IEC (R20) Diameter	Copper Diameter	Overall Diameter (Grade 1)		Wire Area		dc Resistance	Current Capacity*
	(mm)	(in)	(in)	(mm)	(in^2)	(mm^2)	(Ω/1000 ft)	(A)
10	2.500	0.102	0.107	2.72	0.008 15	5.50	1.00	10.38
11	2.240	0.091	0.096	2.431	0.006 47	4.17	1.26	8.230
12	2.000	0.081	0.086	2.172	0.005 13	3.31	1.58	6.530
13	1.800	0.072	0.076	1.943	0.004 07	2.79	2.00	5.180
14	1.600	0.064	0.068	1.737	0.003 22	2.08	2.53	4.110
15	1.400	0.057	0.061	1.711	0.002 56	2.01	3.18	3.260
16	1.250	0.051	0.055	1.389	0.002 03	1.31	4.02	2.580
17	1.120	0.045	0.049	1.247	0.001 61	1.04	5.06	2.050
18	1.000	0.040	0.044	1.118	0.001 27	0.82	6.39	1.620
19	0.900	0.026	0.039	1.003	0.001 01	0.65	8.05	1.290
20	0.800	0.032	0.035	0.897	0.000 80	0.52	10.1	1.020
21	0.710	0.029	0.032	0.805	0.000 64	0.41	12.8	0.810
22	0.630	0.025	0.028	0.721	0.000 51	0.32	16.2	0.640
23	0.560	0.023	0.026	0.650	0.000 40	0.26	20.3	0.510
24	0.500	0.020	0.023	0.582	0.000 32	0.21	25.7	0.400
25	0.450	0.018	0.021	0.523	0.000 25	0.16	32.4	0.320
26	0.400	0.016	0.018	0.467	0.000 20	0.13	41.0	0.253
27	0.355	0.014	0.016	0.419	0.000 16	0.10	51.4	0.202
28	0.315	0.013	0.015	0.376	0.000 13	0.080	65.3	0.159
29	0.280	0.011	0.013	0.340	0.000 10	0.065	81.2	0.128
30	0.250	0.010	0.012	0.305	0.000 079	0.051	104	0.100
31	0.224	0.008 9	0.011	0.274	0.000 062	0.040	131	0.079
32	0.200	0.008 0	0.010	0.249	0.000 050	0.032	162	0.064
33	0.180	0.007 1	0.009	0.224	0.000 040	0.026	206	0.050
34	0.160	0.006 3	0.008	0.198	0.000 031	0.020	261	0.040
35	0.140	0.005 6	0.007	0.178	0.000 025	0.016	331	0.031
36	0.125	0.005 0	0.006	0.163	0.000 020	0.013	415	0.025
37	0.112	0.004 5	0.005 7	0.145	0.000 016	0.010	512	0.020
38	0.100	0.004 0	0.005 1	0.130	0.000 013	0.008	648	0.016
39	0.090	0.003 5	0.004 5	0.114	0.000 009	0.006	847	0.012
40	0.071	0.003 1	0.004 1	0.104	0.000 008	0.005	1080	0.010

*Based on a current density of 1275 A/in^2 (1000 circular mils/A).

Note: IEC R20 is an intermediate wire gauge eventually to be replaced by IEC R40.

Table F–2 Wire Table — IEC R40

Copper Diameter (mm)	Maximum Overall Diameter — Insulated Wire Grade 1 (mm)	Grade 2 (mm)	Grade 3 (mm)	Current Capability* (A)
2.650	2.730	2.772	2.811	11.66
2.360	2.438	2.478	2.516	9.25
2.120	2.196	2.235	2.272	7.46
1.900	1.974	2.012	2.048	6.00
1.700	1.772	1.809	1.844	4.80
1.500	1.570	1.606	1.640	3.74
1.320	1.388	1.422	1.455	2.89
1.180	1.246	1.279	1.311	2.31
1.060	1.124	1.157	1.188	1.87
0.950	1.012	1.044	1.074	1.50
0.850	0.909	0.939	0.968	1.20
0.750	0.805	0.834	0.861	0.93
0.670	0.722	0.749	0.774	0.702
0.600	0.649	0.674	0.698	0.598
0.530	0.576	0.600	0.623	0.467
0.475	0.519	0.541	0.562	0.375
0.425	0.466	0.488	0.508	0.300
0.375	0.414	0.434	0.453	0.234
0.335	0.372	0.391	0.408	0.186
0.300	0.334	0.352	0.369	0.149
0.265	0.297	0.314	0.330	0.117
0.236	0.267	0.283	0.298	0.092
0.212	0.240	0.254	0.268	0.075
0.190	0.216	0.228	0.240	0.060
0.170	0.194	0.205	0.217	0.048
0.150	0.171	0.182	0.193	0.037
0.132	0.152	0.162	0.171	0.029
0.118	0.136	0.145	0.154	0.023
0.106	0.123	0.132	0.140	0.019
0.095	0.111	0.119	0.126	0.015
0.085	0.100	0.107	0.114	0.012
0.075	0.089	0.095	0.102	0.009

*Based on a current density of 2.11 A/mm^2 (1000 circular mils/A).

Skin effect is the apparent increase in wire resistance when high-frequency ac currents are passed through them. A wire's real resistance when involving losses within a switching power supply is given in Equation F.1.

$$R_{\text{total}} = R_{\text{DC}} + R_{\text{AC}} \tag{F.1}$$

R_{AC} is the result of multiplying the below ratio with the dc resistance for a round copper wire such as round magnet wire. The equation below is the percent of increase of the ac resistance over the dc resistance for a single strand of round copper wire in open air.

$$\frac{R_{AC}}{R_{DC}} \approx 0.47 \cdot \pi \cdot d \cdot \sqrt{f} \tag{F.2}$$

where d is the diameter of the wire in centimeters.
f is the frequency of the fundamental frequency of the current waveform in hertz.

As one can see, the larger-diameter wires suffer a much more rapid degradation in ac resistance with increasing frequency than do smaller-diameter wires. So it is advantageous to use multiple strands of smaller wires instead of one large diameter wire. The ac current density of the smaller wires (>30 AWG) can actually be pushed to two to three times the assumed current density used in the charts because their surface area to cross-sectional area ratio is much greater.

The ac resistance increase due to skin effect given above should be considered as a minimum. When wires are placed next to one another and placed in layers within a transformer, the near field magnetic effects between wires further crowd the current density into even smaller areas within the wire's cross-section. For instance, when wires are wound next to one another, the current is pushed away from the points of contact along the surfaces of the wires to areas orthogonal to the winding plane. When layers are placed on top of one another the inner layers show much greater degradation in apparent resistance than do the outermost layers.

For estimation purposes, using the above resistance numbers will be sufficient.

References

Ben-Yaakov, Sam and Gregory Ivensky, *Passive Lossless Snubbers for* High Frequency PWM Converters, Seminar 12, APEC 99.

Balogh Laszlo, *Practical Considerations for MOSFET Gate Drive Techniques in High Speed, Switch-mode Applications*, Seminar APEC99. March 1999.

Brown, Marty, "Laying Out PC Boards for Embedded Switching Supplies," *Electronic Design*, December 6, 1999.

Brown, Marty, *Practical Switching Power Supply Design*, Academic Press, 1990.

Carsten, Bruce, *Avoiding the EMI Accident*, Seminar APEC99, March 1999.

Carsten, Bruce, *Switchmode Design Techniques Above 500 kHz*, High Frequency Power Conversion Conference, May 1986.

Carsten, Bruce, *High Frequency Losses in Switchmode Magnetics*, High Frequency Power Conversion Conference, May 1986.

Li, Alan, Brij Mohan, Steve Sapp, Izak Bencuya, and Linh Hong, *Maximum Power Enhancement Techniques for SuperSOT-6 Power MOSFETs*, Fairchild Ap Note AN1026, April, 1996.

Martin, Robert F., *Harmonic Currents*, Compliance Engineering—1999 Annual Resources Guide, Cannon Communications, LLC, 103–107.

Venable, H. Dean, *The K Factor: A New Mathematical Tool for Stability Analysis and Synthesis*, POWERCON March 1983.

Index